Rationality and Reality

STUDIES IN HISTORY
AND PHILOSOPHY OF SCIENCE

VOLUME 20

RATIONALITY AND REALITY

Conversations with Alan Musgrave

Edited by

COLIN CHEYNE
University of Otago,
Dunedin, New Zealand

and

JOHN WORRALL
London School of Economics,
London, UK

Springer

A C.I.P. Catalogue record for this book is available from the Library of Congress.

ISBN-10 1-4020-4206-X (HB)
ISBN-13 978-1-4020-4206-X (HB)
ISBN-10 1-4020-4207-8 (e-book)
ISBN-13 978-1-4020-4207-8 (e-book)

Published by Springer,
P.O. Box 17, 3300 AA Dordrecht, The Netherlands.

www.springer.com

Cover: Photograph of Alan Musgrave used with kind permission of Gudrun Perin, Guelph, Canada

Printed on acid-free paper

Printed in the Netherlands.

TABLE OF CONTENTS

TABLE OF CONTENTS

ACKNOWLEDGEMENTS

The editors are indebted to Robert Nola, David Papineau and Stephen Gaukroger for their invaluable assistance and advice, and to Alan Musgrave for his enthusiastic support. We are also grateful for the secretarial assistance of Chris Stoddart, Sally Holloway, and especially that of Kate Anscombe, who cheerfully and efficiently carried out the bulk of the manuscript preparation.

We originally planned this book as a tribute to Alan Musgrave to mark his planned retirement from the Philosophy Department at the University of Otago in 2005. Subsequently this retirement theory was refuted, so this stands as a tribute to his work so far. We did not want it to be a standard festschrift; hence the idea of critical essays with Alan having the right of reply. We would like to thank all the contributors (including each other) for their cooperation in making this the testament to Alan's standing in the profession, both in Australasia and worldwide, that we believe it to be.

NOTES ON CONTRIBUTORS

HANS ALBERT is Professor Emeritus of Sociology and Philosophy of Science at the University of Mannheim, Germany. He is the author of *Treatise on Critical Reason*, and *Between Social Science, Religion, and Politics*, as well as many other books and articles on social science, economics, philosophy, and religion.

ANDREW BARKER is Professor of Classics at the University of Birmingham, UK. He is the author of *Greek Musical Writings* (2 volumes), *Scientific Method in Ptolemy's Harmonics*, and several other books and many articles on ancient Greek music and musical theory.

ALAN CHALMERS is an Adjunct Senior Research Fellow in the Philosophy Department at Flinders University, Australia. He is the author of *What Is This Thing Called Science?* and *Science and Its Fabrication*, and articles on the history and philosophy of physical science.

COLIN CHEYNE is a Senior Lecturer in the Department of Philosophy, University of Otago, New Zealand. He is the author of *Knowledge, Cause, and Abstract Objects*, and articles on epistemology and philosophy of mathematics.

MARK COLYVAN is Professor of Philosophy at the University of Queensland, Australia. He is the author of *The Indispensability of Mathematics*, co-author of *Ecological Orbits*, and he has published articles on philosophy of logic, decision theory, philosophy of science, and philosophy of mathematics.

GREGORY CURRIE is Dean of the Faculty of Arts at the University of Nottingham, UK and a member of the Philosophy Department. His most recent book is *Arts and Minds*. He is working on a project on narrative representations of agency.

VOLKER GADENNE is Professor of Philosophy and Theory of Science at the University of Linz, Austria. He is the author of *Philosophie der Psychologie* and articles on philosophy of science.

NORETTA KOERTGE is a Professor Emeritus in the Department of History & Philosophy of Science, Indiana University, USA. Her early work addresses problems arising out of Popper's methodology. Recently she has edited anthologies that criticize postmodernist and feminist accounts of science, such as *Scientific Values and Civic Virtues*.

DEBORAH MAYO is Professor of Philosophy and Economics at Virginia Tech, USA. She is the author of *Error and the Growth of Experimental Knowledge* which received the 1998 Lakatos Prize award, and was a Director of a NEH Summer Seminar in 1999 on Induction and Experimental Inference.

ROBERT NOLA is a Professor of Philosophy at The University of Auckland, New Zealand. He has recently authored *Rescuing Reason*, co-authored *Philosophy, Science, Education and Culture*, and co-edited *After Popper, Kuhn and Feyerabend: Recent Issues in the Theory of Scientific Method.* His current work is a book on scientific method.

GRAHAM ODDIE is a Professor of Philosophy and Associate Dean of Humanities and Arts at the University of Colorado at Boulder, USA. He is the author of *Likeness to Truth*, and *Value, Reality and Desire*, as well as numerous articles on metaphysics, epistemology and ethics.

STATHIS PSILLOS is Associate Professor in the Department of Philosophy and History of Science at the University of Athens, Greece. His book *Causation and Explanation* has received the British Society for the Philosophy of Science President's Prize. He is also the author *Scientific Realism: How Science Tracks Truth* and co-editor of the forthcoming *Routledge Companion to the Philosophy of Science*.

HOWARD SANKEY is Associate Professor in the Department of History and Philosophy of Science at the University of Melbourne, Australia. He has written on incommensurability, rationality and scientific realism. His publications include *The Incommensurability Thesis* and *Rationality, Relativism and Incommensurability*, as well as several edited volumes.

MICHAEL REDHEAD was Professor of History and Philosophy of Science at the University of Cambridge, UK, specializing in the Philosophy of Physics. He was Vice-President of Wolfson College Cambridge. He won the Lakatos Prize in 1988, and is a Fellow of The British Academy. He is currently Co-Director of The Centre for Philosophy of Natural and Social Science at the LSE.

JOHN WORRALL is Professor of Philosophy of Science at the London School of Economics, UK. A former editor of *The British Journal for the Philosophy of Science*, he is the author of numerous articles in philosophy of science and is currently completing a book called *Reason in 'Revolution': A Study of Theory-Change in Science*.

COLIN CHEYNE

INTRODUCTION

Alan Musgrave's philosophical credo is encapsulated by the title of his 1993 book *Common Sense, Science and Scepticism*. But to say that Musgrave believes in common sense, science and scepticism is pretty uninformative. After all, many of his philosophical opponents are likely to claim as much. However Musgrave has a distinctive position on these matters in which they are closely interrelated, and this position has interesting and wide-ranging ramifications, encapsulated in their turn by the title of his 1999 *Essays on Realism and Rationalism*.

For Musgrave it is simply common-sensical to believe that an external world exists independently of the workings of our minds and that the aim of science is to discover what we can about that world, even though we cannot hope to have certain knowledge of it. Musgrave's robust common sense about the external world seems to have clashed with antirealist philosophies from the start. He tells of how as an undergraduate he was 'both fascinated and repelled by Berkeley's idealism. All that ingenuity wasted on a crazy view' (Nola 1995, p. 31[1]). Consequently, he has devoted his philosophical life, in one way or another, to uncovering the errors in Berkeley's arguments, and to detecting and denouncing any arguments that threaten to lead to similar conclusions.

Thanks to his clear-sighted view of what realism about a mind-independent world actually entails, Musgrave has not been short of targets. He has shown a remarkable ability to detect antirealist leanings, even those to which their authors may be oblivious. For example, according to Musgrave, Hilary Putnam's internal realism is no realism at all, rather it is 'Kant generalised and relativised'.

If the external world is truly mind and language independent, then, on Musgrave's view, any serious investigation of that world (science, in particular) is, and ought to be, a robustly realist enterprise. But when philosophers have turned their attention to science, the lure of antirealism has proved strong. Instrumentalism, pragmatism, constructive empiricism, social constructivism, conceptual idealism, *et alia*—all are antirealistic views of one kind or another. Constructivism or conceptual idealism is especially obnoxious because it is pre-Darwinian. Darwin tells us that human beings are the product of a pre-existing world containing unobserved creatures such as dinosaurs and unobservable entities such as quarks. Constructivism reverses this process, making the pre-historic world and its denizens the products of human beings and their minds or words or concepts or whatever.

[1] I have drawn on Robert Nola's interview with Alan Musgrave throughout this Introduction.

C. Cheyne & J. Worrall (eds.), *Rationality and Reality: Conversations with Alan Musgrave,* 1–6.

Scepticism enters the picture when Musgrave considers what it is that attracts philosophers down the idealist path. He believes that they are motivated by the desire for certain knowledge. Idealist philosophies are vain attempts to rule out the possibility of sceptical scenarios, such as Descartes' evil demon or Putnam's brain-in-a-vat. Musgrave takes the (in his opinion) common-sense view that we are obviously fallible epistemic agents; our best efforts pursued to the limit could yield false theories. But we can take comfort from the common-sense fact that we have no reason to think that any of those sceptical scenarios are true. After all, they, and sceptical scenarios in general, are not directed at our beliefs, but rather at attempts to establish those beliefs.

However, one person's common sense is another's absurdity. To claim that the world existed before we did and that we can make mistakes is surely uncontroversial. The same, it would seem, can be said for simple arithmetic. But Musgrave considers belief in the existence of abstract objects an offence against common sense. Lacking causal powers and a location in space and time, abstract objects are 'weird' entities. Since numbers are paradigmatically abstract, Musgrave denies their existence. But if numbers do not exist, then arithmetic, taken at face-value, is false. Musgrave may be right, but mathematical realists will feel justified in claiming that, in this instance, they have common sense on their side. Musgrave's allegiance to common sense can also come under pressure when he is confronted with a possible counterexample to one of his positions. He is rather fond of responding that he will face down the objection by 'biting the bullet'. But surely there is a limit to how much bullet-biting one can engage in whilst keeping enough of one's teeth to defend common sense.

Alan Musgrave's insistence that common sense and scientific realism should be taken seriously and that any hint of idealism be forthrightly eschewed, prompted colleagues at Virginia Tech (he was there as Visiting Professor) to present him with a farewell gift—a sign reading "BEWARE OF THE mad DOG realist". He proudly displays it in his office. But he sometimes points out that his realism is not all that extreme; that 'lap-dog' realist might be more appropriate. After all, he doesn't claim, as some scientific realists appear to, that all current theories of science are true or ought to be believed. Nor is he a realist about abstract entities. Even so, the canine reference is apt. There is something of the blood-hound in the way he sniffs out and tracks down any whiff of idealism, and something terrier-like about the way he holds on once he has his philosophical teeth into an antirealist opponent. (Musgrave's realism has also been labelled 'Coronation Street Realism' by a Kantian opponent. Musgrave, an erstwhile Mancunian, happily accepts this label.)

Musgrave's take on common sense, science and scepticism also gives rise to his distinctive rationalist epistemology. The critical methods of common sense and science cannot, he believes, establish the truth of our hypotheses about the world. But our realist hypotheses are rational or reasonable insofar as they survive our best attempts to fault them, and insofar as they best explain the phenomena. We may reasonably believe a falsehood. If the state of the critical discussion changes, and we find a reason to think our belief false, it will no longer be reasonable to believe it. But what we have found is that our belief is false, not that we were unreasonable to believe it. Truth is a desideratum on reasonable belief, not a defining condition of it. What goes for truth and belief, also goes for reliability and method. We may reasonably follow an unreliable method of forming beliefs. If we find a reason to think our method unreliable, it will no longer be reasonable to follow it. But what we have found is that our method was unreliable, not that we were unreasonable ever to have followed it. Reliability is a desideratum on reasonable method, not a defining condition of it. This avowedly internalist position is Musgrave's version of Popper's critical rationalism. All resting, he claims, on common sense.

One more aspect of Musgrave's philosophical work should be mentioned. His demand for clarity and his detestation of obscurantism and philosophical 'word salads'. The clarity and force of his own prose speaks for itself. The dense, jargon-clotted prose of much recent continental and post-modernist philosophy particularly draws his ire. But he is equally averse to analytic writings that are mired by complex technicalities. Not that he thinks that technicalities can always be dispensed with. His view is that any worthwhile philosophical position can be, and should be, introduced in a few clear sentences, and supporting arguments outlined as valid arguments with two or three premises. 'Bells and whistles' can be added later, if need be. And when it comes to illustrative examples, especially in the philosophy of science, the simpler the better. Billiard balls colliding rather than a pair of entangled electrons entering a Stern-Gerlach magnet. On one occasion a speaker was struggling in the face of Musgrave's demands for further clarification. 'I'm sorry,' he said, 'I can only try to make it clearer with a mickey-mouse example.' 'Don't apologise,' said Musgrave, 'I prefer mickey-mouse examples.'

***

Alan Musgrave was born in 1940, into a working-class family in Manchester, England. Success in the 11+ examinations enabled entry to a Grammar School education, and a scholarship then led to acceptance at the London School of Economics in 1958, where he intended to study law. But he was told that he was unlikely to succeed in the legal profession without the backing of a rich father, which he did not have. Then a letter arrived from the LSE saying that any student accepted for any course of study could switch to a new course, Combined Honours in Philosophy and Economics. So he switched having only a vague idea of what economics is, and no idea about philosophy.

From the beginning Alan was fascinated by philosophy, being particularly struck by the lectures of Karl Popper, which he attended regularly, although he never actually spoke with Popper during his undergraduate years. His first tutor was Joseph Agassi. From Alan's account, Agassi adopted the 'sink-or-swim' approach and offered little guidance, although his comment on Alan's first piece of work, 'Full of jargon—write simply', appears to have had a life-long impact.

Imre Lakatos replaced Agassi in Alan's final year. If anything, Lakatos provided even less assistance, apparently writing Alan off because he knew little mathematics and physics. Neverthelesss, the young Musgrave did not 'sink'. His examination results prompted Lakatos to persuade him to do graduate research, rather than to embark on his intended career as a schoolteacher. He enrolled for a PhD with Popper as his supervisor. The benign neglect continued—Popper refused to help him choose a topic. However, Lakatos did encourage and assist him to remedy his illiteracy in mathematics and physics. In return for this 'private tuition' Alan assisted with the preparation of Lakatos's 'Proofs and Refutations' for its first publication in 1963-64.

It was another two years before Alan embarked on his thesis proper. In the meantime, in 1962, he married as well as becoming Popper's research assistant. Two years later he was appointed to a temporary assistant lectureship, but a couple of years after that when his wife became pregnant he decided to seek something more permanent. He applied for and was offered a position in Scotland. Before he could take it up, Lakatos intervened once again and persuaded the Director of the LSE to offer Alan a tenurable lectureship there.

In 1965 he was involved in the organisation of the famous International Colloquium in the Philosophy of Science held in London. This provided the opportunity for him to meet such luminaries as Carnap, Tarski, Kuhn, Quine, Kreisel, Mostowski, and Bernays. He recalls attending a paper on the foundations of set theory by Mostowski which he found mostly unintelligible. It was followed by heated but even more unintelligible discussion. Then Tarski intervened. His contribution to the discussion was couched in simple, easily understood language ('baby-talk' in Alan's words) and clearly illuminated the issue. This incident had a lasting impression on him, and generations of participants at Otago philosophy seminars have subsequently been grateful for Alan's interventions couched in 'baby-talk'. Following the conference, Alan had the opportunity to co-edit with Lakatos the fourth volume of the proceedings, which became the influential best-seller *Criticism and the Growth of Knowledge*.

In 1969 Alan successfully completed his PhD thesis, 'Impersonal Knowledge—a Criticism of Subjectivism in Epistemology'. By this time he had become disillusioned and depressed by the student troubles at the LSE as well as the infighting within the department. And getting by on a lecturer's salary meant living with his wife and two children in a small house in North London with little opportunity to enjoy the cultural offerings of London-life. So when a job offer with better prospects came up he jumped at it.

Otago University in Dunedin, New Zealand was looking for a new Professor of Philosophy. Because Popper was well-known there, having been at Canterbury University (then Canterbury College) from 1937 to 1945, he was approached for suggestions. He recommended Alan. When Alan flew out to New Zealand for the interview it was the first time he had left the UK and his first aeroplane flight. He was offered the Chair and took up the job in 1970.

The Otago Chair had been held by some distinguished predecessors, including J, N. Findlay, D. D. Raphael, John Passmore and J. L. Mackie. Although Alan was only 29 years old at the time, he was not the youngest occupant. Findlay had been the same age and Duncan Macgregor, the foundation professor in 1871, was two years younger. However, Alan's has been the longest reign. He has been the Head of the Department for more than 35 years, apart from brief periods of occupancy by Bob Durrant, Greg Currie, Andrew Moore and Colin Cheyne.

Alan quickly established himself as a lively and popular teacher. His lectures were remarkable for their clarity and erudition, laced with jokes and gossip about famous philosophers and scientists, and delivered in his Lancashire accent with an obvious enthusiasm for the subject matter. Alan's approach to teaching contrasts sharply with his experiences at the LSE. He is unfailingly helpful and encouraging to all interested students, often dismaying his colleagues with the time and trouble he takes with 'cranks' and 'lame-ducks'. But if he is sometimes overly-generous to less able students, he sets very high standards for those who show promise. As a result the best philosophy graduates from Otago have established themselves with distinction in a variety of professions throughout the world. These include the philosophers Jeremy Waldron, Graham Oddie and Tim Mulgan, and Pamela Tate, Solicitor-General of Victoria, Australia.

Alan also set about strengthening and enhancing the department's research culture. This he achieved by dint of his encouragement and enthusiasm, rather than heavy-handed prescription. A weekly seminar for the presentation of papers by department members and visitors has run continuously during term time for more than 30 years; a record matched by few other departments in the university. Attendance by staff and graduate students is expected, and any absence is likely to be queried. Regular contributions by members of the department are expected.

The seminars are notable for vigorous and lengthy discussion. It is here that Alan's detestation of obscurantism often comes to the fore. If Alan doesn't understand the paper or suspects that the students do not understand, he can be relentless in his demand for clarity. One suspects that some visitors, particularly in the past, have seen their visit to this remote part of the philosophical world as an opportunity to relax and explore New Zealand's spectacular scenery. But there has been nothing relaxing about the reception of their seminar paper. Nevertheless, the trickle of visitors has become a steady stream, and many distinguished philosophers have returned, sometimes for extended periods. Sir Karl Popper visited in 1972 and it was at the department seminar that Pavel Tichy famously demolished Popper's definition of verisimilitude.[2]

With Alan at the helm, research publications from the department increased in quality and frequency. A number of young scholars established themselves there. Some (such as the logician Pavel Tichy) stayed, while others (including Greg Currie and Paul Griffiths) moved on to prominent positions elsewhere. So it should have been no great surprise (although it apparently was to some) that Otago Philosophy did well in New Zealand's recent Performance Based Research Fund assessments (similar to the UK's RAE). In fact the department's score was the highest of all departments over all disciplines in the entire country.

Throughout his period at Otago Alan has continued to publish. Although mostly focussed on defending critical realism and critical rationalism, his papers have covered a wide range of topics, from ancient Greek astronomy to recent theories of economics, from scientific explanation to theories of truth, and from psychologism to the problem of induction. His introductory lectures on epistemology formed the basis of his *Common Sense, Science and Scepticism* (1993) and sixteen of his articles were collectively published as *Essays in Realism and Rationalism* (1999).

***

[2] See Svoboda, Jespersen & Cheyne 2004, p, 27 for further details of the incident.

The papers in this volume are presented in the same order that Alan Musgrave responds to them. He explains the rationale for that ordering at the beginning of his paper (p. 293). When the authors of the papers received the first draft of Alan's resonses they mostly reacted in one of two ways. They either expressed delight that Alan had agreed with them or they expressed delight that he had seen fit to disagree with them in his characteristically forthright manner. Alan's generous support, tempered with astute criticism, has been a source of inspiration for many philosophers over many years. Long may it continue to be so.

REFERENCES

Lakatos, I. (1963-64) 'Proofs and Refutations.' *British Journal for the Philosophy of Science* 14: 1-25, 120-39, 221-45, 296-342.

Lakatos, I. & Musgrave, A. (eds) (1970) *Criticism and the Growth of Knowledge* (Proceedings of the International Colloquium in the Philosophy of Science, vol. 4). London, Cambridge University Press.

Musgrave (1993) *Common Sense, Science and Scepticism: A historical introduction to the theory of knowledge*. Cambridge, Cambridge University Press.

Musgrave, A. (1999) *Essays on Realism and Rationalism.* Amsterdam: Rodolfi.

Nola (1995) 'Interview with Alan Musgrave' *Metascience (New Series)* 7: 122-33.

Svoboda, V., Jespersen, B. & Cheyne, C. (eds) (2004) *Pavel Tichy's Collected Papers in Logic and Philosophy.* Prague & Dunedin: Filosofia & University of Otago Press.

GREGORY CURRIE

WHERE DOES THE BURDEN OF THEORY LIE?

1. OBSERVATIONS AND STATEMENTS ABOUT OBSERVATIONS

Discussing the supposed theory-ladenness of observation, Alan Musgrave says:

> The slogan [observation is theory laden] cannot be literally true. If anything is 'theory laden" it cannot be observation but rather statements made on the basis of observation. Observation is simply an act that humans and other creatures perform, a special kind of event or process occurring in the nervous system of humans and other creatures. How can an act or event or process be 'theory-laden'? (1983, p. 46)

I deny that there is a conceptual or metaphysical error in the idea that observation is theory laden. Of course, 'theory laden', as applied to experience or to anything else, must be some sort of metaphor; nothing can literally be laden with theory. If we stick with the idea that it is observation that is at issue here, the metaphor points us in two directions. One is towards the idea that theory influences observation. There is a sense in which this is uncontroversially true, since the theories that people have are among the things that influence their decisions about where to look; if you don't look at something, you won't have the relevant experience. This cannot be what is at issue in discussions of theory-ladenness. We can distinguish between looking, which is an action, and perception, which is what happens when you look and is not an action. The controversial claim about the influence of theory on perception is this: two people with equivalently functioning perceptual systems but possessed of different theories may have perceptions with different contents when they look at the same thing. This is not the error, described by Musgrave as "fashionable rubbish" (p. 47), according to which people with different theories see different things when they look in the same direction. For the content of perception is a matter of the way perception represents the world as being, not a matter of the way things are in that region of the world where we are looking. Two people may have distinct perceptual contents when looking at the same bit of the world, either because at least one of them is misperceiving, or because, while neither is misperceiving, their perceptual contents reflect different aspects of what is there.

The second direction in which the metaphor of theory-ladenness points us is towards the idea that the elements of cognition – the elements out of which cognitive states like beliefs are built – are also the elements of perception. On this view

C. Cheyne & J. Worrall (eds.), Rationality and Reality: Conversations with Alan Musgrave, 7–18.

perception has conceptual content, just as beliefs have. To have a belief is to be in a state, the content of which is a structured entity. The constituents of this structure are concepts, and possession of this belief requires of the subject possession of those concepts. Believing that owls fly requires that I have the concepts owl and flying, and having a perception with the content there is a flying owl requires this also. So perception can represent the world as being a certain way to me only if I have a concept of it being that way.[1] That, anyway, is the view. On this view, what your senses can represent the world as containing depends on the concepts you possess. So once again, two people with functioning perceptual equipment, presented with the same distal layout may yet be in perceptual states with distinct contents, because they do not possess the same concepts.

2. OBSERVATION AND ITS ROLE IN EPISTEMOLOGY

As we have seen, Musgrave takes theory-ladenness to be an essentially linguistic phenomenon. He says that the theory-ladenness of observation is properly understood as the claim that 'in describing what we observe we must use terms which also figure in our general theories' (p. 54). This, he claims, entails that we always face the possibility that some future event will undermine the statement about what we have observed, and so the statement has the character of a conjecture. But I shall continue to question the idea that theory-ladenness cannot be thought of as a property of perceptual experiences, but must be thought of instead as a property of statements about experience. I am not denying that there is a sense of theory-ladenness that applies to statements; I am neutral on that question. But I am denying that such a sense, if there is one, is the only legitimate sense.

The idea of theory-ladenness, as Musgrave explicates it, provides the link that gets us to fallibilism: the view that "all statements [including statements that describe observations] are conjectural" (p. 53). This seems to me an odd characterisation. After all, the traditional epistemological issue is one concerning the fallibility of experience, or as it is sometimes put, of the senses. If theory-ladenness and fallibility are properties only of statements, fallibilism is consistent with the following view: the senses never deceive us; the problem is just that we sometimes misdescribe our experiences when we formulate sentences descriptive of them. But Musgrave himself, in reminding us of the "ancient lineage" of the problem he is interested in, remarks that "The ancient sceptics disputed the epistemological claim that by using our senses we can come to be absolutely sure of the truth-values of observation statements" (p. 45). The problem is centrally with the reliability of the senses themselves, and only peripherally with language. And it is significant that Musgrave finds it difficult to stick to the linguistic version of the thesis. He says "It is not seeing (or perceiving or observing) which might be theory-laden, but rather seeing (or perceiving or observing) that something is the case" (p. 47). But if *seeing* is an

[1]. This is roughly the formulation of Peacocke (1983, p. 7), who has since changed his mind on this topic (1992, Chapter 3).

event or process, then *seeing that* is surely also an event or process. It certainly isn't a linguistic entity.

Musgrave's aim, in the paper from which I have quoted, is to provide a limited and to some extent revisionary defence of some views of Popper, who has himself insisted that epistemologists stop talking about experiences and talk instead about statements: it is, says Popper "the logical connections between scientific statements which alone interest the epistemologist" (1959, p. 99).[2] But there are signs that Popper misunderstands the claim that experiences, and not just statements about experiences, are important for epistemology. He says that the view that statements can be justified not only by statements but also by perceptual experiences as "psychologism" (p. 94), and suggests that psychologism comes about because people mistakenly think that feelings of conviction can help to establish an hypothesis—a view which, he says, no sensible person would take concerning the validity of a logical inference (p. 98). But the idea that experiences are important objects of epistemological attention does not depend on the claim that feelings of conviction have anything to do with whether something counts as knowledge or not. Here Popper may be confusing experiences with something like *sensations*, conceived of as purely phenomenological states or, as Evans puts it, states "intrinsically without objective content" (Evans 1982, p. 123). But experiences essentially have *representational content*: they represent the world as being a certain way. It is the content of experience that matters to epistemology. It is this content which creates the possibility that an experience may provide a rational basis for the assertion of a statement describing some state of affairs. The epistemological problems connected with experience are those concerning the nature and determinants of its representational content, and the relation of that content to the contents of our beliefs. 'Feelings of conviction' have nothing to do with it.

An objection would be to say that, according to me, it is not the experience itself, but the content of the experience that matters from the point of view of knowledge. So the experience drops out of the picture. But I am not claiming—and nor, I think, is any traditional epistemologist claiming—that it is contents alone that serve to justify, or rationally warrant, our assertions. I am claiming that experience is capable of playing a justificatory role in epistemology because of its content, and hence that some particular experiences—namely those with the right kinds of contents—do justify some assertions. What matters is not content alone, but the content's being the content of an experience.

Problems about the content of experience and the relations between experience and belief are currently receiving a great deal of attention in philosophy, from the two different directions I have already mentioned. Attention comes first of all from a group of philosophers and cognitive scientists interested in whether experience is cognitively impenetrable or, more generally, encapsulated. It was this notion of cognitively impenetrability that I had in mind when I spoke earlier of experience being (or not being) independent of theory. In the broadest sense, any belief we have

[2] This is one aspect of what I take to be a disastrous turn in Popper's thought: the idea of epistemology 'without a knowing subject'. See Currie (1989).

counts as a theory.[3] Cognitive impenetrability is, roughly speaking, the property a system may have of delivering representations which are unaffected by the subject's cognitive states, of which beliefs are an example. It seems fair to say that a cognitively impenetrable system will deliver representations which are not, at least in one sense, theory laden. Encapsulation is a related but more general notion: a system is encapsulated if it delivers representations which are not affected by the representations made available by any other system at any level. Thus encapsulation entails cognitive impenetrability, but not vice versa.

The second direction from which attention comes is from philosophers interested in the question whether experience has conceptual content. Thus McDowell argues that it is only if experiences have contents which are adequately characterisable using concepts possessed by the subject of the experience that an experience can constitute a reason for making a judgement. Otherwise, experiences must be regarded as 'brute' and, while they may be causes, can never be reasons (McDowell 1994).

3. COGNITIVE PENETRATION AND CONCEPTUAL CONTENT

How are these two issues connected? It would simplify things if we could show that, as a matter of necessity, a system is cognitively penetrated just in case it delivers outputs with conceptual content. In that case, any results we could establish concerning systems with one of these features could be asserted of systems with the other feature. Call this claim N(CP if CC). Consider one half of this claim: N(If CC then CP). Bill Brewer argues that a system might be cognitively impenetrable yet one that delivers representations with conceptual content. So he denies this half. His reason is that there is a sense of 'imagining that all swans are white' according to which the content of the imagining is as fully conceptual as that of believing that all swans are white, but one can imagine this without believing it (Brewer 1999, p. 176). However, this is not enough to establish the lack of entailment between 'X is a representation with conceptual content' and 'X is the deliverance of a cognitively penetrable system.' The question is not whether one can imagine that all swans are white without believing that they are: we all agree that this is true. The question is whether you can imagine this without drawing on any beliefs at all. It seems to me that the answer must be no, for the following reason. It is only beliefs, and not imaginings, that have the kind of world-relatedness necessary for concept possession. Beliefs are states which are apt to be caused by certain distal events, and to bring about certain kinds of behaviours in response to those events; beliefs connect us to the world. Imaginings cannot do this, for even in ideal circumstances it is not the case that imagining P is generated in response to it being true that P, and generates P-appropriate behaviour. For concepts of things in the world, we are dependent on belief.[4] It is only when we are in possession of a suitable stock of belief-generated concepts that we can form imaginings of various kinds. In particular, in order to

[3] There are, of course, much narrower notions of theory, but this one is familiar; see e.g. Jackson (1997).

[4] Or rather, beliefs and concepts are interdependent in ways that imaginings and concepts are not.

imagine that P we must have a set of beliefs adequate to provide us with the concepts that would appear in a canonical description of P. The source of our concepts must be our beliefs, in the sense that, while it is possible for there to be a creature with beliefs but no imaginings, a creature with imaginings but no beliefs is not possible. In that case, the power of a system to generate representations (of whatever kind) with conceptual content depends on that systems access to the subject's beliefs. So N(If CC then CP).

How about the other half: N(If CP then CC)? Can we say that the deliverances of a cognitively penetrated system must have conceptual content? Here is an argument for denying this. We need to distinguish questions about the determination of perceptual content, and questions about the constitution of perceptual content. The claim that perception is cognitively penetrated is a claim about determination: it is the claim that, in order to fix the content of the subject's perceptions we need to hold fixed, not merely facts about the direction of looking, but also facts about the subject's beliefs. The claim that perceptual content is conceptual is the claim that the subject cannot have a perceptual content unless he or she possesses the concepts that would be deployed in a canonical description of that content.[5] There does not seem to be any reason why, if beliefs determine perceptual content, they should also transmit to it their own distinctive kind of content.

However, it might be objected here that I have ignored an important aspect of the idea of cognitive penetrability. This idea is originally due to Pylyshyn, according to whom a process is cognitively penetrable if it is rationally sensitive to the semantic content of its inputs.[6] This suggests a distinction between a system being influenced by certain inputs, and its being *rationally* influenced by them: a more demanding condition. Taking over this idea, we might say that in order to show that perception is cognitively penetrated it is not enough to show that beliefs are among the determinants of perceptual content; one must show that the contents of perception display rational sensitivity to the contents of the beliefs. In order to show that a subject's perceptual system is cognitively penetrated by the belief that P, we would have to show that the perception is rationally intelligible in the light of the belief. If the subject's perceptual content is *there is a cat in front of me*, and the relevant belief is *there is a cat in front of me*, I suppose we will agree that this is a case of genuine cognitive penetration. But if the belief were *Goldbach's conjecture is false*, it is difficult to see any rational connection with the perception. The case where perception and belief have identical contents (the case where the subject 'sees what he believes is there' as we might rather misleadingly put it) is an obvious case where the content of the belief renders intelligible the perception. But there are other kinds of cases which, while hard to systematise, certainly exert an influence on our judgement. It has been said that susceptibility to the Muller-Lyre illusion is the result of having acquired beliefs about the distances and sizes of things as typically seen in the architectural environments of western societies; the arrowheads are apt to

[5] Here I follow Cussins (1990).

[6] See Pylyshyn (1981). Cognitive penetration is defined in another way by Stich and Nichols in the course of arguing against the simulation theory (Nichols et al. 1996, p. 46). Their sense is not relevant to this discussion.

be interpreted as interior and exterior angles, with different implications for distance and hence for length.[7] Here, beliefs about angles, edges and distances in certain environments are said to affect the contents of our perceptions involving lines on a two dimensional surface. Regardless of whether this hypothesis is, in the end, supported by the evidence, it certainly is a plausible candidate for explaining the perceptual illusion. Its plausibility consists in its seeming intelligible to us how such beliefs are rationally related to such perceptual states, though I think we would have a hard time articulating principles underlying the idea of rational relatedness. If the claim was that people were subject to the Muller-Lyre illusion just in case they believed that polar bear liver is poisonous, or that Curzon was Viceroy of India, it would have no such intuitive appeal. We can see similar kinds of relatedness exemplified in claims to the effect that perception is influenced by desire; perhaps people are more likely to see things as food if they are hungry, or to see things as dangerous if they are afraid.[8] Whatever their truth, such claims offer at least the promise of intelligibility.[9]

If we now insist that cognitive penetrability involve this idea of rational sensitivity to content, then it follows that the outputs of a cognitively penetrated system must have conceptual content if they have content at all. For the test of the claim that a subject is deploying concepts is, exactly, the subject's capacity to make his or her thinking conform to norms of rational inference.[10]

I think we need to distinguish two kinds of cognitive penetrability. The first, stronger kind, conforms to Pylyshyn's demand for rational sensitivity. The second, weaker kind, requires only that the system have beliefs among its determinants.[11] My claim is that strong cognitive penetrability, but not weak cognitive penetrability, entails the requirement of conceptual content. How important is the notion of weak cognitive penetrability? The claim that perception is weakly cognitively penetrated would certainly be a challenge to the idea that perception constitutes a neutral common ground for scientists of different theoretical persuasions. Indeed, it could be argued that this kind of cognitive penetrability would be more worrying than strong cognitive penetrability would be. After all, if perception is cognitively penetrated in ways that respect constraints of rational intelligibility, then we have some hope of identifying, tracking and correcting for the resulting biases. But if all we can say is that belief tends to influence the content of perception, without our

[7] See e.g. Churchland (1988). See also Gregory (1998), pp. 150-1.

[8] See Pylyshyn (1999, Section 1) for examples of this kind from the 'new look' psychology of Jerome Bruner and others.

[9] If the human mind has a track-record of seeking intelligibility where none is to be found, this is one reason to be suspicious of claims about theory-dependence. See Currie and Jureidini (2004).

[10] This holds even if, as I shall argue later on, we revise our notion of what conceptual content is.

[11] Interpretationalists will worry at this point, holding that the conditions for the possession of belief are the conditions for the attribution of belief, where attribution is made in the light of normative constraints. How, they will, ask, could we ever be in a position to assert that it was the belief that P that was causing the proneness to perceptual illusion, when there is no rationalizing connection between the two? Surely a situation could arise where (i) we have all the reasons an interpretationalist could want to attribute the belief that P to a group of subjects and (ii) find that just those subjects are prone to the illusion. Why would anyone then deny that it was possession of that belief that is causing, or at least a partial cause of, the illusion?

being able to identify, in terms of rational connectedness, the kinds of beliefs likely to influence a given perceptual content, then bias will be very hard to avoid indeed. On the other hand, it is true, I think, that the kinds of cases which people have actually suggested as exemplifying cognitive penetration of perception have been cases of strong cognitive penetrability. Pylyshyn's empirical (and incidentally highly sceptical) survey (1999) offers a very large class of purported cases of cognitive penetration of perception, all of which exhibit, to varying degrees, rational sensitivity.[12] Indeed, it is hard to see how systematic research into cognitive influences on perception could be conducted *without* the assumption that cognition and perception are rationally related; what sorts of cognitive influences would we be looking for otherwise? In that case, cognitive penetrability, as it is likely to be used as a tool for research, entails conceptual content. Putting all this together, my tentative conclusion is that, for relevant cases, we have our equivalence: N(CC iff CP).

On the issue of cognitive penetration I have little more to say: here the situation seems to me very complex, conceptually and empirically. The conceptual difficulty arises because we have no clear antecedent understanding of the point at which perceptual systems deliver their outputs and belief takes over. Pylyshyn, for example, makes it clear that his own claims about the cognitive impenetrability of visual processing are not meant to apply to experience as understood phenomenologically; rather his claim concerns an hypothesised "early visual system", the design of which is to be inferred on the basis of our best overall theory of perception and cognition (1999, Section 7.2). And Fodor, who sometimes sounds like a confident advocate of the 'perception is theory-neutral' view,[13] exhibits considerable uncertainty about exactly what it is that is cognitively impenetrable: on some occasions even perceptual belief is included.[14] At other places, it isn't perception that is theory neutral, but some early part of perceptual processing: more or less Pylyshyn's conclusion.[15] For the rest of this paper I'll focus on the issue of conceptual content.

[12] Thus it has been claimed that the fusion of random dot stereograms is improved by prior information about the nature of the object. And chess masters' rapid visual processing and good visual memory for chess boards manifests itself only when the board consists of familiar chess positions and not when it is a random pattern; this is said to support the idea that it is the system of classification that they have learned which allows masters to recognize and encode a large number of relevant patterns.

[13] See especially where Fodor, taking as his targets Churchland, Goodman, Kuhn and Hanson, claims that 'arguments against the possibility of drawing a principled observation/theory distinction have been oversold' (Fodor 1984, p. 120).

[14] 'there is a class of beliefs that are typically fixed by sensory/perceptual processes, and... the fixation of beliefs in this class is... importantly theory neutral' Fodor (1984, p. 120). It is clear that the beliefs in question are perceptual beliefs. But later Fodor says 'the fixation of perceptual belief is the evaluation of such hypotheses in the light of the totality of background theory' (Fodor 1984, pp. 135-6). This latter way of putting it sounds better: perception represents the lines in the Müller-Lyer illusion as the same length, but because of background knowledge we do not form the belief that they are the same. The claim that perception is theory-neutral is not the claim that perceptual belief is theory-neutral. At most, it is the former claim that Fodor's argument entitles him to.

[15] 'The point of perception is the fixation of belief, and the fixation of belief is a... process... that is sensitive... to what the perceiver already knows. Input analysis may be encapsulated, but perception surely is not' (Fodor 1983, p. 73). So the argument goes: perception contributes to belief fixation; belief fixation is cognitively penetrable; so perception is cognitively penetrable. The principle

4. THE CONTENT OF EXPERIENCE

One way to put Popper's view about the proper concerns of epistemology is in terms of an image due to Wilfred Sellars, and revived by John McDowell: that experience lies outside the space of reasons.[16] Popper is not alone in holding this view. Another version of the view is that of Donald Davidson, who says that "nothing can count as a reason for holding a belief other than another belief" (Davidson 2001, p. 141). Popper and Davidson disagree about something here, since Davidson thinks that beliefs belong to the space of reasons while Popper wants to exclude even them. But they agree, apparently, that experiences do not belong to the space of reasons. This is the starting point for McDowell's attempt to rehabilitate experience: he wonders how Davidson can make room for the idea of rational constraint on belief from outside. McDowell's own solution is to bring experience into the space of reasons by seeing it as possessing the same kind of content as is possessed by belief: conceptual content. In my view this is a mistake; we can account for the reason-giving potential of experience without insisting that experience has only a kind of content available only to a possessor of the relevant concepts.[17] Let me explain.

In order for experiences to constitute reasons for judgement, must those experiences have conceptual content? This is the standard formulation of the question. McDowell and Brewer claim that the answer to it is yes. Others say no.[18] Brewer argues, for example, that a reason for judging (like a reason for doing anything) must be a reason *for* the subject.[19] While we can say that there was a reason why Fred should have done this, even though Fred did not have that reason available to him, the question we are interested in here does not concern such reasons. It concerns the possibility that a subject may *have* an experience as a reason for making a judgement; such a reason has to be a reason *for him*. Brewer claims that this can happen only if the reason in question has conceptual content. For it is only when the subject conceptualises the content that it is a reason for him. It is not enough that what we are given as the reason for judging is related to the subject's mental state 'by the theorist in some way'; the proposition must be the content of the

operating here—what contributes to an outcome which is cognitively penetrated is itself cognitively penetrated—would have us conclude that everything is cognitively penetrated. Since perception is cognitively penetrated (by the argument above) anything that contributes to perception is cognitively penetrated (by principle implicit in the argument above), so even the lowest level of perceptual processing (retinal stimulation, for example) is cognitively penetrated. The principle is wrong: what makes a system cognitively impenetrable is just the independence of its outputs from belief; what is done with those outputs later on is nothing to the question. If Fodor wants to agree with the top-down theorists that perception is cognitively penetrable, he should not cite the contribution of perception to belief fixation as a reason for this.

[16] Sellars (1956, p. 76). Page reference is to the reprint, as a book, by Harvard University Press, 1997. See also McDowell (1994, p. 5).

[17] I say 'only' (first occurrence) here because the dispute is really between those who think that experience has only conceptual content and those who deny this, some of whom, e.g. Peacocke, (1992), think that experience has contents of both kinds. As will become apparent, I reject this way of putting the issue.

[18] See e.g. Evans (1982), Peacocke (1992), Crane (1992).

[19] Brewer (1999, Chapter 5). See also McDowell (*ibid*, p. 140): 'nothing can count as a reason for holding a belief except something else that is also in the space of concepts.'

subject's mental state 'in a sense which requires that the subject has all of its constituent concepts.'[20]

I suggest a different way of looking at the matter. We can think of the content of experience as itself nonconceptual, but as providing a potential reason for judgement on the part of a subject who is equipped to conceptualise that content. Thus someone who judges that P on the basis of an experience is someone with a reason so to judge only if he judges that P on the basis of his articulating the content of the experience, and in the process of so doing, deploys the concepts that appear in a canonical formulation of that content. It is crucial to this proposal that the experience and the judgement have the very same content; that is what makes the experience the (potential) ground of the judgement. But how can the experience and the content have the very same content if the content of the one is nonconceptual and the other conceptual? I take 'conceptual content' to be a misleading term: content is itself not conceptual. To say that this or that state is a state with conceptual content is best taken as a way of saying that the subject is in a state which is (i) contentful and (ii) one which the subject could not be in unless he or she possessed the relevant concepts. What happens in the case of experientially based judgements is this: the subject has an experience with the content P; possessing the relevant concepts—those sufficient to articulate the content of P—he or she then judges that P. The experience and the judgement have exactly the same content.[21] The difference between them is that the subject can have the experience without possessing the concepts, but cannot make the judgement without possessing the concepts. An experience functions as an opportunity: an opportunity for someone, with the right conceptual equipment, to make a judgement which will thereby, in normal circumstances, be justified. In taking up the opportunity, the subject may judge to be true exactly that which his experience informs him of. The difference between the perception and the judgement is not in the nature of their contents (for the contents are the same) but in what is required of the subject in order to be in these states. To have an experience with the content P one does not need any concepts, but to believe that P one needs concepts—those concepts which one would display mastery of in satisfying what Evans calls the generality constraint. Thus someone who believes that Socrates is a philosopher must be in a position to think the thought that S is a philosopher, for any singular concept S which he or she possesses, and to think the thought that Socrates is an F, for any general concept F which he or she possesses (Evans 1982, Section 4.3).[22]

What sort of content is it, then, that is the common property of perception and of belief? There are several proposals for such a content. But I suggest, rather tentatively,

[20] Brewer (1999, p. 152). This argument is presented briefly in Sedivy (1996).

[21] Strictly speaking the condition of sameness of content is too strong; all that is required is that the content of the experience be such as to entail, or at least rationally mandate the content of the judgement. What is required for this is that it is at least possible that the content of the experience and the content of the judgement be the same. Anyone who believes that the content of the experience is *of a different kind* from that of the judgement must deny this.

[22] Evans offers one way of explaining the role of concepts in thinking; there are others. See e.g. Crane (1992). Crane also advocates the view that perception has nonconceptual content. For comment on Crane's proposal see Currie and Ravenscroft (2002, Chapter 5).

that it is content individuated by possibility. In other words, we should think of the contents of perceptions and of beliefs (and indeed statements as well) as functioning to narrow down the range of live possibilities for the subject. As we experience more, form more beliefs, and hear more assertions, we progressively refine our picture of which state of affairs, out of all those that are possible, is actual; we never, of course, remotely approach a situation where we fix on a particular world (a particular maximal state of affairs) as actual. This is the sort of view of the contents of beliefs that Robert Stalnaker has been urging upon us for many years (Stalnaker 1984 & 1999). It is remote from conceptions of content according to which concepts are constituents of content.[23] On the individuation-by-possibility approach, content is a set of ways the world might be. Indeed Stalnaker recently suggested that we should agree with McDowell that experiences and beliefs have the same kind of content, but disagree with him to the extent that we say that the content is nonconceptual. Stalnaker puts it like this:

> Let us grant (without looking too hard at what this means) that states of belief and judgement are essentially conceptual—states and acts that require the capacity to deploy concepts, and manifest the exercise of this capacity. That does not by itself imply that the concepts that subjects deploy and are disposed to deploy when they are in such states or perform such acts are thereby constitutive of the contents that are used to describe the states and acts. (1998, p. 352)

Stalnaker does not put forward this idea as a solution to the problem of saying how a nonconceptual content for perception would make it possible for us to justify experientially based judgements. But it seems to me to do this job rather well.[24]

One objection to this proposal is as follows. I insist that perceptions and beliefs have contents of the same kind, and indeed that some perceptions have the same specific content (type) as some beliefs. And I claim that in both cases the contents are nonconceptual; concepts are not constituents of these concepts. But I also claim that possession of concepts is essential for the possession of beliefs, but that this is not essential for the possession of contentful perceptual states. How can this be? The only reasonable answer is to say that concepts are constituents of beliefs but not of perceptions. But this answer is not available to me.

My answer to the objection is to insist that we distinguish two questions:

1. What are the constituents of the content of a given kind of mental state?

and

2. What are the possession-conditions for a mental state of that kind?

Standardly, these two questions get very closely associated answers for the case of belief: the constituents of the contents of beliefs are concepts, and concepts are what you need in order to possess states of that kind. But on my view there are kinds of mental states which (i) are contentful; (ii) do not have concepts as the constituents

[23] As is Peacocke's account in terms of scenario content (Peacocke 1992, Chapter 3).

[24] It is time to make good an omission. In Currie and Ravenscroft (2002) we briefly suggested the idea described above, without referring to Stalnaker (1998), which we did not know about, but certainly should have known about, at the time of writing.

of their contents, and (iii) can be possessed by a subject only if he or she possesses certain concepts. Beliefs are states of this kind; perceptions are not. The reason that a believer must possess certain concepts is not that the concepts go to make up the contents of the belief, but because being a believer requires certain kinds of facility with that content—inferential facility for example—that is constitutive of concept possession. In general we do not expect that differences between kinds of mental states that show up as differences of functional role will be reflected in differences of constitution. Take belief and desire. These are distinct kinds of states, with distinct functional roles; the belief that P is apt to play a role in theoretical reasoning that the desire that P is not apt to play. Yet it is generally agreed that the belief and the desire have the same kinds of constituents. In that case it does not seem unreasonable to say that beliefs and perceptions differ in functional role, but not in their constituents. [25]

REFERENCES

Brewer, B. (1999) *Perception and Reason*. Oxford: Clarendon Press.

Churchland, P.M. (1988) 'Perceptual Placticity and Theoretical Neutrality: a Reply to Jerry Fodor.' *Philosophy of Science* 55: 167-187.

Crane, T. (1992) 'The Nonconceptual Content of Experience.' in T. Crane (ed.) *The Contents of Experience. Essays on Perception*. Cambridge, Cambridge University Press.

Currie, G. (1989) 'Frege and Popper: Two Critics of Psychologism.' in C. Gavroglu, Y. Goudaroulis and P. Nicolacopoulos (eds) *Imre Lakatos and Theories of Scientific Change*. Dordrecht: Kluwer.

Currie, G. & Jureidini, J. (2004) 'Narrative and Coherence.' *Mind & Language* 19 (4), 409-427

Currie, G. & I. Ravenscroft (2002) *Recreative Minds: Imagination in Philosophy and Psychology*. Oxford: Oxford University Press.

Currie, G. & K. Sterelny (2000) 'How to Think About the Modularity of Mindreading.' *Philosophical Quarterly* 50: 145-167.

Cussins, A. (1990) 'The Connectionist Construction of Concepts.' in M. Boden (ed.) *The Philosophy of Artificial Intelligence*. Oxford: Oxford University Press.

Davidson, D. (2001) *Subjective, Intersubjective, Objective*. Oxford: Clarendon Press.

Evans, G. (1982) *The Varieties of Reference*. Oxford: Oxford University Press.

Fodor, J. (1983) *The Modularity of Mind. An Essay on Faculty Psychology*. Cambridge, MA: Bradford Books, MIT.

Fodor, J. (1984) 'Observation Reconsidered.' *Philosophy of Science* 51: 23-43.

Gregory, R. (1998) *Eye and Brain*. 5th edn. Oxford: Oxford University Press.

Jackson, F. (1997) 'All That Can be at Issue in the Theory-Theory Simulation Debate.' *Philosophical Papers* 28: 77-96.

McDowell, J. (1994) *Mind and World*. Cambridge, MA: Harvard University Press.

Musgrave. A. (1983) 'Theory and Observation: Nola versus Popper.' *Philosophica* 31: 45-62.

Nichols, S., Stich, S., Leslie, A. & Klein, D. (1996). 'Varieties of Off-Line Simulation.' in P. Carruthers & P. K. Smith (eds) *Theories of Theories of Mind*. Cambridge: Cambridge University Press.

Peacocke, C. (1983) *Sense and Content*. Oxford: Clarendon Press.

Peacocke, C. (1992) *A Study of Concepts*. Cambridge, MA: MIT press.

Pylyshyn, Z. W. (1981) 'The Imagery Debate: Analogue Media versus Tacit Knowledge.' *Psychological Review* 88(1): 16-45.

Pylyshyn, Z. W. (1999) 'Is perception continuous with cognition?' *Behavioral and Brain Sciences* 22: 341-65.

[25] Thanks to Colin Cheyne and John Worrall for the invitation to honour a friend and former colleague. Thanks also to Bill Fish and Nick Jones for discussion of these issues; neither, I think, agrees with me about content.

Sedivy, S. (1996) 'Must Conceptually Informed Perceptual Experience involve Nonconceptual Content?' *Canadian Journal of Philosophy* 26: 413-431.

Sellars, W. (1956) 'Empiricism and the Philosophy of Mind.' in H. Feigl and M. Scriven (eds) *Minnesota Studies in the Philosophy of Science*. Minneapolis: University of Minnesota Press.

Stalnaker, R. (1984) *Inquiry*. Cambridge, MA: MIT.

Stalnaker, R. (1998) 'What might Nonconceptual Content be?' in E. Villanueva (ed) *Concepts*. Atascadero: Ridgeway Publishing Company.

Stalnaker, R. (1999) *Context and Content*. Oxford: Oxford University Press.

COLIN CHEYNE

TESTIMONY, INDUCTION AND REASONABLE BELIEF

Many, perhaps most, of our beliefs about the world are acquired on the basis of testimony. Not only do we often believe what we are told, but if we were never to accept the word of others, then our belief sets would be extremely meagre. The contents of many of our testimonial beliefs are empirical generalisations, such as the philosophers' old favourite: "All swans are white". Empirical generalisations are, of course, at the heart of the problem of induction. But discussions of the problem of induction seldom acknowledge the role that testimony plays in the acquisition of beliefs. In this paper I examine Alan Musgrave's version of the (Popperian) critical-rationalist solution to the problem of induction; in particular, its application to testimonial beliefs. I conclude that critical rationalists have a problem devising an epistemic principle for testimonial belief that does not either allow any belief-content to be reasonably believed or makes reasonable testimonial beliefs unattainable.

1. THE PROBLEM OF INDUCTION

Does Karl Popper's critical rationalism provide a solution to the problem of induction? Alan Musgrave believes that it does. Many others are not so sure. Indeed, many others are sure that it does not.

According to Musgrave (1999, p. 315), the problem of induction, as Popper sees it, is to avoid the conclusion of the following argument from Hume:

> We do, and must, reason inductively.
> Inductive reasoning is logically invalid.
> To reason in a logically invalid way is unreasonable or irrational.
> Therefore, we are, and must be, unreasonable or irrational.

In order to by-pass exegetical issues, Musgrave suggests that we call this argument a 'Humean' argument. For similar reasons, let us call the solution that Musgrave discusses a 'Popperian' solution. For the most part I shall confine my discussion to the particular version of the solution discussed by Musgrave in his 1999 paper 'Critical Rationalism'. Although he has discussed this Popperian solution in a

C. Cheyne & J. Worrall (eds.), Rationality and Reality: Conversations with Alan Musgrave, 19–29.

number of earlier publications, his introductory remarks to the 1999 paper have the air of a frustrated author making one further attempt to convince his critics. So we have grounds for supposing that it provides a reasonably definitive version of the Popperian solution according to Musgrave.[1]

The Popperian solution aims to avoid the conclusion of the Humean argument by rejecting the first premise. But it seems quite easy to reject the first premise. To reason inductively is to argue, for example, from the fact that all experienced swans are white to the conclusion that all swans are white. Perhaps we do reason in this way. But we need not. We might, instead, ignore the evidence and conclude that all swans are green. We might even attempt to justify such a belief by appealing to the principle that one ought to believe whatever seems most appealing, or whatever one would most like to be true, or whatever the bible says. Unfortunately, to acquire beliefs in this way is clearly unreasonable, so rejecting the first premise on this basis does not avoid the irrationalist conclusion of the Humean argument. What the Popperian needs to do is not just to deny that we must reason inductively but to demonstrate that there is a reasonable alternative to inductive reasoning.

Inductive reasoning takes us from beliefs about the immediate evidence of our senses to beliefs that transcend that evidence. It is evidence-transcendent beliefs that are at the heart of the problem of induction. Implicit in the Humean first premise is the notion that inductive reasoning is the only potential starter as a rational means of acquiring evidence-transcendent beliefs. Because inductive reasoning is invalid and, hence, unreasonable, it follows that evidence-transcendent beliefs are unreasonable. Why is inductive reasoning the only starter as a rational means of acquiring evidence-transcendent beliefs? Well, the beliefs at issue are contingent beliefs about the way the world is, so it would seem that the only rational support for such beliefs must come from evidence about the way the world is. But that sort of evidence, the evidence of our senses, is, by definition, not evidence-transcendent. We appear to have no choice but to proceed from non-evidence-transcendent beliefs to evidence-transcendent beliefs. But to so proceed is to reason inductively and, thus, to be unreasonable.

With the claim that evidence-transcendent beliefs are unreasonable in place, the Humean can then restate the original argument as:

Evidence-transcendent beliefs are unreasonable.
We do, and must, have evidence-transcendent beliefs.
Therefore, we are, and must be, unreasonable.

The Popperian task becomes that of avoiding this irrationalist conclusion by demonstrating how we may acquire evidence-transcendent beliefs that are reasonable without resort to inductive reasoning. Critical rationalism is offered as the solution.

I have merely sketched the argument that Musgrave employs to take us from the first Humean argument to the role of critical rationalism in avoiding irrationalism,

[1] It is not, however, Musgrave's final word. Another, shorter article appeared in 2004. So far as I can see that paper contains nothing that affects the points I make here.

since I do not intend to question that argument here. But one important point must be noted: the ambiguity of the term 'belief'.

> It can refer to the thing believed or the content of the belief, and it can refer to the mental act or state of believing that thing or content.... Talk of the reasonableness or otherwise of a belief inherits this ambiguity. Is it the belief-content which is reasonable or otherwise, or is it the person's believing that content? (Musgrave 1999, p. 320)

The critical rationalist argues that it is believings, not belief-contents, that are reasonable or unreasonable. I agree. In this paper I use the terms 'belief' or 'believing' to refer to a mental act or state of believing, reserving the term 'belief-content' for the content of a belief. So, for example, the term 'evidence-transcendent belief' refers to a believing with an evidence-transcendent belief-content.

2. THE CRITICAL-RATIONALIST SOLUTION

The critical rationalist's task is to demonstrate how we may acquire reasonable believings that have evidence-transcendent belief-contents. The critical-rationalist solution may be seen as a particular version of a general form of solution to the problem of induction. This general form may be stated as follows:

(GS) If person A is in epistemic situation S then A's believing that P is reasonable.

A particular solution then consists in supplying an epistemic principle according to which there are situations S such that A's believing that P is reasonable for some evidence-transcendent P. That a solution to the problem of induction may take this form, rather than consist in a vindication of inductive reasoning, is an important insight. Once we see that a solution to the problem may take this form, our task becomes that of supplying an appropriate epistemic principle. According to the Popperian solution, the epistemic principles of critical rationalism are just what we are looking for. But note that if we are dissatisfied with the principles offered by the critical rationalist we are not thereby thrown back to the task of vindicating inductive reasoning. Rather, we may continue the search for more satisfactory epistemic principles.

I turn now to the epistemic principles of critical rationalism as stated by Musgrave (1999, pp. 348-49).[2] There are four principles:

(CR) *A*'s non-perceptual believing that *P* is reasonable (at time *t*) if and only if *P* is that hypothesis which has (at time *t*) best withstood serious criticism from *A*.

(E) *A*'s perceptual believing that *P* is reasonable (at time *t*) if and only if *P* has not failed to withstand criticism from *A* (at time *t*).

[2] I have made minor modifications for stylistic consistency.

(T) *A*'s testimonial believing that *P* is reasonable (at time *t*) if and only if *P* has not failed to withstand criticism from *A* (at time *t*).

(ADL) If *A* reasonably believes *H* (at time *t*), and *A* has validly deduced *P* from *H* (at time *t*), and *P* has not failed to withstand serious criticism from *A* (at time *t*), then *A* reasonably believes *P* (at time *t*).

An initial problem with this formulation is that believings that arise from testimony or from a valid deduction are non-perceptual believings and, according to the first principle (CR), are reasonable *only if* their contents have best withstood serious criticism. This contradicts principles T and ADL which only require that such belief-contents have not failed to withstand criticism. According to CR the content of a testimonial believing must withstand serious criticism in order to be reasonable, but according to T it need not. Similarly for inferential believings with respect to ADL. This contradiction can be avoided by restricting the application of CR to non-perceptual, non-testimonial and non-inferential believings. No doubt this is what Musgrave intended. For him, perceptual, testimonial and inferential believings are epistemically privileged in virtue of the manner in which they are acquired. They are reasonable *unless* they happen to have failed the test of criticism. This is in contrast to all other, less privileged, believings, which are not reasonable *until* they have passed the test of serious criticism. I suggest that we reformulate CR as:

(CR) *A*'s non-perceptual, non-testimonial or non-inferential believing that *P* is reasonable (at time *t*) if and only if *P* is that hypothesis which has (at time *t*) best withstood serious criticism from *A*.[3]

3. TESTIMONY AND REASONABLE BELIEF

Now the critical rationalist's idea is that CR provides us with the means to solve the problem of induction. If I hypothesise that all swans are white and then subject this hypothesis to serious criticism which it withstands, then, according to CR, it is reasonable for me to believe that all swans are white. Since this provides at least one way of aquiring a reasonable belief which has an evidence-transcendent content, the problem is solved. So far, so good, as long as we accept CR as a correct epistemic principle. But what the critical rationalist may not have noticed is that Principle T by itself will do the trick. Suppose you tell me that all swans are white and suppose that your claim has not failed to withstand criticism from me, then by Principle T it is reasonable for me to believe that all swans are white. If T is a correct epistemic principle, then testimony can provide me with reasonable evidence-transcendent beliefs. Principle T is all that we need to solve the problem of induction.

But wait, there is more! Principle T does not place any restrictions on the *content* of a testimonial belief. (At least, not on the content as such. There is only a

[3] If there should be further 'privileged' methods of belief acquisition, then CR may be further adjusted accordingly. Memory is a likely candidate.

restriction on the relation between content and believer, namely that the content not have failed to withstand the believer's criticism.). So the content could be evidence-transcendent, as in my example. It could even be a contradiction or a logical falsehood. Musgrave (1999, pp. 338-39) discusses such cases and explicitly allows that believing a contradiction or logical falsehood may be reasonable. Furthermore, in one of his examples it is testimony that underpins a reasonable but contradictory belief. The example concerns a sign on a ship that says 'SHIP'S BARBER: I am the man who shaves every man on the ship who does not shave himself'. By reading such a sign a passenger acquires the reasonable testimonial belief that there is a ship's barber who shaves every man on the ship who does not shave himself. This reasonable belief has a contradictory belief-content and it remains reasonable (one supposes) so long as the believer is unaware that it is contradictory.

Now if Principle T allows for the possibility of someone's reasonably believing contradictions then, given the right circumstances, virtually any belief-content may be reasonably believed. Rules of logic may be reasonably believed, including false rules of logic. If a logician assures a student that affirming the consequent is a valid form of argument and this claim does not fail to withstand criticism by that student, then it is reasonable for the student to believe that affirming the consequent is a valid form of argument. Now if it is reasonable for someone to believe that affirming the consequent is a valid form of argument, then surely it would be reasonable for that person to use this form of argument when reasoning. A similar conclusion may be drawn if someone were to be assured that reasoning inductively is valid or cogent. Indeed, this assurance is more likely than the former one, given that many expert logicians and philosophers have argued that inductive reasoning can be valid or cogent. It follows that it may be reasonable to argue inductively and, given that affirming the consequent and induction are (in fact) invalid, then it also follows that it may be reasonable to reason invalidly. This, of course, contradicts the third premise of the Humean argument, namely, 'To reason in a logically invalid way is unreasonable'.

If we accept Principle T and apply it along the same lines as Musgrave (1999) then we arrive at the conclusion that it may be reasonable to have evidence-transcendent beliefs, to reason inductively and to reason invalidly. The problem of induction is not so much solved as blown away!

One response to this conclusion might be to see it as a *reductio* of Principle T, since any principle that solves the long-standing problem of induction so easily must be false. But what, exactly, is wrong with Principle T? As it stands, it appears to be rather weak, perhaps improbably so. According to T, as long as you refrain from criticising what you are told, your testimonial beliefs are reasonable. That cannot be right.

Suppose that I have heard talk of black swans in the past but ignore this when someone tells me that all swans are white. My believing what I have been told would not, on the face of it, be reasonable. But has the content of my belief failed to withstand criticism? Lacking a fuller account of what constitutes criticism, it is difficult to say. What this example does suggest that having a reason to doubt tends to undermine the reasonableness of a belief. I suggest strengthening (or clarifying) Principle T accordingly:

(T′) *A*'s testimonial believing that *P* is reasonable (at time *t*) if and only *A* does not have sufficient reason to doubt *P* (at time *t*).[4]

This leaves open the question as to what constitutes sufficient reason. However, any sensible account of sufficient reason we settle on, will still allow us to solve the problem of induction with ease. We can imagine cases in which I have little or no reason to doubt an informant's evidence-transcendent claim, in which case, by T′, my testimonial believing of that claim will be reasonable. Once again, the problem of induction is solved without an appeal to CR. And, once again, a belief that inductive reasoning is reasonable may not be unreasonable, from which it appears to follow that inductive reasoning may be reasonable.

Note that there are two distinct ways in which we may doubt what we are told. We may have evidence that the content of the claim is false (evidence of the existence of a black swan, for example) or we may have evidence that our informer is untrustworthy (either because of insincerity or incompetence). Incorporating this distinction into T′ gives:

(T′) *A*'s testimonial believing that *P* is reasonable (at time *t*) if and only *A* does not have sufficient reason to doubt *P* or to doubt the trustworthiness of the testifier (at time *t*).

Once again, this makes no difference to the claim that T′alone provides a solution to the problem of induction. There is no reason to suppose that the conditions imposed by this revised T′ may not easily be met when someone assures me that all swans are white or that induction is reasonable.

But talk of the reliability of informants does draw attention to something that has been overlooked so far, namely, what Hume himself had to say about testimony. According to Hume (1748/1975, pp. 111-116), we are only justified in believing testimony if we reasonably believe that the testifier is trustworthy. We must assure ourselves of a testifier's trustworthiness before we accept the testimony.[5] This is a stronger requirement than that of T′. It requires a further reasonable belief, not just the absence of doubt. Incorporating Hume's requirement yields:

(T′′) *A*'s testimonial believing that *P* is reasonable (at time *t*) if and only *A* does not have sufficient reason to doubt *P* (at time *t*) and *A* reasonably believes (at time *t*) that the testifier is trustworthy.

Now to believe that someone is trustworthy is to believe something that is evidence-transcendent. Just as 'All swans are white' concerns unobserved swans, 'Person B is a sincere and competent testifier' concerns unobserved actions, including future actions. If this further evidence-transcendent belief is itself a testimonial belief then

[4] This revision accords with Musgrave's gloss on principle T, 'Trust what other folk tell you, unless you have a specific reason not to.' (1999, p. 349)

[5] Other interpretations of Hume are possible, e.g., that we need only establish the reliability of testimony in general or of various classes of testifiers. These interpretations lead to similar conclusions.

a regress threatens. We can escape the regress only by having a non-testimonial belief in the trustworthiness of some testifier or other. T′′ rules out the possibility of our aquiring a reasonable belief based on testimony alone. Having a reasonable testimonial belief depends on our having a reasonable non-testimonial belief that the testifier is trustworthy, or that a testifier to the trustworthiness of the original testifier is trustworthy, or…. So the problem of induction cannot be solved simply by appealing to T′′. Maybe the critical-rationalist principle CR has a vital role to play in the solution of the problem of induction after all. In order to reasonably (and non-testimonially) believe that a testifier is trustworthy, the proposition that the testifier is trustworthy must withstand serious criticism.

However, we are not out of the woods yet. Firstly, it is not clear that Hume's restriction is reasonable and it is certainly not clear that Musgrave would accept it. The demand that we ought to seriously criticise the trustworthiness of others before we accept their word deviates significantly from our actual practice. Secondly, even if it is a reasonable restriction, it does not rule out the possibility that it may be reasonable to reason inductively. Suppose the trustworthiness of a testifier does withstand my serious criticism and that testifier then assures me, with much impressive argumentation, that inductive reasoning is reasonable, then, according to T′′, it will be reasonable for me to believe that inductive reasoning is reasonable.

4. THE CRITICAL RATIONALIST RESPONDS

What might critical rationalists say in response to these observations? One response might be to agree that the Popperian solution's epistemic principle for testimonial belief does allow that there are circumstances in which inductive reasoning may be reasonable. Perhaps the problem of induction need not be the bug-bear that has apparently loomed so large in the history of modern philosophy. But this response suggests that we might take a Pascalian approach to inductive reasoning. Recall Pascal's recommendation to those who hanker after theistic belief but find that reason cannot help. Adopt a religious life, chant prayers and masses along with the believers, and belief in God will follow naturally. Likewise, if you can't avoid inductive reasoning but are troubled by its apparent irrationality, then surround yourself with inductivists, pay close attention to their prognostications, and a reasonable belief in the reasonableness of induction will surely follow.

Another response might be to claim that I have misunderstood the problem of induction or that the problem has been mistated in the versions above. Note that both versions contain premises and conclusions which employ the first-person plural:

We do, and must, reason inductively.
We are, and must be, unreasonable or irrational.

We do, and must, have evidence-transcendent beliefs.
We are, and must be, unreasonable.

Who are 'We'? Perhaps 'We' refers to those of us who believe that inductive reasoning is logically invalid and that to reason in a logically invalid way is unreasonable or irrational. So, perhaps a solution to the problem of induction requires an account of how *we* may acquire evidence-transcendent beliefs that are reasonable without abandoning *our* beliefs about induction and validity. If so, then our believing in the reasonableness of induction is ruled out. However, our acquisition of reasonable evidence-transcendent beliefs on testimony is not ruled out, and if something like T′ is the appropriate epistemic principle, then our acquisition of such beliefs by testimony alone is not ruled out. CR still appears to be superfluous to a solution of the problem of induction. (And we are still left with the apparently unsatisfactory conclusion that *others* may reasonably reason inductively.)

A third way of responding might be to point out that Musgrave's version of the Popperian solution is overly ambitious. The problem is to demonstrate how we may acquire evidence-transcendent beliefs that are reasonable. A solution to this problem only requires the provision of a sufficient condition for the acquisition of such beliefs. Musgrave's version purports to provide a complete set of necessary and sufficient conditions for reasonable belief. But if it is, indeed, reasonable to believe that hypothesis which has best withstood one's serious criticism then the following condition is all that is required to solve the problem of induction:

(CR′) *A*'s believing that *P* is reasonable (at time *t*) if *P* is that hypothesis which has (at time *t*) best withstood serious criticism from *A*.

So the critical rationalist may claim that an account of what is required for our testimonial beliefs to be reasonable need not be part of the solution. But I'm not so sure. CR′ leaves us with the problem of how we may avoid irrationality in our everyday lives. In our day-to-day lives we do, and must, have evidence-transcendent beliefs, but we cannot possibly subject them all to serious criticism. We must obviously rely, *inter alia*, on the testimony of others. Without a condition for reasonable testimonial belief, this 'modest' critical-rationsalist solution to the problem of induction still leaves us with what we may call 'the problem of testimony'.

Critical rationalists may concede this but claim that that is not their problem. Further, they may claim they have, at least, solved the problem of induction as far as scientific knowledge is concerned. Science, they may say, is much more epistemologically demanding than everyday life. CR′ sets a sufficiently high (but achievable) standard for reasonable (induction-free) scientific belief. What constitutes an appropriate standard for everyday life is another matter, and one for which they are not offering an answer.

The problem with this latter response is that everyday beliefs (including testimonial beliefs) appear to play an ineliminable role in serious scientific criticism. Could I seriously criticise the hypothesis that all swans are white without relying on reports from others? And it goes much deeper than that. Consider, for example, any instruments that may be employed as an aid to criticism. We must believe that they will do what their manufacturers say they will. Even if we insist on making them ourselves, we must have evidence-transcendent beliefs about the materials we choose, and so on. So critical rationalists taking this line must either provide an epistemic principle for reasonable testimonial belief or an account of serious criticism which does not rely on testimonial belief. I seriously doubt that the latter is possible.[6] As we have seen, the former will almost inevitably provide a solution to the problem of induction for which a critical-rationalist principle is superfluous.

5. POPPER ON TESTIMONY

It may be appropriate, at this point, to consider what Popper himself says about testimony. The issue arises in the context of his discussion of the sources of knowledge (Popper 1969, pp. 21-30). On the one hand, he argues that no source of knowledge (including testimony and tradition) has ultimate authority; but, on the other hand, any source is welcome, as long as the claim itself is open to critical examination. His position is anti-foundationalist and anti-authoritarian. '[W]e do not test the validity of an assertion or suggestion by tracing its sources or its origin, but we test it, much more directly, by a critical examination of what has been asserted.' (pp. 24-25).

This suggests that Popper might not approve of Musgrave's supplementary epistemic principles, since they privilege testimony, perception and inference as sources of knowledge or belief. However, Popper also asserts that 'the most important source of our knowledge…is tradition' (p. 27), and notes, with respect to an assertion about the Prime Minister's whereabouts, that 'reasonable people might simply accept the answer "I read it in *The Times*"' (p. 22). This latter point seems to be in with accord with Principle T′: we may reasonably believe what we are told unless we have reason to doubt it. But it does not follow that Popper believes that testimony or any other source should be specifically privileged. Rather, he argues strongly that whatever applies to one source, applies to all. And, at least in some passages, what applies to all sources seems to be the condition that Musgrave applies only to testimony, perception and inference. '[I]f we are doubtful about an assertion, then the normal procedure is to test it, rather than ask for its sources; and if we find independent corroboration, then we shall often accept the assertion without bothering at all about sources' (Popper 1969, p. 23). In other words, critical rationalism becomes something like:

(CR*) A's believing that P is reasonable (at time t) if and only either A does not have sufficient reason to doubt P (at time t) or *P* is that hypothesis which has (at time *t*) best withstood serious criticism from *A*.

Now I am not concerned here with exegesis. After all, Popper wrote little on what constitutes reasonable belief, and was somewhat cautious on the issue as to which hypotheses should or should not be believed or accepted. 'I do not demand that

[6] See Coady (1992) for an extended exploration of the way that our epistemic resources are inextricably enmeshed with testimony.

every scientific statement must *have in fact been tested* before it is accepted. I only demand that every such statement must be *capable* of being tested' (Popper 1968, p. 48). But if we take seriously his argument that no source of knowledge ought to be especially privileged, then critical rationalism will consist of one basic condition. What we have on the table is either the rather lax CR* or a much stricter:

(CR**) A's believing that P is reasonable (at time t) if and only *P* is that hypothesis which has (at time *t*) best withstood serious criticism from *A*.

The contrast between CR* and CR** is similar to that between T′ and T′′. CR* makes reasonable belief so easily available that the epistemically naïve or reckless may, particularly with the assistance of testimony, reasonably believe pretty much anything, including the claim that inductive reasoning is reasonable, while CR** puts reasonable belief beyond the reach of most, if not all, of us. If serious criticism requires prior reasonable beliefs (and it is difficult to see how criticism based on unreasonable beliefs could be serious), then a regress threatens and CR** cannot provide a solution to the problem of induction.

6. CONCLUDING REMARKS

Our exploration of the place of testimonial belief in a critical-rationalist solution to the problem of induction leaves us with the following dilemma. Either critical rationalism is so demanding that reasonable beliefs (particularly those with evidence-transcendent contents) are beyond us and the problem of induction is unsolved; or critical rationalism is rendered so undemanding by its epistemic principle for testimonial belief that the problem of induction is trivially solved. It may be that the problem that I have highlighted is not especially related to critical rationalism or the problem of induction. Rather it is a more general problem concerning epistemic principles for testimonial belief. Is there a correct epistemic principle which does not either allow any belief-content to be reasonably believed or put reasonable testimonial beliefs (perhaps all reasonable belief) beyond us?

In any good conversation there comes a time when it is approriate to pause in order to hear the response of fellow conversants. That time has doubtless passed. I look forward to Alan Musgrave's response. Past experience suggests to me that I will not be disappointed, and that I will acquire beliefs more reasonable and closer to the truth than those I now hold.

REFERENCES

Coady, C.A.J. (1992) *Testimony*. Oxford: Clarendon Press.

Hume, D. (1748) *An Enquiry Concerning the Human Understanding*, in Nidditch, P.H. & Selby-Bigge, L.A. (eds.) (1975) *Hume's Enquiries* 3rd edn. (Oxford: Clarendon Press).

Musgrave, A. (1999) 'Critical Rationalism' in his *Essays on Realism and Rationalism*. Amsterdam: Rodolfi.

Musgrave, A. (2004) 'How Popper [Might Have] Solved the Problem of Induction.' *Philosophy* 79: 19-31.
Popper, K. (1968) *The Logic of Scientific Discovery* 2nd edn. London: Hutchinson.
Popper, K. (1969) *Conjectures and Refutations: The Growth of Scientific Knowledge* 3rd edn. London: Routledge and Kegan Paul.

JOHN WORRALL

THEORY-CONFIRMATION AND HISTORY

1. MUSGRAVE ON 'LOGICAL' VERSUS 'HISTORICAL' ACCOUNTS OF CONFIRMATION

There are very many topics in philosophy of science on which Alan Musgrave and I see eye to eye. So it has not been easy to do the decent Popperian thing and pick a (friendly) fight with him. However, thinking again about his influential (1974) paper on theory-confirmation ('Logical versus Historical Theories of Confirmation') solved my problem. Despite having some of its heart in some of the right places, both the argument of that paper and the position it ends up endorsing are, I believe, importantly off-beam. In this paper I shall explain why and clarify what I think is the correct account of the issue that he addressed. I shall finally take the opportunity to contrast my views on confirmation with those of Deborah Mayo (see in particular her 1996); Mayo was herself indebted to Alan Musgrave's paper and has developed her own influential account of the issues it raises. Although Alan's paper was published in 1974, the problem it faces has not been given a satisfactory resolution—at least not one that has met widespread acceptance. It remains very much a live issue within current philosophy of science.[1]

Musgrave begins his paper with a sharp formulation of the prediction versus accommodation issue: is there some epistemic premium on *predictive* success? That is, does a theory obtain, *ceteris paribus*, more confirmation from a piece of evidence that it correctly predicts than it does from an 'otherwise equivalent' piece of known evidence that it correctly entails?

He takes it that a 'purely logical' account of confirmation must answer this question negatively. Any such account sees confirmation as entirely based on the logical relationships between the theory, T, and the piece of evidence, e, at issue; and hence must entail, whatever the details of the logical relationships it highlights, that the question of whether or not e was already known to hold, or was already in 'background knowledge' however construed, when T was proposed is entirely

[1] For example, Musgrave's views are one of the starting points for the very recent paper on prediction and accommodation by Chris Hitchcock and Elliott Sober (2004).

C. Cheyne & J. Worrall (eds.), Rationality and Reality: Conversations with Alan Musgrave, 31–61.

irrelevant to confirmation. All logical accounts have their difficulties —in particular, in Musgrave's view, they supply no satisfactory answer to the 'paradox of confirmation.'[2]

An historical (or more accurately—as he allows—a 'logico-historical') account, on the other hand, sees confirmation as a relationship, not just between T and e, but also a third variable: 'background knowledge', b. All variants of the historical view entail that T fails to be confirmed by any e that is in b, even if T (of course, in conjunction with appropriate initial conditions and auxiliaries) entails e. All variants of this account do indeed have an historical element on Musgrave's view: the answer to the question 'does e confirm T?' may very well be different in two different historical epochs, because these will be characterised by different states of background knowledge.

But exactly which evidential results should be taken to be in 'background knowledge' and hence fail to be possible confirmers of new theories, according to this historical approach? Musgrave distinguishes three versions of the approach, characterised, as he sees it, by three different answers to this question.

According to the first answer—which produces 'the strictly temporal view'—background knowledge contains 'all the relevant experimental results, hypotheses, *etc.*, which are "known to science" when [the] theory [in question] was proposed' (*op. cit.*, p. 8). This entails that a theory T is *only* confirmed by facts that were unknown at the time of T's initial proposal and cannot be confirmed by any evidence that was already known to hold when T was first articulated. Musgrave points out *both* that this suggestion flies in the face of quite clear-cut intuitions about some particular cases (e.g. that the General Theory of Relativity (hereafter: GTR) was confirmed by getting right the already well-known details of the precession of Mercury's perihelion) *and* that it seems difficult to discern any convincing general rationale for giving such a crucial role to purely temporal considerations.

On the second version—the 'heuristic view'—the relevant background knowledge for assessing the confirmation of theory T is, Musgrave takes it, restricted to those known facts and results that *were involved in the development* of T. This gives scope for the recapture of some of the intuitive judgments about particular cases: GTR may be confirmed by the details of Mercury's orbit, for example, provided that those details played no role in the construction of GTR (as indeed they did not). However, it is not clear, suggests Musgrave, that this account has any convincing rationale, and, in any event, it is altogether too person-relative: '[i]f different scientists take different routes to the same theory, then the evidential support of that theory as proposed by one of them might be different from its evidential support as proposed by the other.' (p. 14) And he regards this—entirely reasonably, it would seem—as in effect a *reductio ad absurdum* of the account.

Musgrave is inclined to endorse the third variant ('for my money it is the best version of the historical approach to confirmation' (*op. cit.*, p. 19)) This holds that the relevant 'background knowledge' for T consists only of the 'touchstone theory'

[2] Since I am one of those (like Hempel himself) who do not believe that *there is* a 'paradox of confirmation' (red herrings do confirm 'all ravens are black' just not very strongly!), this cuts no ice with me.

for T—in effect T's most plausible current rival. A theory T is then confirmed by any correct piece of evidence e that it entails provided that e is not also entailed by its 'touchstone' T′. Clearly there will in general be two types of such evidence: evidence that contradicts the touchstone T′ and evidence on which T′ is simply silent. On this account, GTR is confirmed by getting the details of Mercury's perihelion correct, since its rival, Classical Physics (hereafter: CP), gets those details wrong. Ditto with the Special Theory (STR) and, say, the Michelson-Morley result.[3] On the other hand, neither STR nor GTR is confirmed by any correct observational result that CP already correctly entails, even if it entails that same result in as straightforward and 'natural' a way as does CP.

2. A CLARIFICATION AND A PROBLEM WITH MUSGRAVE'S CLASSIFICATION

The main purpose of this paper will be to argue that, despite its neatness and intuitive appeal, Musgrave's whole classificatory scheme is off-beam: I shall argue for what might look like a modified version of Musgrave's second variant of the historical view, but also show that the account I favour is, when properly understood, logical rather than the historical! However a couple of detailed points about Musgrave's classification should be made beforehand.

First, a clarificatory point: Musgrave's approach, along with much of the subsequent literature (including my own contributions)[4], is focussed on one particular aspect of the general issue of confirmation—the impact of general observational or experimental results on deterministic theories that entail them. Of course this does not exhaust all confirmational issues—in particular those concerning stochastic or probabilistic theories. Although investigators such as Deborah Mayo and, more recently, Christopher Hitchcock and Elliott Sober in their (2004) have developed accounts that attempt to cover both deterministic and probabilistic cases, I think that there are problems with these accounts and will continue throughout this paper to concentrate (at least very largely) on the particular type of issue outlined. Was Fresnel's theory of diffraction better confirmed by the (novel) 'white spot' result than it was by getting the already known details about straightedge diffraction right? Was Einstein's theory better confirmed by the (novel) prediction of light-bending than it was by accounting for the already known facts about Mercury's perihelion? And so on. I recognise, of course, that statistical issues lie hidden here: real experiments and observations always show a certain amount of variation and the issue of how we get from real data to general observational or experimental results of the kind we are considering itself involves statistical considerations. Nonetheless I shall ignore these issues for current purposes and just take as a starting point the fact that certain general results

[3] Of course these judgments depend on *which particular versions* of classical physics we are considering (it is for example well known that Dicke and others eventually produced a version of CP that does yield the correct account of Mercury's motion)—therein lies much of the tale that will unfold in this paper.

[4] See my (1985) and especially (2002).

have been accepted as evidence, leading on to the question of the extent to which various deterministic theories that entail that evidence are confirmed by it.

Secondly, there are immediate questions about the completeness (or perhaps aptness) of Musgrave's classification: where, in particular, does the currently most widely held account of confirmation—personalist Bayesianism—figure within his scheme? I suppose that intuitively most philosophers of science would regard the Bayesian theory as the archetypically 'logical' approach to confirmation. Musgrave sees the logical approach as an aspect of 'modern logical empiricist orthodoxy' (*op.cit.*, p. 2) and Bayesianism certainly seems to be what eventually became of that orthodoxy, even to the extent to its being explicitly adopted by Carnap in his later years.

Bayesians standardly measure the support that evidence e lends to theory T by the difference between T's 'prior' probability, p(T), and its 'posterior' probability in the light of e, p(T,e). This appears to make Bayesian confirmation a two-place relation and hence indeed to make Bayesianism a 'logical' account on Musgrave's characterisation. However, Bayesians insist that all probabilities are *implicitly* relativised to background knowledge,[5] in fact to the background 'knowledge' of a particular Bayesian agent, where background knowledge, at least in the most straightforward account, consists simply of everything that the agent takes as evidence, ahead of the time at which we are considering the question of whether, and to what extent, the particular piece of evidence e confirms the particular theory T.

This relativisation to what a particular agent takes to be background knowledge is one—comparatively under-emphasised—source of the enormous (and in my view clearly unacceptable) subjectivity in the Bayesian approach: what a Bayesian 'agent' counts as evidence and hence puts in background knowledge is purely a matter for the agent, no less than are her 'priors'. Moreover, far from a Bayesian agent being required to justify every change in her degree of belief in a theory by appeal to the principle of conditionalisation, such an agent is entirely free (so far as the constraints of rationality are concerned) to feel at any stage that the epistemological earth has moved, that her background knowledge has changed and hence that an (in principle quite unconstrained) reassignment of all probabilities is called for. So for example there is nothing in pure Bayesian theory to rule that the following scenario involves anything that is counter-rational: a 'scientific' creationist begins with a very high prior for creationism and a very low prior for Darwinism; conditionalising on the accumulating evidence in approved Bayesian manner, however, leads her posterior for creationism to become steadily smaller and her posterior for Darwinism steadily greater; next however she receives a (perhaps further and powerful but ineffable) message from God or elsewhere that leads her to revise all her erstwhile judgments and to call for a new round of assignments of priors in the light of a radically revised background 'knowledge'; this new assignment of priors sees creationism back at a very high level and Darwinism back at a very low one. Of course we would all suspect the sincerity of such a creationist and the Bayesian trades in real, rather than

[5] Most deny that this dependence should be captured by explicitly conditionalising on background knowledge hence producing an absolute confirmation measure—see below.

merely *alleged*, degrees of belief—but were there such a sincere Creationist, the Bayesian could raise no objection to her (surely in fact irrational) belief-dynamics.

However, although such a sudden change in (personal) background knowledge is permitted in theory, when Bayesianism is applied *in practice*, things are generally made to look altogether more sensible and objective: it is quietly assumed that everyone will have the same background knowledge; that this is gradually augmented with extra material that everyone regards as evidence;[6] and that no sudden shifts of the kind just envisaged in fact occur.

Bayesianism appears, then, in practice at least, to be a version of Musgrave's historical approach—judgments about the impact of evidence e on theory T are made relative to, or in the light of, background knowledge (in practice implicitly assumed to be general amongst all competent agents at any given historical epoch). And indeed, on the most straightforward, 'natural' construal Bayesianism would fall squarely into the first, 'purely temporal' camp.

On this most natural construal, the background knowledge that the Bayesian sees as relevant for assessing the impact of some result e on T would consist of everything that is accepted as evidence (that is, assigned probability one) ahead of the question being raised of what impact particular piece of evidence e has on theory T. Suppose we are, then, asking about the confirmation of T at a time when some evidence e is already known (say that we are interested in the impact of the evidence about the precession of Mercury's perihelion on the GTR when that theory was first proposed). The fact that e is already known and accepted as evidence, entails that e will already at that time be part of background knowledge and hence that $p(e) = 1$. But this in turn immediately implies—as has been heavily emphasised under the name of the 'problem of old evidence'—that e cannot Bayesian-confirm T: if $p(e) = 1$, and T entails e, then it straightforwardly follows that $p(T,e) = p(T)$ and hence that there is, on the Bayesian account, no confirmation. As previously noted, this implication flies in the face of a number of intuitively firm judgments of confirmation in particular cases (which is why, of course, it is known as the *problem* of old evidence).

However, some Bayesians, such as Colin Howson and Peter Urbach (see their 1994), have insisted that the old evidence problem is based on a misunderstanding of the approach. If e is the evidence whose confirmational impact is under consideration at time t, then, if e is already known, that is, accepted as evidence and hence as part of background knowledge B at t, the correct background against which to make the confirmational judgement is not B itself, but rather, so to speak, $B - \{e\}$: the relative complement of B with respect to $\{e\}$, that is, the background knowledge that you 'would have had at time t, had you not known e but all else remained the same.'[7] It is, as I have argued elsewhere (especially in my (2000a)), extremely tricky (to say the very least) to make coherent sense of this counterfactual judgment. For

[6] Of course on pure personalist Bayesianism, 'evidence' (really 'evidence for the agent') is anything that the agent comes to assign probability one! (This is another massive and comparatively underemphasised source of subjectivism in the account.)

[7] See also the burgeoning literature on both Bayesian and non-Bayesian 'belief revision'.

current purposes, however, let's assume that its admitted intuitive appeal should override any formal difficulties—where does this alternative construal of the relevant background place Bayesianism within Alan Musgrave's scheme?

The answer, I think, is 'outside of it'. This version of Bayesianism *is* 'historical' in the sense that historically varying background knowledge plays a role in confirmation: it makes it entirely possible that the answer to the question 'Does e support T and if so to what extent?' may be different for different historical epochs, because of the differing content of background knowledge. But Musgrave assumes that for all versions of the historical approach, this variability will rest on the question of whether or not the piece of evidence, e, at issue is itself a part of the relevant background knowledge. On this alternative construal of Bayesianism, on the contrary, the evidence whose confirmational impact we are interested in is automatically 'subtracted' from background knowledge before the Bayesian formulas are applied. (Hence Colin Howson (see, for example, his 1990) in particular believes that Bayesianism, when properly construed, makes both the issue of when some evidence was discovered, and that of whether or not it was used in the construction of some theory, entirely irrelevant to confirmation.) The historical nature of confirmation on this alternative Bayesian view depends instead on the (ineffable) way in which background knowledge informs judgments of 'prior' probability.

3. A PROBLEM WITH MUSGRAVE'S PREFERRED VERSION OF THE HISTORICAL THEORY

It is not, then, clear that Alan Musgrave's classification scheme covers all accounts of confirmation that currently deserve serious attention and more problems in this regard lie ahead. But let's return for the present to operating within his scheme, and consider the merits of his own preferred alternative version of the 'historical account'. This asserts, remember, that the relevant background knowledge, membership of which prevents an observational or experimental result from confirming some theory T, is supplied by T's 'touchstone theory'—its most plausible current rival. This 'touchstone account' implies that GTR, for example, cannot be confirmed by any empirical result that is already entailed by CP.

But this is surely an extraordinarily counterintuitive judgement and hence not one that any sensible account would, on reflection, want to endorse. Scientists will, naturally, be especially interested in the question of whether GTR, for example, is *better* confirmed, obtains *greater* empirical support from the total evidence, than CP, and this will direct particular attention to those pieces of evidence that are entailed by Einstein's theory but not also by the classical one. But this is an issue of *extra* empirical support, not empirical support *simpliciter*. Assuming at least that both theories yield some piece of data in a 'natural' (non *ad hoc*) way (as is the case, for example, with the accounts they give of the precession of the equinoxes), then surely the reasonable judgement is that *both* CP *and* GTR are confirmed by the phenomenon—this is why the precession of the equinoxes, unlike, say, the precession of Mercury's perihelion, is irrelevant to the *comparison* of the degrees of evidential support of the two theories.

I can see no general rationale for Musgrave's preferred alternative and it certainly leads to any number of intuitively extremely awkward consequences. GTR entails the correct details of the precession of the equinoxes and it does so in as straightforward, natural, non-*ad hoc* way as does CP. Why on earth, then, should it fail to provide any confirmation for GTR just because there is another theory that also gets the phenomenon correct? Or consider what the account says not about the confirmation of the newer theory in some case of inter-theoretic rivalry but about the confirmation of the *older* one. Presumably, once GTR has been articulated, the precession of the equinoxes ceases to be a possible confirmation for CP too—since GTR now becomes the classical theory's 'touchstone' no less than vice versa. This means that while Newton's theory was confirmed by the precession of the equinoxes in, say, 1900 (when its 'touchstone theory' was what? Galileo's (very partial) mechanics? or Aristotle's more comprehensive but hopeless system?), by 1914, when nothing relevant had changed either in the theory or (of course) in the phenomena, it was no longer confirmed by those phenomena because a new theory, GTR, had arisen that equally well entailed a correct description of them. (I suppose that the 'touchstone' theorist could claim, alternatively, that the right way to judge the empirical support gained by a theory is by always taking as background knowledge that theory's chief rival *at the time it was introduced*. But this alternative is worse, *much* worse than the original. For one thing, it would disqualify the account as an historical one on Musgrave's terms—the question of a theory's confirmation by e eternally carries with it the historical context of that theory's initial articulation and hence the question becomes ahistorical! More importantly, the alternative yields even more counterintuitive results than the initial suggestion. Admittedly the alternative would have the intuitively pleasing consequence of allowing Newton's theory to retain its support from phenomena such as the precession of the equinoxes or the existence of Neptune even after the articulation of a rival that equally adequately explains them. But at the same time it would of course, all too readily, yield the judgment that, as well as there being empirical phenomena (like the Michelson-Morley experiment, and the precession of Mercury's perihelion) that support GTR but not CP, there are also phenomena (like the precession of the equinoxes or the existence of Neptune) that support CP but not GTR (because CP supplies a more demanding 'touchstone' for GTR than it itself had faced when first articulated). And all this, despite the fact that GTR entails correct descriptions of these phenomena too in a 'natural' non *ad hoc* way just as CP does![8] Surely the right judgment, as I suggested earlier, is that these phenomena support *both* theories and hence drop out of the equation when it comes to *comparing* the empirical supports enjoyed by the two theories.)

[8] Of course, as Kuhn liked to emphasise (see in particular his 1977), it does sometimes happen—especially early on in the development of some new theory—that different pieces of evidence point in different directions: e better supporting the new theory T′, while e′ better supports the older theory T. But it *clearly* cannot be the right judgment in general, that once some evidence has (fully) confirmed T, it always supplies a reason for preferring it.

We saw that Musgrave castigates the other two alternative construals of the historical approach as lacking any obvious rationale, but, as we have just now seen, his own preferred version certainly does no better. This surely should make us reflect again on the general underlying claim: Why should 'background knowledge' *in any form* be a factor in empirical support? There is of course, as just remarked, an obvious rationale for taking background knowledge to be a factor in *increased* support: if we are interested in why new theory T′ is *better* supported than its earlier rival T, then results that T has already either predicted or adequately explained will be in background knowledge and hence *may* drop out of the equation [9]—if e already confirms T then the fact that it also confirms T′ *may* provide no reason to prefer T′ over T. But it seems difficult to see why the fact that an empirical result e is already in background knowledge *in any sense* should by itself totally rule out e as support for some newly proposed theory T′, in the *non-comparative* sense of support.

Alan Musgrave's residual Popperianism leads him to claim that a justification for giving background knowledge this central role might be developed by considering which bits of evidence do or do not supply a proper *test* of the theory concerned. The suggestion is that for some reason results already in background knowledge at the time of T's proposal cannot provide a test of T. But why should this latter claim be true? If we already know that e holds rather than some alternative result of the experiment or observation it describes, then of course the fact that it turns out that some new theory T entails e rather than any of the alternatives will not have us on the edges of our seats wondering if the theory might turn out to be refuted by this particular experiment or observation. In that sense there is no test from old data. But why should that sense have the slightest epistemic relevance? The new theory is by no means *a priori* guaranteed to entail correct descriptions of all the phenomena equally well dealt with by its predecessor. (Indeed if Popper's account of new theories as 'bold conjectures' were true, it would be a miracle if this happened in a field where the old theory had had any considerable degree of empirical success). Still less is there an *a priori* guarantee that the new theory will get right all known phenomena, whether or not dealt with successfully by its predecessor. And indeed few, if any, theories do get *all* known phenomena correct (at least when first proposed). There is, then, a clear sense in which such a theory *was* tested by the already known data: it might have entailed different data that contradicts that actually recorded, but in fact it did not. If a theory might perfectly well have got some already known phenomenon wrong, but in fact got it right, then it seems perverse to rule ahead of time that this success fails to count as surviving a 'test', and so cannot yield any degree of empirical support for that theory.

Alan Musgrave's preferred solution of the prediction versus accommodation problem is, I claim, wrong; and, as so often happens in philosophy, this is because he has got the problem wrong.

[9] The fact that it only *may* drop out of the equation is important: if T provided only an *ad hoc* accommodation of e, while T′ genuinely predicts e (in the non-temporal sense, see *below*) then, on the account that I favour, e may, on the contrary and far from dropping out of the equation provide an important reason for preferring T′ over T.

4. THE REAL PROBLEM: PREDICTION VERSUS ACCOMMODATION

The problem is not whether new evidence counts more than old—it doesn't (at any rate it doesn't just because it's new). The problem is *adhoc*ness (indeed the real problem is perhaps seeing that the *adhoc*ness problem is the *only* problem in this area).

In the early 19th Century, the classical wave theory of light predicted the results of various diffraction or interference experiments. Intuitively these results told very strongly in favour of this theory against its then rival—the emission or corpuscular theory of light. Yet, as we would expect on Duhemian grounds, the emissionists by no means immediately surrendered. Duhem emphasised that single 'isolated' theories such as the corpuscular theory have no empirical consequences of their own, but achieve them only when conjoined both with specific assumptions (answering the questions: what velocities do the light-corpuscles have? and what masses? most importantly, what forces are they subjected to in particular circumstances?) and with further auxiliary and instrumental assumptions. It follows that there is always logical leeway for holding onto the central theory in the light of experimental 'anomalies' and looking to modify either a specific or auxiliary assumption. 18th and 19th century corpuscularists duly obliged—some postulated, for example, a force of diffraction, exercised on the light-corpuscles as they pass the edges of any 'gross' opaque object; others considered the possibility that the fringe phenomena that wave theorists attributed to interference and/or diffraction were in fact physiological phenomena. Although in this case it was never achieved, it clearly has to be possible in principle for the emissionists to have given themselves an expression for the 'force of diffraction' with so many (initially free) parameters that any given particular fringe phenomenon (or finite set of such phenomena) could have been accommodated. Certainly by appealing to (unknown) physiological facts about vision an entirely cheap corpuscularist 'explanation' was suggested at the time and could have been developed in some detail.

Or consider another case where this sort of dodge definitely works. ('Works' in the sense that it produces a theory that yields the accommodated data, not of course in the sense that it produces a scientifically respectable theory that does so.) The fossil record looks like strong confirmation of the Darwinian theory of evolution. (Of course the situation is less straightforward in this example because that theory does not actually deductively entail any particular aspect of the fossil record, but this is inessential to the point at issue.) As is well known, however, it is trivially easy for the 'scientific' creationist to 'match' this success. All that she needs to do is follow Gosse and assert that God decided, when creating the Universe in 4004 BC, to include some pretty pictures in some rocks that look awfully like the marks of the skeletons of now extinct organisms but are in fact *just* pretty pictures, and to include some buried bone-*like* objects that seem to fit together to form the skeletons of impressive and now extinct creatures but are in fact just artefacts, and so on. She will thus create a version of 'scientific' creationism that entails the correct facts about the (now alleged) 'fossil record', but clearly it would be absurd to hold that this requires us to abandon the view that this record supports the Darwinian theory over its rival.

There is a long tradition in science of deeply engrained distrust of such *ad hoc* moves. We surely require an account of the confirmation of theories by evidence that underwrites the judgement that the interference effects continued to give more empirical support to the wave theory in the early 19th century even once it had been indicated that emissionist accounts of those effects could be constructed, and similarly underwrites the judgment that the fossil record continues to give good empirical reason to prefer the Darwinian theory even after creationists have availed themselves of the 'Gosse dodge'. But how *exactly* are we to capture these judgments within a generally defensible account of confirmation?

The obvious initial suggestion is to say that no theory can be confirmed by evidence that it has simply accommodated in this *ad hoc* way, where the advocates of the theory have taken the evidence at issue as given and *used it* to produce a specific version of their favoured theory that yields that evidence. At least when the notion is used liberally, these are all exercises in parameter-fitting. The idea behind the 'diffracting force' emissionist account of fringe-phenomena was to start from a very complicated expression for that force as a function of the distance from the diffracting object (allowing this to be attractive at some distances and repulsive at others) and then use particular fringe measurements to fix those parameter values so that the required phenomena are entailed. Similarly, the Creationist's general theory—that God created the Universe in 4004 B.C. 'essentially' as it now is—effectively gives the Creationist a whole series of 'free parameters' that specify how *exactly* it was that God chose to create the universe: if you observe particular patterns in some rocks, then that specifies one part of God's creation, this 'parameter' value is tied down on the basis of the observation and this, unsurprisingly, produces a specific theory that entails the observed data—the theory being of course that God created the Universe, not just any old how, but in particular with these patterns in these rocks.

The positive side of the account would then be that a theory is confirmed by any piece of data a correct description of which it entails, provided that the evidence was not used in the construction of the specific version of the theory that entails it, *whether or not* the data was already known. There appears to be, then, an important methodological distinction between accommodation and prediction in the general sense in which it is generally used in science (meaning simply that some evidence follows from a theory without having needed to be accommodated within it)[10].

The most straightforward way to capture this difference would, of course, be by ruling that theories are confirmed only by predictions (understood as not requiring novelty) and not at all by accommodations. This amounts, it would seem, to the

[10] Here for example is an especially clear passage from French's excellent textbook on Newtonian Mechanics: '[L]ike every other good theory in physics, [the theory of universal gravitation] had predictive value; that is, it could be applied to situations besides the ones from which it was deduced. Investigating the predictions of a theory may involve looking for hitherto unsuspected phenomena, or it may involve recognising that an already existing phenomenon must fit into the new framework. In either case the theory is subjected to searching tests, by which it must stand or fall.' (French 1971, pp. 5-6)

'heuristic account' as Musgrave characterises it—namely, the version of the historical approach to confirmation which identifies those results that belong in background knowledge and hence cannot confirm the theory as those that have been used in the construction of the theory. The 'heuristic account', as so construed, is also sometimes known as the 'no double use' or 'use novelty' rule.[11]

I shall in fact argue, first, that this 'most straightforward way' of underwriting the prediction/accommodation distinction is altogether too straightforward to be true; and secondly, that the correct way to underwrite the distinction and hence arrive at the correct account of confirmation of the sort here at issue produces a view that cannot properly be regarded as a version of Musgrave's historical approach. However, it should be noted that even in its most straightforward form, the 'no double use' account seems to have some immediate attractions. *First,* it accords with a range of intuitive judgments about particular cases (one such is the precession of Mercury's perihelion and the GTR) where 'old evidence' is taken to provide strong support for a theory: provided that the facts about Mercury's orbit were not involved in the construction of its explanation within GTR, then there is no reason, on this account, to deny that those facts support GTR. And *secondly* (and of course relatedly) the account relegates the time-order of theory and evidence *in itself* to what it should be—namely, a complete historical irrelevance (what possible *general* justification could there be for old evidence always to count less? why give such an epistemic role to what may have been a mere historical accident?).

However, despite these attractions, the 'no double use' rule has been alleged to face at least two fundamental objections of its own. The objection that Musgrave himself cites, as we already noted, concerns the fact that it seems to make theory-confirmation an unacceptably relativistic (enquirer-relative) affair:

> If different scientists take different heuristic routes to the same theory, then the evidential support of that theory as proposed by one of them might be different from its evidential support as proposed by the other. In short, Zahar's ['heuristic' or 'no double use'] view makes confirmation a person-relative affair. (*op. cit.*, p. 14)

An even more frequently voiced criticism of this view is that, just like the purely temporal view that it attempts to replace, it flies in the face of deeply held intuitions about particular cases. Nickles, Mayo, Howson and others[12] have all pointed to cases in which evidence e was used in the construction of some theory T and yet where e was taken to provide (strong) support for T. As Colin Howson, for example, claimed (*op. cit.*, p. 231) 'counterexamples abound to' the idea that evidence used in the construction of a theory cannot be used in its support, and indeed 'can be invented ad lib'. In the next section, I address this second objection—therefore cunningly renamed 'objection one'. I then show how to develop the idea underlying the no double use rule so to produce an account that escapes objection one, and then I will show how the developed view also overcomes Musgrave's objection (now 'objection

[11] See for example Nickles (1987)

[12] Nickles *op.cit,* Mayo (1996) and Howson (1990)

two').[13] Finally, I will show how this developed view is not properly regarded as a version of the historical approach as Alan Musgrave construes it.

5. OBJECTION ONE AGAINST THE 'HEURISTIC' VIEW: USED DATA SOMETIMES (STRONGLY) CONFIRMS

Allan Franklin once gave a seminar talk at the LSE under the title '*Ad hoc* is not a four letter word'. Underneath the (multiple) surface correctness of this title, there lies a somewhat deeper but no less correct point: scientists entirely legitimately use data all the time in the construction of their theories. If, to take the most clear-cut case, general theoretical considerations leave open the value of some important parameter, then how else would a scientist tie down that parameter's value *except* by using data? The only other alternative that seems open would be to conjecture a value and then test (and then re-conjecture when the test is failed as it almost inevitably will be, and then re-test…)—but this attempt to find a needle in a (generally infinitely large) haystack would be madness. Here is one extremely simple but canonical instance of the systematic use of data in theory-construction.

Suppose a mid-19th Century scientist already accepted the *general* wave theory of light—the theory that light from any particular source consists of waves of some wavelength or other transmitted through the luminiferous aether. This general theory does not specify the wavelength of any particular kind of monochromatic light—say light from a sodium arc. The scientist would like a more detailed theory that *does* specify that wavelength. Rather than attempt to conjecture a value, she would 'deduce' the specific theory, involving the specific value of the wavelength, 'from the phenomena'. She would look for some consequence, e, of her general theory T, where e characterises some observable magnitude (fringe separation in some particular experiment, say) as a one-to-one function of the wavelength. She would perform the experiment using light from a sodium arc, measure the magnitude at issue—here, the fringe separation (call the result of this measurement e′)—and infer to a more specific theory T′. So, for example, subject to a couple of idealisations, it follows from the general wave theory that, in the case of the famous two-slit experiment, the (observable) distance X from the fringe at the centre of the pattern to the first fringe on either side is related to (theoretical) wavelength λ, via the equation $X/(X^2 + D^2)^{1/2} = \lambda/d$ (where d is the distance between the two slits and D the distance from the two-slit screen to the observation screen—both of course observable quantities). It follows analytically that $\lambda = dX/(X^2 + D^2)^{1/2}$. But all the terms on the right hand side of this last equation are measurable. Hence particular observed values will determine the wavelength (within of course some small margin of experimental error), and so determine the more specific theory T′, with the parameter that had been free in T now given a definite value—again within a margin of error. Far from being scientifically questionable, this is, to repeat, entirely standard (and patently legitimate) scientific procedure.

[13] My treatment here follows and builds upon that given in my (2002)—actually written for a conference in 1999.

Several of the most celebrated episodes from the history of science involve using data (often anomalous data for an earlier theory) to construct a new theory. For example, Adams and Leverrier used the data from Uranus's orbit that had proved inconsistent with the initial Newtonian account essentially as follows. They took it that the basic Newtonian theory (of mechanics plus universal gravitation) was correct, and then worked backwards from the initially anomalous Uranian data to figure out what assumptions would have to be made about a further trans-Uranian planet, such that, when that further planet's gravitational interaction with Uranus was taken into account (along of course with the gravitational interaction with the sun and the other, already known planets), the overall Newtonian theory would ascribe the correct orbit to Uranus. This manoeuvre, as is well known, led to the discovery of Neptune—one of Newtonian theory's greatest successes and indeed one of the most impressive confirmations of any theory in the history of science.

So how, in the light of facts like these, could anyone have defended the 'heuristic account' of confirmation, committed, as it seems to be, to the view that evidence used in the construction of a theory *can never* confirm it? In the specific case from optics that I just sketched, there is a very clear sense in which e, the fringe data used in the construction of the more specific wave theory T′ supports that theory: *given that* the general theory T has already been accepted, e *deductively entails* T′, and what better support could there be than deductive entailment?

Colin Howson likes to emphasise a still more general sort of case—standard statistical examples such as the following (see again his 1990). We are given that an urn contains only black and white balls though in an unknown (but fixed) proportion; we are prevented from looking inside the urn but can draw balls one at a time from it. Suppose that a sample of size n has been taken (with replacement) of which k have been found to be white. Standard statistical estimation theory then suggests the hypothesis that the proportion of white balls in the urn is $k/n \pm \varepsilon$, where ε is calculated as a function of n by standard confidence-interval techniques. The sample evidence is the basis here of the construction of the particular hypothesis, and surely also supports that particular hypothesis at least to some (good) degree—the evidence for the hypothesis just *is* that a proportion k/n of the balls drawn were white.

Deborah Mayo cites and analyses in more detail the same case and also cites the following 'trivial but instructive example' (1996, p. 271). Suppose one wanted to arrive at what she characterises as 'a hypothesis H' about the average SAT score of the students in her logic class. She points out that the 'obvious' (in fact uniquely sensible) way to arrive at H is by summing all the individual scores of the n students in the class and dividing the result by n. The 'hypothesis' arrived at in this way would clearly be 'use-constructed'. Suppose the constructed 'hypothesis' is that the average SAT score for these students is 1121. It would clearly be madness to suppose that the data used in the construction of the 'hypothesis' that the average SAT score is 1121 fails to support that hypothesis. On the contrary, as she writes (*ibid.*):

> Surely the data on my students are excellent grounds for my hypothesis about their average SAT scores. It would be absurd to suppose that further tests would give better support.

Exactly so: the data provide not just excellent, but, short of some trivial error, entirely *conclusive* grounds for the 'hypothesis'—further tests are entirely irrelevant. (This is precisely why it seems extremely odd to talk of a 'hypothesis' at all in these circumstances—a point to which I will return *below* in my more extensive consideration of Mayo's views.)

6. RESPONSE TO OBJECTION ONE: TWO SORTS OF CONFIRMATION

Does the admission that these sorts of 'deductions from the phenomena' (such as the deduction of the specific version of the wave theory T′ from the general wave theory T plus fringe data e) provide clear-cut cases of theories that *are* supported by data used in their construction spell the end for the heuristic account of confirmation?

To start to see that the answer is 'no', consider again the 'Gosse dodge' within 'scientific' Creationism, or indeed any of the other standard cases of blatantly *ad hoc* moves in defence of a theory that have been cited in the literature.[14] In all these cases, the specific theory is 'deduced from the phenomena'—meaning, as always, of course deduced from the phenomena *plus already accepted general principles*.[15] 'Deduction from the phenomena' is a very powerful technique in the case where the necessary general principles are indeed *generally* accepted and therefore, presumably, themselves have strong evidence in their favour. But what if, on the contrary, the necessary 'background principles' are not universally accepted as based on sound evidence, but instead accepted only by some group or other, one with its own particular axe to grind?

If you *were already convinced* of the general Creationist claim that God created the Universe 'essentially' as it now is in 4004 B.C., then the data that your irritating Darwinian supporters insist on calling the 'fossil record' do of course deductively Those data thus give you not only good but 'essentially' *conclusive* reason to accept this particular version of the general theory that you already accepted on other entail[16] the more specific version of your theory that says that part of God's creation was some pretty pictures in the rocks and buried bone-like artefacts, and so on.

[14] Another favourite example that I and others have used elsewhere is provided by Immanuel Velikovsky's famous theory that a large chunk of Jupiter broke away and careered towards the Earth, orbiting it on a series of occasions before (somehow or other) settling down to a quieter life as the planet Venus. Velikovsky saw these close encounters with this 'comet' as the explanation for 'events' 'recorded' in the Old Testament—such as the parting of the Red Sea and the fall of walls of Jericho. Velikovsky recognised that other contemporary record-keeping cultures ought, in that case, to have recorded cataclysms on a similar scale, since such amazing effects of the 'comet' were unlikely to have been confined to the particular area of the Middle East covered by the Old Testament scribes. He found one or two (arguable) confirmations, but several altogether more clear-cut apparent refutations. But Velikovsky rose to the task, arguing that in the cultures that otherwise kept records the cataclysmic events associated with the 'comet' had proved *so* cataclysmic that 'collective amnesia' had set in there. Of course he read off which particular record-keeping cultures had suffered from this unfortunate complaint precisely by noting which ones had no records of suitable cataclysms.

[15] 'Deduction from the phenomena' was of course Newton's preferred method of theory-construction (and of avoiding 'hypotheses'). For Newton on deduction from the phenomena and references to the literature, see my (2000b).

[16] It is admittedly only a more or less deduction—it would be a valid deduction only if the Creationists assumed that the world now is exactly as it was when god created it, but of course even they have to admit that there has been some change (hence the 'essentially' as it now is).

grounds. In this regard the case is surely no different from the (intuitively more scientifically respectable) case of the early 19th Century optical scientist, who, being already convinced of the general wave theory, deduces from the phenomena the more specific version with specific wavelengths for light from particular monochromatic sources: in this latter case too, *given* that she accepts the general wave theory, T, the fringe data, e, give her (in this case entirely) conclusive reason to accept the particular version of the theory T′, involving a now fixed value of an initially free parameter.

But the natural reaction to the Creationist/Gosse dodge case is surely that while the 'fossil record' data may indeed give you reason, let's say conclusive reason, to adopt the particular Gossefied version of Creationism, this is an ineliminably conditional judgment—the evidence gives you absolutely no reason to have adopted the general Creationist view in the first place. If you are going to be any sort of Creationist at all, then this data gives you as solid a reason as could be for being a Gosse-dodge-Creationist, but it gives you absolutely no reason to be any sort of Creationist at all! There is no reason to think that the general underlying theory itself obtains any empirical support just because the specific version of it entails the correct empirical data.

What is sauce for the goose is sauce for the gander. Exactly the same judgment is valid in the (intuitively scientifically respectable) wave theory case: the fringe data, e, give you solid (indeed conclusive) reason to believe T′ (the wave theory with a specified wavelength for monochromatic light from a sodium arc), *provided* that you have already accepted the general wave theory (with free parameter), but give you absolutely no reason to accept the general wave theory in the first place. Both in this—seemingly legitimate—case and in the, apparently illegitimate case of the Gosse dodge, the correct judgment seems, then, to be twofold: *first* that, if the general underlying theory is taken as given, then if e is used in the construction of a specific version of that general underlying idea, e gives very strong (perhaps conclusive) support for the specific theory; however, *secondly*, there is no support from that evidence for the general, underlying theory itself.[17]

The difference between the two cases seems clearly to be that while there were *other, independent* empirical reasons for taking the general wave theory of light seriously, there are no such reasons in the case of 'scientific' creationism. There was already good reason to accept the general wave theory with the free parameter,

[17] A similar remark also applies to Colin Howson's statistical examples: so long as the basic theory or 'model' is given (basically in his urn case, that we are dealing with a 'Bernouilli process' with fixed, but unknown parameter p (the proportion of white balls in the urn)), then the evidence that k/n of the sampled balls were white gives support (in this case of course not conclusive) for the specific theory that estimates p as lying in the interval $k/n \pm \varepsilon$. But that data gives no conceivable reason for having greater faith in the idea that this is the correct model. (Indeed this is not an issue that would normally even arise in that case.)

ahead of any measurement of fringe distances with light from the sodium source. Hence, when evidence e turns out deductively to entail the specific theory T′ (complete with filled-in value for the wavelength of light from the sodium arc) *given* T, we can 'discharge the antecedent' and infer that e gives us (of course some, defeasible) reason to accept T′ *full stop*. In contrast, in the Gosse dodge case, exactly because there is no independent reason to accept the underlying general Creationist account, the fact that the fossil record entails the Gosse dodge variant of Creationism, justifies *only* the *conditional* judgment that e gives us reason to accept the Gosse dodge variant to the extent that (but only to the extent that) we already have reason to accept the general theory.

But how exactly can these general underlying theories earn their independent empirical support, as, if the line I am defending is correct, in some cases they must do? After all, the Duhem problem is exactly posed by the fact that such general theories do not have directly checkable empirical consequences of their own. All empirical tests of the wave theory of light, for example, are tests of the general wave theory *plus* particular assumptions. It seems, then, that if the dual approach to confirmation that I am outlining is to be at all coherent, there must be a contrast class to the sorts of cases we have considered so far. That is, there must be some empirical tests, the results of which not only confirm the specific version of the theory that entails the results of those tests, *but also* confirm the underlying general theory (despite the fact that that general theory does not entail those results on its own). It must be the case, in other words, that scientists do sometimes take it that the empirical success of some particular version of a general theory gives good reason to accept the general theory itself—and in particular good reason to seek to develop another specific theory for a different field of phenomena based on that same general theory.[18]

Certainly this seems to be an actual feature of scientific practice: for example, the discovery of Neptune seems to have been regarded as a success not just for the particular Newtonian model of the universe (now involving Neptune), but also for Newtonian gravitational theory itself. Similarly, returning to optics, both the (new) white spot result and the (long known) straightedge diffraction experiments were taken to support not only Fresnel's specific wave theory of diffraction that entailed them, but also the general theory of light as waves in an elastic medium on which it was based. Hence these phenomena, although following only from the specific wave theory of diffraction, were taken as providing good reason to develop another specific theory based on the same general elastic medium wave theory to deal with the quite separate phenomena of polarisation and crystal optics. (See my 1989.)

My claim, then, is that scientists do not restrict themselves simply to judgments of the conditional kind that we highlighted—judgments to the effect that, *against the given background* of some general framework theory, some piece of evidence e gives strong support to some specific version of the general theory. They *also*

[18] This in practical terms seems to me the main work that confirmational judgments do in the development of science.

sometimes see the general framework theory as empirically supported. Yet, as Duhem showed us, such support must always be achieved, not directly, but *via* specific versions of the general theory (*i.e.* not the general theory alone but that theory plus some further assumptions). Some, but not all, types of empirical success must somehow spread from the particular theory that directly enjoys them to the underlying general theory.

What kinds of empirical success turn this second and stronger confirmational trick? The answer, I think, is two kinds, of which the more straightforward is the following. A scientist starts with some general theory T, uses e to fix some parameter in T and thus creates (by 'deduction from the phenomena') the more specific theory T′; T′then goes on to make some *further* independent prediction e′. If e′ is experimentally verified then this confirms not only the specific theory T′ but also the underlying more general theory T. This is exactly what happens in our first wave theory case: once the parameter corresponding to the wavelength of light from a sodium arc has been fixed using the fringe distances in the two-slit experiment, the more specific theory thus created can then go on to be directly tested in *other* experiments using light from the same source (notably the single slit diffraction experiment). (It is standard to talk of 'overdetermination' of parameter-values in such cases: the initially free parameter could be fixed using *any one* of a range of experimental results and the specific theory with fixed parameter would then proceed to entail the rest of that range of results.)[19]

A more significant episode in the history of the wave theory illustrates the same lesson. The result of the experiment of Fresnel and Arago—that the interference fringes in the two slit experiment disappear when the light from the two slits is oppositely polarised through the interposition of suitably oriented quartz plates—more or less forces the wave theorist to adopt the view that the wave motion in light occurs *at right angles* to the direction of propagation, rather than *along* the direction of propagation, as previously believed. ('More or less' because you can deduce the specific tranverse wave theory from the general theory (light is some sort of wave in a medium) plus the Fresnel-Arago result, only if you add some further extra assumptions, that are, however, entirely 'natural'.[20]) The Fresnel-Arago result then very strongly confirms the tranverse version of the wave theory in the first (conditional) sense—if you have already accepted the general wave theory then the

[19] Alan Musgrave too highlights the importance of independent testability and independent evidence (*op. cit.*, p. 6) But he takes it that the idea that scientific theories require not just testability, but *independent* testability to be accepted is captured by his favoured third variant of the historical approach: T is independently testable through any of its empirically checkable consequences that are not also consequences of its 'touchstone' T′. But as we are now seeing the really important idea is not one involving a comparison between theories, instead a single theory is independently tested by any piece of evidence that it makes a prediction about, provided that evidence was not 'written into' the theory in advance.

[20] Light waves could for instance in principle have *both* a transverse *and* a longitudinal component. However the fact that this Fresnel-Arago result (along with others) shows that any longitudinal component could have no observable effect means that simplicity dictates it be rejected. (It is in this particular sense, rather than any nebulous general way, that simplicity judgments play an important role in science.)

result shows (pretty well) that tranverse waves are what you *must* plump for. However, the fact that—having in effect deduced the transverse wave version from the Fresnel-Arago result—that experimental result can in turn be deduced from the transverse version of the theory would clearly give anyone unconvinced of the general wave theory no further reason to adopt it. But Hamilton saw that the transverse wave theory made predictions about the wave surface in particular types of birefringent crystal and hence about certain phenomena in crystal optics that are quite independent of the initial Fresnel-Arago result; and these predictions were successfully tested by Lloyd. These crystal optics results represent exactly the sort of *independent* evidence that, unlike the Fresnel-Arago result, *does* support not only the particular theory that entails it but also the general underlying approach—they *do* give the unbeliever extra reason to adopt the wave theory approach in general.

Finally, in the famous Newtonian case, using the (initially anomalous) data from Uranus's orbit to fix (in fact, in this case, re-jig) a parameter about the number of other planets affecting that orbit produces a theory that turns out to entail an independently checkable prediction about the existence of a further (and hitherto unrecognised) planet. Confirmation of this prediction in the form of the discovery of Neptune supports not only the specific version of Newtonian theory, partially created from the Uranian data, but also the general Newtonian theory itself. So the 'prediction' of the Uranian data gives only the first, conditional sort of support for the specific Newtonian model, while observations of Neptune yield the stronger kind of support that reaches the general theory by 'confirmational osmosis'.

These cases, then, exhibit the *first* type of stronger confirmation—*independent* evidence. The second type is equally important. This sort of confirmation (again: of the general underlying theory, rather than of some specific theory, *given* the general underlying theory) is provided in cases in which, roughly speaking, some prediction 'drops out of' the basic idea of the theory. Here's an example.

The explanation of the phenomena of planetary stations and retrogressions within the Ptolemaic geocentric theory is often cited as a classic case of an *ad hoc* move. The initial geocentric model of a planet, Mars, say, travelling on a single circular orbit around a stationary Earth, predicts that we will observe constant eastward motion of the planet around the sky (superimposed, of course, on a constant apparent diurnal westward rotation with the fixed stars); this is directly refuted by the fact that Mars' generally eastward (apparent) motion is periodically interrupted by occasions when its gradually slows to a momentary halt and then begins briefly to move 'backwards' in a westward direction, before again slowing and turning back towards the east. The introduction of an epicycle of suitable size and the assumption that Mars moves around the centre of that epicycle at a suitable velocity while the whole epicycle itself is carried around the main circular orbit (now called the deferent) leads to the correct prediction that Mars will exhibit these stations and retrogressions. Although not as straightforward as normally thought, this case surely is one that fits our first, entirely conditional, kind of confirmation—if you *already accept* the general geocentric view, then the phenomena of stations and retrogressions give you very good reason to accept (and in that sense they strongly confirm) the particular version of geocentricism involving the epicycles.[21] However the fact that stations and retrogressions are 'predicted' (better: entailed) by the specific version of geocentricism with suitable epicyclic assumptions gives absolutely no further reason to accept (and so no support for, or confirmation of) the underlying basic geocentric (geostatic) claim.

The situation with Copernican heliocentric (or again, better, heliostatic) theory and planetary stations and retrogressions is, I suggest, entirely different. According to the Copernican theory we are, of course, making our observations from a moving observatory. As the Earth and Mars both proceed steadily eastward around the sun, the Earth, moving relatively quickly round its smaller orbit, will periodically overtake Mars. At the point of overtaking, although both are in fact moving consistently eastward around the sun, Mars will naturally *appear*, as observed from the Earth, to move backwards against the background of the fixed stars. Planetary stations and retrogressions rather than needing to be explained *via* specially tailored assumptions ('having to be put in by hand' as scientists sometimes say), drop out naturally from the heliocentric hypothesis. Copernican theory, in my view, genuinely *predicts* stations and retrogressions even though the phenomena had been known for centuries before Copernicus developed his theory. (I am talking here about the qualitative phenomenon not the quantitative details which, as is well known, need to a large extent to be 'put in by hand' by both theories—and courtesy of multiple epicycles in Copernicus no less than in Ptolemy.[22])

The way that Copernicus's theory yields stations and retrogressions may, indeed, seem to be *so* direct that it challenges Duhem's thesis: doesn't the basic heliocentric hypothesis on its own, 'in isolation', entail those phenomena? This is a general feature of the sort of case I am trying to characterise: the way that the confirming phenomenon 'drops out' of the basic theory appears to be so direct that scientists are inclined to talk of it as a direct test of just the basic theory, in contradiction to Duhem's thesis. But we can see that, however tempting this judgment might seem, it cannot be literally correct.

No theory T, taken 'in isolation', can deductively entail any result e, if there is any assumption A which is both self-consistent and consistent with T and yet which together with T entails not-e. So in the case we are considering, if the basic Copernican theory alone entailed stations and retrogressions, then there would have to be no possible assumption consistent with that basic heliocentric claim that, together with it, entailed that there would be no stations and retrogressions. But

[21] This is often thought of as the archetypically *ad hoc* move (epicycles are almost synonymous with ad hoccery). However the Ptolemaic move does produce an independent test (and indeed an independent confirmation) but not one that, so far as I can tell, was ever recognised by any Ptolemaist. It follows from the epicycle-deferent construction that the planet must be at the 'bottom' of its epicycle and hence at its closest point to the Earth exactly at retrogression. But this, with other natural assumptions, entails that the planet will be at its brightest at retrogression—a real fact, that can be reasonably confirmed for some planets with the naked eye. (Of course even had it been recognised, this test would not have been reason to continue to prefer Copernicus over Ptolemy, since, as will immediately become apparent, the former too entails—in an entirely non *adhoc*—way that the planet is at its nearest point to the Earth at retrogression.)

[22] See, for example, Kuhn (1957)

there *are* such possible assumptions. Suppose for example that the earth and Mars are orbiting the Sun in accordance with Copernicus's basic theory. Mars happens, though, to 'sit' on an epicycle, but only starts to move around on that epicycle when the Earth is overtaking Mars and does so in such a way as exactly to cancel out what would otherwise be the effects of the overtaking (that is, the station and retrogression). Of course this is a monstrous assumption—but it is both internally consistent and consistent with the basic heliocentric view. The existence of this assumption implies that, contrary to first impressions, Duhem's thesis is not refuted in this case: the heliocentric hypothesis *alone* does not entail the phenomena.

However those first impressions and the monstrousness of the auxiliary necessary to 'prevent' the entailment of stations and retrogressions both reflect just how 'natural' the extra assumptions are that are necessary for heliocentricism to entail the phenomena. All that needs to be assumed, in addition to the basic idea that Mars and the Earth are both orbiting the sun, is that the Earth (which has an observably smaller average period) moves relatively quickly round its smaller orbit and hence periodically 'laps' Mars. (Many philosophers—including both Duhem and Quine themselves—have been overimpressed by Duhem's arguments. There is nothing in those arguments that favours 'holism' in any serious sense, nor that contradicts the idea that some predictions require *fewer* auxiliary assumptions than others.)

A similar case is again provided by the classical wave theory of light. Fresnel's account of diffraction is so natural within the context of the general idea that light consists of periodic motions transmitted through an elastic medium, that he was led to suggest that no auxiliary assumptions are involved:

> I am … going to show that one can give … a general theory [of diffraction] within the system of waves *without the aid of any secondary hypothesis,* by depending on the Huygens principle and that of interferences, which are one and the other consequences of the fundamental hypothesis. (Fresnel 1819, pp. 282-3; my translation and emphasis.)

However, without going into the details, the same message applies here as in the heliocentric theory. The 'direct test' or 'no auxiliary needed' view cannot be literally correct but it is easy to see why Fresnel claimed it was—the great plausibility of the claim reflects the naturalness of the auxiliary assumptions that were in fact necessary.

So in summary, the *real* heuristic view of confirmation that emerges from this consideration of objection one and that I want to defend is as follows:

Two types of confirmation need to be distinguished. *First* a purely intra-paradigm or intra-research programme judgement—e supports specific theory, T′, *relative to* a given general theoretical background T. The most straightforward case is where e, in conjunction with T, deductively entails T′. Even manoeuvres that are patently *ad hoc* (in the pejorative sense) produce specific theories that are confirmed in this (ineliminably) conditional sense. The *second* type of confirmation, unlike the first, produces support not only for the specific theory that entails the phenomenon at issue, but also for the general underlying theory which does not. There are in turn two cases in which this second type—call it 'unconditional support'—is produced: (i) cases of independent evidence (e entails T′ modulo T, but then T′ turns out also to predict e′ which is experimentally verified) *and* (ii) cases where e 'drops naturally out' of T (or, if you like, where the T′ that really entails e is the 'natural version' or 'natural extension' of the underlying general theory T).

I am confident that this dual account of confirmation captures all the intuitive judgments that have been cited in this debate, *both* those used to support the heuristic account or 'no double use' rule in its original formulation *and* those used by critics of that view as originally formulated. Is that all that can be said in its favour or can the heuristic account also be given a plausible general rationale?

The justification of the first (conditional) sort of judgement of confirmation is surely straightforward. If e deductively entails the specific theory T′, given the more general theory T, then e confirms T′ for anyone who already accepts T in the clear sense that it supplies conclusive reason for also accepting the more specific theory T′ (and, in cases of 'near deduction', e supplies a very strong reason for accepting T′, given that the background general theory T is already accepted). This first sort of confirmation in a clear-cut way 'passes the confirmational buck': e, in these cases, demonstrates that you ought to have exactly as much (or, in the 'near deduction' case, almost as much) confidence in T′ as you have in T (despite, of course, the greater content of T′). From outside the 'paradigm', this sort of confirmation shows that T and T′ are, given the evidence e, epistemically inseparable—they stand or fall together.

As for the justification of the second, unconditional and hence more powerful, sort of confirmation, here, for all philosophers' fancy talk, we are, I think, just thrown back on the basic, intuitive 'no miracles' consideration (despite feeling its force, I have always thought that 'no miracles *argument*' was an overly flattering description). The two types of case—of independent evidence and evidence that 'drops out of the basic idea'—that are identified by my account as producing this type of confirmation are exactly the sorts of case that elicit the no-miracles response: 'surely it would be a miracle if the theory could have such evidence in its favour and yet be somehow entirely off-beam?' We are, of course, from the point of view of deductive logic, as always with ampliative inference, committing some version of the fallacy of affirming the consequent. That is why we need to be circumspect about the conclusion to be drawn. This conclusion should not, of course, be that the theory is true (the history of science would soon put paid to that conclusion), nor yet I think, even in an intuitive sense, 'approximately true', but rather 'along the right lines'—probably destined to have its structure preserved, perhaps in approximate or limiting case form, in later successful theories. I do not claim that this is much of a justification; I do believe that it is the only justification we can ultimately give for any account of the confirmation of theories by evidence.

7. OBJECTION TWO AND THE RESPONSE TO IT: SAME THEORY, SAME EVIDENCE, SAME CONFIRMATION

The objection to the 'heuristic' account raised by Alan Musgrave himself—now 'objection two'—was, remember, that the account is unacceptably investigator-relative.

Reformulated to take account of the distinction that I have just now emphasised, the objection goes as follows. Two scientists, A and B, employ two different methods of construction—A uses evidence e, B does not; nonetheless A and B still arrive at the very same theory T; when that theory 'turns out' to entail e, e will confirm—in the strong unconditional sense—theory T as constructed by scientist B, but not as constructed by scientist A (who will, instead, obtain only the conditional sort of confirmation from e). But this, so the objection goes, is surely ridiculous—if they arrive at the very same theory then surely that theory ought to receive the same confirmation from any piece of evidence including e, independently of the way the theory was arrived at. Hence, since the account has a ridiculous consequence, it cannot be correct.

How could the two-scientist story that underlies the objection ever in fact be realised? It cannot be emphasised sufficiently that 'means of construction' is, in the mature sciences at least, *not* a personal notion—finding out about it does *not* require combing through a scientist's personal diaries and the like. It depends instead on the research programmes involved. And these programmes can be articulated and objectively assessed.

The most straightforward way in which two different scientists might take different routes in trying to develop a theory for the same field of phenomena is in fact by pursuing two different research programmes: Biot tried to develop a corpuscularist account of diffraction, Fresnel a wave account; the Ptolemaists tried to develop a geostatic account of observed planetary motions, Copernicus a heliostatic one; and so on. But of course no pair of scientists can possibly arrive at the *very same* theory in such ways (though they might very well, of course, arrive at two different but empirically equivalent theories). The specific theory that scientist A arrives at will of necessity entail the general, 'hard core' theory underlying her research programme, while the specific theory that scientist B arrives at will equally entail the general hard core theory underlying his different research programme—the hard cores of rival programmes are, by definition, inconsistent and so, therefore, are the two specific theories.

This 'two scientist' story, then, can only start to make sense if A and B are working within the same research programme. Again it is important to realise that there is no significant subjective element here: a research programme either supplies a theoretical reason for parameter to have a particular value or it does not. It is, for instance, just a fact that the wave optics programme, to take again my favourite example, supplies no general theoretical reason to fix the parameter corresponding to the wavelength of light from a sodium arc at any particular value (at least within a wide range). A more extensive (but in the 19th century, of course, unavailable) theory involving that wave theory but also an account of the radiation of light from particular sources with particular chemical constitutions, and subject to particular inputs of energy, might conceivably have done so, but the wave theory of light itself—objectively—just does not. Hence no 19th century scientist could see a theoretical justification for taking some particular value of the parameter, and all such scientists needed instead to use the results of experiments to fix the value (or take a blind guess—see *below*). Such a scientist could not have seen a theoretical justification for a particular value of that parameter, because there was no theoretical justification to be seen.

If both scientist A and scientist B work systematically in such a case, then both would need to use data in order to arrive at their more specific theory—it couldn't be that A, say, used data but B purely theoretical considerations in arriving at the same theory, since there are no such theoretical considerations to be considered. The only way that the two-scientist story could get going in such a case would be if one of them, A say, made a blind guess at the value of the parameter left free by theoretical considerations and yet happened, by simple good fortune, to hit on the very same value that B arrived at systematically by using data e. Each starts from the same general theory T, each arrives at the same more specific theory T′, though by different routes. Is not the 'heuristic' approach then forced into the absurdity that T′ as arrived at by systematic scientist B fails to be supported in the stronger sense by the evidence e, while that very same theory as arrived at by unbelievably lucky A *is* supported in that stronger sense by e?

This, admittedly wild, possibility is one that used to exercise Peter Urbach.[23] Once it is realised, however, that we are not appraising scientists but rather theories-in-the-context-of-research-programmes, then any apparent awkwardness here evaporates. The stronger sort of confirmation that I have highlighted is the sort that spills over from the specific theory that entails the relevant data to the underlying general theory or programme. The chief practical impact of such confirmation is to supply confidence in the successful extendability of that same general idea to a different (sub-)field (which will of course mean constructing a different specific theory T′′). Clearly lucky A in the above case has not shown anything relevant in this regard. She has *not* shown that the underlying general idea deserves this sort of support from e, since she has not shown that there are theoretical considerations attached to that general idea tying down the relevant parameter to the particular value, that she merely (and with quite incredible good luck) conjectured. The correct judgement is surely the one supplied by my dual account: (i) that B has shown, while A has not, that T′ is maximally confirmed by e in the conditional sense: B has shown that, since e entails T′, modulo T, if you accept the general theory T you must accept the more specific theory T′; while (ii) A has *not* shown that e supplies 'unconditional' 'stronger' support for T′ in the sense that would spill over to the underlying T. Of course, if it turns out that T′ also yields further so far unconsidered (though actual) data e′, then e′ (unlike e) does provide this stronger unconditional sort of support and it supplies it for T′ as proposed by either A or B. Of course it does: A and B have proposed the same theory!

So far we have considered the case where the underlying research programme gives the theorist no reason why a particular parameter should have a particular value and hence she needs, if she is to work systematically at all, to invoke data. Suppose now, to the contrary, that there *is* a theoretical justification, provided by the research programme concerned, why some parameter λ should have a particular value, but scientist A fails to see that reason. A instead uses evidence e to tie down

[23] See his (1978).

λ, at, say, the value λ_0; thus producing a theory, $T(\lambda_0)$, that, in turn, entails some further, initially unconsidered, evidence e′. Scientist B, on the other hand, sees that her research programme already supplies a theoretical reason why λ should have the value λ_0 and goes *directly* to $T(\lambda_0)$, pointing out that it entails *both* e *and* e′. I cannot see any reason why this scenario couldn't be realised, though I am doubtful that there are any real historical examples. However, if there are such examples, what we ought to say about them again seems entirely straightforward: *not* that the theory as proposed by A is supported (in the stronger, general theory or research programme supporting sense) only by e′, while the very same theory as proposed by B is supported (again in that stronger sense) by both e and e′; instead we would say that scientist B has *shown*, what A simply subjectively failed to recognise, that the theory is supported in this stronger sense by both e and e′. Once it has been realised what the different support judgments I have highlighted are doing—giving merely conditional support against the background of a presupposed general theory or, more interestingly, giving support to that more general theory itself (though *via* a specific representative of it)—then any apparent mystery in this sort of case too disappears.

Could there be, objectively speaking and laying aside random guessing, more than one route within a research programme to the same theory? And would the existence of such multiple routes pose any threat to the theory of support that I have outlined?

I can think of only one such way. And this is an entirely benign case that has already in fact been mentioned. Quite often with powerful scientific theories (as, for example, in the simple wave-theoretic case I sketched above involving the determination of the wavelength of monochromatic light from a particular source) experiments *overdetermine* the value of that parameter in the following sense. The general theory, in this example the general wave theory in which the values of all wavelengths of monochromatic light-sources are free parameters, entails not just one, but a range of formulas, involving the wavelength and measurable quantities *in different experiments*. So for example, alongside the equation cited above linking the wavelength to measurable fringe-distances in the two-slit experiment, the general theory entails another equation linking that wavelength to measurable fringe-distances in the one-slit diffraction experiment. This *does*, then, admit a genuine scientist A/scientist B scenario: A might produce T′ out of T in the way outlined earlier, using the result of the two slit experiment with monochromatic light from a sodium arc to fix the value of the wavelength, and then use T′ to predict the exact outcome of the one slit experiment with light from the same source in quantitative, rather than merely qualitative terms; while Scientist B on the other hand might produce what turns out to be the very same theory T′ on the basis of the result of the one-slit experiment with light from a sodium arc and then use it to predict the quantitative details of the two-slit experiment. (Of course the fact that it turns to be the very same theory is a contingent fact reflecting the predictive power of the wave theory. Scientist A using the two-slit data might have produced T′, while B's use of the one-slit data led to the different T′′ (in fact inconsistent with T′). This would mean that A's theory failed the one slit diffraction test, while B's failed the two-slit test.)

Is this really a problem for the dual account of confirmation I have sketched? Let's call the two-slit fringe data with light from a sodium arc e_1 and the one-slit fringe data using the same light-source e_2. Telling it from the point of view of scientist A, e_1 confirms T' in the conditional sense (it entails T' given the general theory T), while e_2 confirms T' in the stronger sense that spills over to T; from the point of view of scientist B, on the other hand, the roles of e_1 and e_2 are reversed: e_2 confirms T' conditionally, while e_1 supplies the stronger T-involving confirmation. These may be strictly different accounts but they are surely equivalent *modulo* any genuine interest that we would have in making confirmation judgments: each of A and B has shown that the general theory needs to fill in one parameter value on the basis of one piece of data, thus producing a specific theory that gains genuine empirical success from the other piece of data (at least—there may of course be other results that specific theory also correctly predicts). So each scientist shows that there is, so to speak, one unit of genuine, unconditional, general-theory-involving data and hence delivers the judgment that that general theory is ahead in terms of empirical support of any theory (such as the rival emissionist theory in the early 19th century) that *merely* accommodates both pieces of data. (On the other hand if, as was not the case historically, there were still a third theory which, without needing either e_1 or e_2 to fix parameters, entailed both of them 'naturally'—in the way that Copernican theory entails planetary stations and retrogressions—then that third theory would be even better confirmed. Intuitively we would want to say that the score—relative of course to just e_1 and e_2 (judgments might be different in view of the *total* evidence)—would be 'Imaginary theory 2, wave theory 1, emission theory 0'; and this is exactly the score that is delivered by my dual account of confirmation.)

In sum, then, objection two fails: contrary to Alan Musgrave's claim, confirmation is not unacceptably inquirer-relative on the approach that I endorse. It is clearly a desideratum on any account of confirmation that it underwrite the judgement 'same evidence, same theory, same confirmation' and my account underwrites exactly this judgment. (This is in contrast, of course, to the judgement 'same evidence, two *rival but empirically equivalent* theories, same confirmation', on the denial of which this whole approach is based.)

Is the 'heuristic' account, when properly understood, a version of the historical approach?

According to Musgrave's classification, all the accounts that make confirmation dependent, not only on theory and evidence, but also on background knowledge, for that reason make confirmation (at least partly) 'historical'. This is because background knowledge may change over time and so the answer to the (in fact elliptical) question 'does evidence e support theory T?' may be different in different eras—eras that are characterised by different states of background knowledge.

When properly understood, however, the 'heuristic' view I advocate does not have this historical character. It does, certainly, make confirmation a three-, rather than two-place relation. But, although describable in a loose way as making confirmation dependent on background knowledge, in fact this account makes confirmation (or rather *both kinds of confirmation*) depend on evidence e, specific theory T′, and the underlying general theory T. It is not a history-dependent, but rather a research programme-dependent account:

(1) Evidence e confirms$_1$ T′ in the context of the general underlying theory T if the conjunction of e and T entails T′ (or more generally to the extent that e and T entails T′);

while:

(2) Evidence e confirms$_2$ T′ in the context of the general underlying theory T if (i) T′ entails e, and (ii) T′ has been developed out of T in a way independent of e.

There is no question, then, of historical variability in either of the types of confirmation-judgment. Fresnel's wave theory of diffraction was, is, and forever shall be, confirmed (confirmed$_2$) by the 'white spot' result—this result follows from that wave theory of diffraction and gives support to the whole wave programme. Fresnel's specific claim that light waves are transverse rather than longitudinal is, was, and forever shall be, confirmed (confirmed$_1$) by the disappearance of the fringes in the two slit experiment when the two beams are oppositely polarised—this result did, does and ever more shall entail the specific transverse claim given the general idea that light is a wave in an elastic medium.

Of course the historical context changes, because other theories are articulated. Hence the question of whether only one theory of light is confirmed$_1$ by the white spot result may (and indeed of course did) have one answer in 1819 ('yes, Fresnel's') and another answer ('no') in the 1860s, once Maxwell's electromagnetic theory had been formulated. Hence the issue of whether some result provides grounds for accepting a theory as the currently best available in its field quite properly, and obviously, has an historical dimension. But, as I argued earlier in considering Alan Musgrave's own preferred version of the historical approach, it would surely be a mistake to confuse these patently historical issues with the ahistorical one of whether some theory is confirmed by some piece of evidence. The main conclusion of this paper is that there are two types of confirmation—both of them (three-place) 'logical'. These confirmation judgments then feed into the clearly historical issue of which currently available theory is *best* confirmed by the currently available evidence.

8. DEBORAH MAYO AND CONFIRMATION VIA 'SEVERE TESTS'

Finally, I want to try to work out an issue that has troubled me for some time—namely, the relationship between my account and the much-discussed views of Deborah Mayo. She—again taking 'Musgrave's neat analysis of the situation' (*op.cit*, p. 255) as one of her starting points—has developed a theory of confirmation that, amongst other things, claims to account both for the cases in which the heuristic account of confirmation accords with our intuitive judgments about particular cases of confirmation and for those cases in which the heuristic account conflicts with those judgments. Mayo in effect takes the heuristic account to be captured by the 'use novelty' or 'no double use' slogan: 'You can't use the same fact twice, once in its construction and then again in its support.' The view that I have been developing here, as we saw, also rejects this slogan in its straightforward interpretation: claiming instead that you can indeed both use a datum in the construction of a theory and use it to support the constructed theory in the sense of support that is conditional on pre-acceptance of the underlying general theory, but that used data cannot support in the stronger, unconditional sense that spreads from the specific theory that entails the data to general theory underlying that specific theory. Intuitive judgments that were in conflict with the 'no double use rule' are in fact judgments of conditional support; intuitive judgments that conform to that rule are judgments of the stronger, unconditional kind of support.

Both Mayo and I, then, claim to capture the intuitively underwritten judgments in all particular cases of confirmation or support—both those that have been cited in favour of the 'no double use' rule and those that have been cited as refutations of that rule. What, then, is the relationship between Mayo's account and my own: which is better, or are they perhaps just two different ways of saying the same thing?

Mayo's basic line of reasoning is very simple. Hypotheses should gain empirical credit only from passing genuine tests; and the more severe the test, the higher the confirmation or support, if the theory passes it. The defenders of the use-novelty account hold that evidence used in the construction of a hypothesis cannot provide a genuine test of it and hence cannot supply genuine confirmation. Underlying their view, on Mayo's analysis, is the claim that a severe test is one that a theory has a high probability of failing; and hence, since a theory constructed with the help of evidence has no chance of failing the 'test' supplied by e, that the use-novelty view is correct. However plausible this may sound, argues Mayo, it in fact misidentifies the probability that we should be concerned to maximise: a non-severe test is *not* one that has a high probability of being passed by a theory, but rather one that has a high probability of being passed by the theory, *even though the theory is false*. As she puts it 'what matters is not whether passing is assured but whether erroneous passing is' (*op. cit.*, pp. 274-5).

In cases where the heuristic view as originally formulated goes wrong—such as her SAT score 'hypothesis' and standard statistical estimation (where we use, for example, the evidence of the observed relative frequency of white balls in a sample to arrive at a theory about the unknown frequency in the urn)—there is indeed no chance of the (constructed) hypothesis failing the 'test'. However, the chance of the hypothesis having passed the 'test' *if it were false*, is zero (in the case of the SAT score example) or very small (in the confidence-interval case). Concentrating on the more straightforward SAT score case, the 'test' of the 'hypothesis' that the average SAT score of her logic students is 1121—the "test" consisting of taking the individual SAT scores and dividing by the number, n, of students (thus producing evidence e)—is in fact *maximally severe*, according to Mayo: 'since there is no way that such a result can lead to passing H erroneously, H passes a maximally severe test with e.' (*op.cit.*, p. 271)

Part of my response will involve justifying the scare quotes I have placed around 'hypothesis' and especially around 'test' in outlining her view; this will in turn lead to the criticism (which Deborah Mayo herself cites and tries—unsuccessfully—to meet) that the SAT score and statistical estimation cases are not representative of the interesting cases from science. While there is no doubt that Mayo's account and my own are based on a number of shared views and intuitions, my account gets the situation straight whereas her own is somewhat skewed.

As already remarked, it does seem extraordinary to call the assertion arrived at about the average SAT score of Mayo's students an 'hypothesis', and at least equally extraordinary to call the process of adding the individual scores and dividing by the number of students a 'test' of that claim. Of course had someone made a 'bold conjecture' about the average score, then one might talk of the systematic process of working out the real average as a test of that conjecture. But boldly conjecturing would clearly be a silly way to proceed in this case, and, as already remarked, not one that would ever be used in science. As it is, the process of adding the individual scores and dividing by the number of students surely is a *demonstration that* the average score is 1121, not a *'test' of* the 'hypothesis' that this is the average score.

This also points to a real problem in applying Mayo's central justification for all confirmation judgments to this particular case. In the circumstances (and assuming that both the data on the individual students and the arithmetic have been carefully checked) there is *no* chance that the average SAT score is *not* 1121. So we are being asked to make sense of a conditional probability—the probability that the claim about the average score would have passed the test, had it been false—where the conditioning event (the claim's being false) has probability zero; and indeed asked not only to make sense of it but to agree that the conditional probability at issue is itself zero. It is well known, however, that—at any rate in all standard systems—$p(A/B)$ is not defined when $p(B) = 0$. It is true that Mayo wants us to concentrate primarily on intuitive judgments about 'probability' and not on what can formally be justified as genuine probabilities. However I confess that I have no idea what it means in this case, even 'intuitively', to imagine that the average score is *not* 1121, when the individual scores have been added and divided by n and the result *is* 1121!

There is not the same formal difficulty of course in the statistical estimation case, where we can readily make sense of the probability that the estimate is wrong (that is, that the interval systematically arrived at on the basis of the sample data does not in fact include the real population value of interest). However it is intuitively quite wrong to talk of 'tests' in this case too. In the deterministic case, we *measure* a parameter (or demonstrate that that parameter has a certain value); in the stochastic case we *estimate* a parameter. Although apostate, I remain enough of a Popperian to put very little weight, in general, on how we happen to talk, but there seems to be in this particular case a very good reason why we do not talk of tests: despite Mayo's claims, a test of a theory surely must have a possible outcome that is inconsistent with the theory— neither the SAT score process nor the confidence-interval technique could possibly refute the 'theory' that we end up with.

As should be clear from my own positive account, I am far from disputing Deborah Mayo's claims that both measurements and estimations of the value of parameters form important aspects of scientific reasoning. I also agree with her in particular that statistical inference from actual experimental data to the claim that we normally regard as 'the' (generalised) result of the experiment is an important, and relatively underexplored, aspect of the logic of science. However these are *not* aspects of any testing-process in science. The lesson to learn, contrary to Mayo's general view, is that science is not all about tests of theories and so not all about attempts to detect error; some of the important logical relationships between evidence and theory are of a quite different nature. Mayo gets herself into trouble by attempting to produce a 'one size fits all' account—all (let's say) *accreditations* of theory by data are, she claims, the results of tests, once tests are properly construed (that is, construed in line with her account, of course).

The problems that this approach leads to are made still clearer when we consider cases that are more representative of reasoning in science. The feature of the SAT score case that makes it unrepresentative, as already indicated, is that there just is no genuine theory around—the framework is simply given, in particular the relationship between the individual scores and the average score is analytic. In the statistical estimation case, there *is* an underlying 'model'—when drawing balls from the urn, for example, we are assuming that it is Bernouilli process with underlying fixed population parameter p (the fixed proportion of white balls in the urn), but this underlying model is not itself usually thought of as at all conjectural. (We can't see inside the urn, so it *might be* that some demon is constantly changing the proportion of white and black balls—but we just assume that this is not the case and don't look for any experimental confirmation of this assumption.)

The interesting scientific cases of 'deduction from the phenomena', as indicated earlier, on the contrary, all involve a general underlying *theory*. In the simplest case, a specific value of a parameter is deduced from the data, but only *given* an underlying general theory that yields, without any experimental input, some functional relationship between the free version of that parameter and experimental results. This underlying theory, although it may be assumed as 'given' for the then current purposes, is itself clearly a substantive and defeasible assumption and as such stands in need of confirmation from evidence no less than the specific theory deduced from it and the phenomena. The (general) wave theory of light replaced the (general) emission theory and was itself then replaced by the electromagnetic, and later photon theory. The general theory's fortunes are subject to the changing verdict of ever accumulating evidence—we need to take its defeasibility into account.

Consider the case, analysed in detail above, where a specific version of the wave theory T′ with a definite value for the wavelength of light from a sodium arc is deduced from the general wave theory T using evidence e from slit- and fringe-distances in the two slit experiment with light from that source. My account entails that e does confirm T′ in this case—strongly, but in the conditional sense. What does Deborah Mayo's account say? In line with her claims about the SAT score case she too will want to say that e gives some good degree of confirmation to T′ (these are exactly the sorts of *real* scientific cases where the unmodified 'no double-use' rule goes wrong). She will be forced to say that this is because e constitutes a pass for T′ in a test that had relatively little chance of passing it if it were false. However, it is surely clear that this 'test' in fact had every chance of passing T′, *whether it is true or not*. Whatever general theory of light is true, that is, whether or not T is true (and T′ can't be true if T is not) T′ was indeed *bound to* get the fringe distances in the two-slit experiment with sodium light correct—exactly because T′ was fitted to e!

The correct judgment is surely the one delivered by my account: e constitutes no test of T′ but it does tell us something positive about it—namely, that it is *the* specific representative of T (so far of course as this particular detail is concerned). If T′ is not correct (and we have, to repeat, no chance of finding that out from the two slit experimental result though we might, as explained, from other experiments), then neither can the general theory T be correct. These further experimental results (such as the prediction of the outcome of the one-slit experiment with light from the same source) *are,* on the contrary, genuine tests. Genuine tests produce the stronger, non-relativised type of confirmation my account talks about (and it is here that Mayo's 'error probability' intuitions —about it being improbable that the theory should get such a test result right if it is, not false, but something weaker like 'structurally off-beam'—come genuinely into play). Parameter-fixing exercises and other such inferences, on the contrary, are important (indeed crucial) aspects of the scientific endeavour and they carry important information about theories, but they are not *tests* of those theories.

There are hints of possible responses to this criticism in Mayo's book (especially when she discusses the problems posed by the possibility of alternative theories, really alternative theoretical frameworks). But even without going into details it seems clear that her approach is in general barking up the wrong tree. We have here two *quite different* roles for evidence *vis à vis* theory, just as my approach implies. This is exactly why my approach yields two quite different notions: confirmation$_1$ and confirmation$_2$. They are not, as Mayo is trying to make them out to be, simply two different aspects of the one drive—to test theories in, and hence to eliminate error from, science.

REFERENCES

French, A. (1971) *Newtonian Mechanics*. Cambridge, Mass: M.I.T. Press.

Fresnel, A.J. (1819) 'Mémoire sur la diffraction.' reprinted in E.Verdet et al. (eds) (1865).

Hitchcock C. and Sober, E. (2004) 'Prediction Versus Accommodation and the Risk of Overfitting.' *British Journal for the Philosophy of Science* 55: 1-34.

Howson, C. (1990) 'Fitting Theory to the Facts: Probably Not Such a Bad Idea After All.' in C. Wade Savage (ed) *Scientific Theories*. Minneapolis: University of Minnesota Press.

Howson, C. and Urbach P. (1994) *Scientific Reasoning: the Bayesian Approach,* 2nd Edition. Chicago and La Salle: Open Court.

Kuhn, T.S. (1977) *The Essential Tension*. Chicago: University of Chicago Press.

Mayo, D. (1996) *Error and the Growth of Experimental Knowledge*. Chicago: University of Chicago Press.

Musgrave, A. E. (1974) 'Logical versus Historical Theories of Confirmation.' *British Journal for the Philosophy of Science* 25: 1-23.

Nickles, T. (1987) 'Lakatosian Heuristics and Epistemic Support.' *British Journal for the Philosophy of Science* 38: 181-205.

Urbach, P. (1978) 'The Objective Promise of a Research Programme.' in Radnitzsky and G. Anderson (eds) *Progress and Rationality in Science*. Dordrecht: Reidel.

Verdet. E. & Senaramont H. (eds) (1865) *Oeuvres Complètes d'Augustin Fresnel.*
Worrall, J. (1985) 'Scientific Discovery and Theory-Confirmation.' in J.C. Pitt (ed): *Change and Progress in Modern Science*. Dordrecht: Kluwer.
Worrall, J. (1989) 'Fresnel, Poisson and the White Spot: the Role of Successful Prediction in Theory-Acceptance.' in D. Gooding et al (eds) *The Uses of Experiment—Studies of Experimentation in Natural Science*. Cambridge: Cambridge University Press.
Worrall, J. (2000a) 'Kuhn, Bayes and "Theory-Choice".' in R. Nola and H. Sankey (eds) *After Popper, Kuhn and Feyerabend.* Dordrecht: Kluwer.
Worrall, J. (2000b) 'The Scope, Limits and Distinctiveness of Newton's Method of "Deduction from the Phenomena".' *British Journal for the Philosophy of Science* 51: 45-80.
Worrall, J. (2002) 'New Evidence for Old' in P.Gardenførs et al (eds) *In the Scope of Logic, Methodology and Philosophy of Science*. Dordrecht: Kluwer.

DEBORAH G. MAYO

CRITICAL RATIONALISM AND ITS FAILURE TO WITHSTAND CRITICAL SCRUTINY

PART I: THE SEVERE TESTING PRINCIPLE IN THE CRITICAL RATIONALIST PHILOSOPHY

1. INTRODUCTION

> Observations or experiments can be accepted as supporting a theory (or a hypothesis, or a scientific assertion) only if these observations or experiments are severe tests of the theory—or in other words, only if they result from serious attempts to refute the theory, and especially from trying to find faults where these might be expected in the light of all our knowledge. (Popper, 1994, p. 89)

The lack of progress in the neo-Popperian philosophy known as 'critical rationalism' may be traced to its inability to show the acceptability of the fundamental principle underlying the above quote:

Severity Principle (SP) Data **x** count as evidence in support of a hypothesis or claim H, only if **x** constitute severe tests of H—only if data **x** (which are in accord with H) result from serious attempts to refute H.

This failure seems deeply puzzling, given the intuitive plausibility of SP, as in Popper's exhortation above. The problem is hardly limited to critical rationalists. Something like SP is endorsed far more generally in philosophy as well as in science, and yet it has been notoriously difficult to actually cash out what 'surviving serious criticism' demands, and why H's surviving the 'ordeal' is good evidence for H. My focus here is on critical rationalists, and in particular on Alan Musgrave's recent (1999) attempt.

What gives SP its plausible-sounding ring is the supposition that 'H's surviving serious criticism' is being used in the way it is ordinarily meant: roughly, that H has been put to a scrutiny that would have (or would almost certainly have) uncovered

C. Cheyne & J. Worrall (eds.), Rationality and Reality: Conversations with Alan Musgrave, 63–96.

the falsity of (or errors in) H, and yet H emerged unscathed, i.e., that H has survived a highly reliable probe of the ways in which H might be false. However, critical rationalists, as they freely admit, do not have resources to articulate anything like 'reliable error probes', and even deny the reliability of the method they espouse. Despite exhortations as in the epigraph from Popper, critical rationalists only espouse a weaker, *comparativist principle CR*:

(CR) It is reasonable to adopt or believe a claim or theory P which *best survives* serious criticism.

But without being able to say that surviving the critical rationalist's actually affords evidence for P, a 'best surviving' claim may still have been very poorly probed, and thus P may be 'best tested' with **x**, even though **x** actually provides scant evidence for P at all.

So, while we may (and most of us do) accept the intuitive principle that CR is supposed to capture (namely the severity principle SP), we have yet to be given grounds to accept CR as instantiating the intended severity requirement. To simply declare CR is a reasonable epistemic principle without giving evidence that following it advances any epistemic goals is entirely unsatisfactory, and decidedly un-Popperian in spirit. So it does not help for the Popperian to insist 'there is no more rational procedure' than to prefer a hypothesis that is well-corroborated, i.e., that has withstood serious or severe criticism (Popper 1962, p. 51), without demonstrating the existence of testing methods that are actually severe. Yet, far from demonstrating the existence of severe error probes (or whatever one wishes to call them), the critical rationalist feels bound to deny that tests that are severe in the critical rationalist's sense are reliable tools for uncovering errors. The critical rationalist is thus guilty of a kind of 'bait and switch', getting our nod for plausible sounding exhortations, as in SP, but then serving up, not the robustly severe tests we thought we were getting, but 'tests' incapable of doing their intended job.

Granted, Popper invites this problem, due in part to his efforts to distinguish himself from the 'inductivists' of the time. The deductive resources to which Popper limited himself allows neither substantiating a claim to actually *have* a severe test or error probe, nor to say that the probability of P's passing test T is low, given P is false. Now that we know so much more about conducting severe testing in experimental practice than was evident through logical-empiricist blinders, one would have expected this weakness to be remedied by Popper's critical rationalist followers. Surprisingly, it has not been. Even so astute a thinker and upholder of common sense as Alan Musgrave (1999) has recently mounted a defence of CR that he openly concedes is circular, admitting, as he does, that such circular defences could likewise be used to argue for principles he himself regards as 'crazy'. (Why is CR a good rule? Because it is a good rule.) As if this were not bad enough, it turns out we cannot even self-referentially apply rule CR, i.e., we cannot show that CR itself is a 'best tested' rule, because it is demonstrably unreliable (and other methods are not). While Musgrave's full argument is subtle and clever, these concessions, or so I shall argue, radically undermine his goal, as they render his argument no argument at all.

I will expose the series of missteps that have landed the critical rationalist in this untenable position. In Part I, I will argue that the critical rationalist arguments, as urged by Musgrave, themselves rest on the ability to distinguish severe from insevere tests and reliable from unreliable error probes, and thus are self-contradictory when denying the possibility of doing any such thing. In Part II, I will show how to rectify the situation by (a) rejecting the erroneous conceptions of inductive or 'evidence-transcending' inference upon which their sceptical slide is based, and (b) showing how to develop an account with the resources to define and apply severe or reliable error probes.

The main points for which I will be arguing are these:

1. An adequate defence of CR must characterize 'withstanding severe or serious scrutiny', and show it corresponds to classifying claims reliably, which neither Popper nor current day 'critical rationalists' have done.

2. Musgrave's argument that all epistemic principles can only be defended circularly, if they are defended at all, is unsound, and confuses 'self-subsuming' methods with 'self-sealing' (circular) methods.

3. In distinguishing 'crazy' and 'non-crazy' methods, Musgrave must assume a reliable classification scheme, which, if drawn out, already goes several steps further than what is alleged by the critical rationalist.

4. Critical rationalists assume falsely that justifying claim P requires either showing it to be true or probable.

5. A satisfactory articulation of withstanding a severe test can achieve the intended goals, without illicit 'justificationist' or metaphysical inductive appeals.

2. BETWEEN SKEPTICISM AND IRRATIONALISM: THE WEDGE IS NOT ENOUGH

The reason that there are only a dozen odd self-styled 'Popperians' in philosophy of science these days, Musgrave ventures, is that 'Popper's chief contribution to philosophy has still not been understood' (p. 314), in particular, philosophers of science have failed to appreciate 'Popper's critical rationalism and the solution to the problem of induction which it contains.' He sets out to rectify this. The secret, in Musgrave's view, is to appreciate fully the way critical rationalism drives a 'wedge' between skepticism and irrationalism (e.g., 1999, p. 322): we can be skeptics about inductively inferring or warranting hypothesis or claim P, but still regard it as reasonable to believe P, or to put it in Popper's locution, to accept or 'prefer' P.

2.1 The Probabilist's View of Justifying Claims is Rejected

What enables this 'wedge' to be 'driven between skepticism and irrationalism' (p. 322), Musgrave thinks, is the critical rationalist's rejection of the traditional justificationist's principle (J) (p. 321):

(J) A's believing that P is reasonable if and only if A can justify P, that is, give a conclusive or inconclusive reason for P, that is, establish that P is true or probable.

Blithely accepting that an 'inconclusive reason' for P must be understood as assigning P a probability, Musgrave touts a philosophy wherein 'P is reasonable' never means having to say there's even an inconclusive reason for P.[1] Later on, I will come back to question and reject this conception of warranting an 'evidence-transcending' claim, but for now we want to trace out Musgrave's reasoning, and his reasoning is this:

By rejecting principle (J), the critical rationalist is free to hold:

(1) it is or may be reasonable to prefer hypothesis P despite the lack of any warrant for the truth of P.

Quoting Popper: 'although we cannot justify a theory…we can sometimes justify our *preference* for one theory over another; for example if its degree of corroboration is greater' (Popper 1976, p. 104), Musgrave proposes to replace Popper's 'justify our preference for' P with 'show the reasonableness of believing P'.[2] So, Musgrave replaces (1) with:

(1′) it may be reasonable to believe hypothesis P despite the lack of any warrant for the truth of P.

Although my own preference is to avoid talk of beliefs altogether (and (1′) is more contentious than (1)), since Musgrave seems prepared to identify belief in P with adopt P as true (p. 327), I will follow his terminology in this throughout Part I.

2.2 Some 'Unsavoury' Wedges

Musgrave cites, as precedents, examples that help to construe the manner in which one can uphold (1′): 'Pascal's wager gives a reason for believing God exists which is not a reason for God's existence. The pragmatic vindication of induction is a

[1] In speaking of P's truth, there is no realist assumption. That P is true or correct may be cashed out in terms of a specific error P asserts to be absent. It may mean that a given claim is adequate in a number of senses: that an assertion about a genuine effect, a causal factor, or a parameter estimate is correct, possibly with margins of error attached (in quantifiable cases).

[2] The use of 'justify' here presumably does not mean show it has a high probability—even critical rationalists have trouble keeping up the linguistic summersaults their defences require.

reason for believing that nature is uniform which is not a reason for the uniformity of nature' (p.322). This suggests a clearer and less contentious formulation of the thesis in (1'):

(1′) we may have information **x** that shows belief in P to be reasonable, even though **x** does not show P to be true or probable.

By upholding thesis (1′), Musgrave argues, the critical rationalist can concede there are no good reasons for inferring an evidence-transcending hypothesis P, while nevertheless maintaining that it is reasonable to believe that P.

In particular, we may have a method or procedure (M) for classifying claims as reasonable or not. That is:

M: P $\rightarrow$ {reasonable, unreasonable}

Associated with each such M is an epistemic principle (EM):

(EM) It is reasonable to believe P if and only if P is classified as reasonable by method M, i.e., iff P satisfies a criterion set out by method M.

We can readily agree with Musgrave that P may be classified as reasonable to believe by method M, even without there being reasons for regarding P as true or probable, while still demanding that the chosen classification method M have some warrant or justification.[3] He himself claims his two illustrative examples 'are unsavoury ones', though one wishes he had explained why. Does he regard them as unsavoury because in each case the reasons for belief are merely pragmatic and do not supply support for the truth of the claims in question (God's existence, nature is uniform, respectively)? That would seem strange, since the whole point of this exercise is to uphold the idea that data **x** may give perfectly good reasons for believing P although **x** fails to supply reasons that P is the case.

In fact, there are features of these 'unsavoury wedges', at least in their intent, that would seem to offer the kind of strategy that would appeal to a critical rationalist. Their linchpin, after all, is their claim to demonstrate that *whether or not the belief in question is true*, there is a payoff attached to adopting a given attitude with respect to P. At times, Popper himself drops hints along these lines in arguing for CR (e.g., 'if we have made it our task…' (1962. p. 51)). Perhaps their unsavouriness, then, is that they fail to ensure the promised payoff (pragmatic or epistemological)?

It is of interest to note that contemporary statistical hypothesis testing, e.g., Neyman-Pearson (NP) tests, exemplify Musgrave's wedge: tests use data **x** to classify statistical hypotheses 'acceptable' or not, without assigning them degrees of probability; however, they will be regarded as good tests only insofar as it can be

[3] Although I will strive mightily to abide by the language the critical rationalist wants us to adopt, I see no reason to share his fear that using the word 'justification' will force me to adopt enumerative induction. For me it is just a synonym for 'warrant'.

shown they very infrequently classify false hypotheses as true (or true hypotheses false), i.e., they must be shown to be reliable in this sense, namely, they have low error probabilities. Tests with high error probabilities are 'unsavoury'. (I return to this in Part II.)

Thus, merely giving us a 'wedge' (between evidence for the reasonableness of believing in P versus evidence for the truth of P) does not take Musgrave very far. We are led to the question of what if any grounds Musgrave provides for the 'belief-adoption' method M championed by the critical rationalist, What grounds are there that CR is not also unsavoury?

3. M_{CR} IS UNRELIABLE

The critical-rationalist position that Musgrave endorses espouses the following method:

(M_{CR}) P satisfies classification method M_{CR} (at time t) if and only if P has best withstood serious criticism (at time t).

The corresponding epistemic principle is that it is reasonable or rational to prefer or believe the comparatively best tested P. That is, the epistemic principle corresponding to method M_{CR}, which we may write as EM_{CR}, is CR, only now he writes it with a time index:

(EM_{CR}) It is reasonable to believe P (at time t) if and only if P has best withstood serious criticism (at t).

The particular 'wedge' offered by CR, then, is this: "if a hypothesis has withstood our best efforts to show that it is false, then this is a good reason to believe it *but not a good reason for the hypothesis itself*" (Musgrave 1999, p. 322).

Even granting the 'wedge', surely the critical rationalist (of the Popperian or Musgrave stripe) wishes to incorporate certain requirements or demands which must be satisfied before it can be said to be reasonable or rationale to believe P, and surely, then, they must regard CR as embodying a method capable of promoting those aims. The question then is: what aims does M_{CR} achieve, such that it makes sense to adopt this principle?

To sum-up this part, we can grant Musgrave's instantiation of (1′):

(1′) we may have information **x** that P is best tested[4] (by method M_{CR}), even though **x** does not show P to be true or probable.

But we deny **x** shows belief in P to be reasonable if it turns out that M_{CR} is unreliable. That is, we insist on:

[4] We substitute in (1′) as Musgrave directs us to: replacing '**x** shows belief in P to be reasonable' with '**x** shows that P is best tested' (i.e., has best survived the critical rationalist's notion of a severe test).

(2) if with high probability method M_{CR} deems P 'best tested' even if P is false, then the passing result **x** fails to show belief in P to be reasonable.

As we will see, method M_{CR} may deem P 'best tested' even if little or nothing has been done to uncover the ways P can be in error, i.e., even if the 'test' would be regarded as having little or no 'severity' at all. If this is so, then M_{CR} may fail its intended task of capturing the severity principle (SP) with which we began.

3.1 CR Fails to Give a Necessary Condition for Reasonable Belief

Remembering that CR is equivalent to our EM_{CR}, we see that the 'only if' in EM_{CR} is false:

(CR⇒) It is reasonable to believe P (at time t) only if P has best withstood serious criticism (at t).

To satisfy the antecedent, it is required only that there be reasons, **x**, to believe P (at t), and by Musgrave's 'wedge', we need not expect that **x** supplies reasons in support of P's truth. So, **x** might be reasons of prudence, pragmatics, or any number of things. Consider for example information **x**:

(**x**) evidence from medical trials shows a high correlation between tolerating the treatment given for disease D and adopting an optimistic belief that one can make disease D vanish by will.

Proposition P is that one can make disease D vanish by will. Evidence **x** may make it reasonable for a patient with D to believe P, even where **x** does not constitute any evidence that P has withstood serious criticism or any kind of criticism. Indeed, P may have failed tests (suppose no patient has ever been able to will disease D away). But it may be prudent to believe P in order to better tolerate the treatment. Or it may be reasonable to believe P if there is evidence **x** that one will otherwise be killed, even where **x** is not evidence that P has survived any kind of criticism or probe of P's falsity. In fact, P may be known to be false.

Thus, CR as an 'if and only if' claim is plainly false: having withstood serious criticism is not a necessary condition for reasonable belief, as Musgrave understands the 'wedge'. But since it is the 'if' claim that seems mostly to be doing the work for Musgrave, we can put this qualm aside for now. However, the 'if' clause, on which Musgrave's argument depends, is also highly problematic.

3.2 Having 'Best Withstood Criticism' is not Sufficient for Reasonable Belief

I will argue that it is also false to claim that:

(CR⇐) If P has best withstood serious criticism (at time t), then it is reasonable to believe P (at t).

It is very important, in evaluating CR to consider, not what we would ordinarily mean by surviving serious criticism, because, as already said, this assumes tests with capabilities that the critical rationalist has no intention of supplying. To begin with, the comparative nature of the rule entitles P to receive the 'best-tested' medal, even if poorly tested, it may be the first ever tested, or slightly less poorly tested than an existing rival! But such a comparative principle of testing is highly unreliable (Mayo 1996). I return to this in Section 10.

In deeming M_{CR} unreliable, I mean that P may be the best-tested so far without P having been probed in the least, and thus it would seem that this does not suffice for it to be reasonable to believe P. (Even if there are other, non-evidential reasons to believe P, e.g., pragmatic considerations, this is no thanks to the antecedent being satisfied.) Why then do critical rationalists settle for comparativist method M_{CR} when principle SP is non-comparative? Presumably, it is felt that the comparatively-best-tested principle is all that can be demanded if the principle is to be applicable. But this just underscores the fact that 'best-tested' in the critical rationalist's sense, need not mean well-tested at all (else the non-comparativist principle SP would be retained).

3.3 Popper on Severe Testing and Corroboration

This comparativism was clearly embraced by Popper. According to Popper, hypothesis P best survives test T with data **x** so long as:

(i) P entails (or otherwise 'fits') **x**

and

(ii) **x** is not predicted, or is counterpredicted, by P's existing rival(s).[5]

As Popper's critics observed from the start (e.g., Gruenbaum 1978) satisfying (ii) does not warrant stronger claims such as:

(ii′) **x** would not be expected were P false

or

(ii′′) there is a low probability of **x**, given that P is false

[5] As Musgrave has elsewhere noted, Popper's condition (ii) may be construed even more weakly, allowing it to be satisfied even if existing alternatives to P say nothing about the phenomenon in **x**. P may pass all the tests that rival(s) P′ do, even if all are silent about certain results.

although at times Popper suggested it did. The reason is that (as Popper was aware) 'P is false' includes the disjunction of all possible hypotheses or claims other than P that would also 'fit' or accord with **x**—the so-called 'catchall hypothesis'—including those not even thought of. Existing data **x** would be just as probable were one of the catchalls true, and P false.

Therefore, P may be the 'best-tested' hypothesis so far, even without P's having been probed especially well at all. So long as P is not falsified, even if no alternative to P exists, P would, on this requirement, count as 'best-tested', or so it seems. But why should it be reasonable to believe in the first hypothesis put forward, say, to account for a phenomenon? Or believe in a full blown theory when only a small portion has been tested? (Mayo 2002a, Laudan 1997).

The intuition behind the severity demand is that mere accordance between **x** and P—mere survival of P—is insufficient for taking **x** as genuine evidence for P. Such survival must be something *that is very difficult to achieve* if in fact P deviates from the truth (about the phenomena in question). The intuition is sound, but Popperian logical computations between statements of hypotheses and data never gave us a way to characterize severity adequately. (In part II, I shall describe an account that enables the needed characterization.)

Popper himself seemed to concede that the various formal definitions C(P,**x**) he proffered were only *potential* measures of the degree to which **x** corroborates P: in order for it to genuinely measure corroboration, Popper claimed, **x** would have to *actually* be the result of a severe test, a notion which was perhaps beyond formalization.

> In opposition to [the] inductivist attitude, I assert that C(P,**x**) must not be interpreted as the degree of corroboration of P by **x**, unless **x** reports the results of our sincere efforts to overthrow P. The requirement of sincerity cannot be formalized—no more than the inductivist requirement that **x** must represent our total observational knowledge. (Popper 1959, p. 418. I substitute his h with P and e with **x** for consistency with Musgrave's notation.)

The important kernal of rightness here is that these inductive logics make it too easy to find evidence in support for hypotheses *without satisfying the requirement of severity.* Unfortunately, Popper's computations suffered from just this weakness.

3.4 How Might Musgrave Respond?

Now Musgrave might respond in two ways:

(a) He might maintain that he (and other critical rationalists) do or would go beyond Popper by fleshing out the demand that '**x** report the results of our sincere efforts to overthrow' or find fault with P.

But how? I doubt he would be satisfied with some sort of subjective or psychologistic 'sincerity' requirement that could not be intersubjectively checked (Musgrave 1974b). Musgrave has, after all, been a long-time proponent of one or another 'objective' novelty requirements, and he might maintain that this allows him to exclude problematic cases. For example, he might classify under 'not sincerely trying to find fault with P' cases where P has been deliberately constructed to account for given data **x**, and no independent evidence for P exists (Musgrave 1974). But the novelty requirement Musgrave endorses, 'theoretical novelty',[6] boils down to Popper's comparatively best-tested requirement (3.3); and, as noted, this fails to provide tests that are actually severe and is, moreover, neither necessary nor sufficient for good evidence (see Mayo 1996).

Finally, even if one granted a given test was a severe and reliable probe of errors, we would still be in need of an account of how to obtain evidence **x** that P has actually *withstood* this test. This a non-trivial task that (as Popper admits) demands evidence of a 'reproducible' or reliable effect, not merely 'non-reproducible single occurrences' (Popper 1959, p. 86). (The mere perceptual claims that Musgrave (1999) is prepared to accept so long as they are not known to fail scrutiny will hardly do.) Musgrave's remark that 'existing critical rationalist literature goes a good way to provide [a theory of criticism]' (ibid. p. 323) must remain a mystery: one finds nothing approaching such a thing in that literature.

Indeed, the whole 'secret' to critical rationalism, as he sees it, is that it escapes the demanding task of developing an account of severe testing that can be shown to be reliable.

(b) Musgrave might, in this vein, insist he cares nothing for the reliability of method M_{CR}.

'Critical rationalists' Musgrave tells us 'deny that the process they commend is reliable' (p. 346). But this will not do. It is one thing to deny one is 'commending' a method as reliable*; it is quite another to be confronted with the blatant unreliability of a method and yet deny that this matters.* Further, since accounts of severe testing exist which are not unreliable, it follows that the critical rationalist's testing method M_{CR} itself fails to survive even moderately severe scrutiny! (More on this later.)

4. THE FALSITY OF THE ALLEGED NECESSITY OF CIRCULAR DEFENCES OF EPISTEMIC PRINCIPLES

Musgrave does spend considerable effort addressing the question of how to defend the critical rationalists' epistemic principle, but remarkably, he does not take up concerns such as those I have just raised. Notably, he does not seem to think he has to. In handling the question, "Why is P's being best-tested (as the critical rationalist understands this) a reason for believing P?" he allows he can do no better than simply repeat the epistemic principle under question!

[6] Musgrave also propounded a notion of 'deductive novelty' which demanded being able to identify if data **x** were required as premises in constructing P (Musgrave 1989). For a discussion of the relationship between novelty and severity see Mayo 1991, 1996.

> There is nothing more rational than a thorough and searching critical discussion. Such a discussion may provide us with the best reason there is for believing (tentatively) that a hypothesis is true—though not, of course, with a conclusive or inconclusive reason for that hypothesis. (p. 324)

Buying the traditional inductivists' (probabilistic) notion of 'a reason' for a hypothesis ((J) above), Musgrave is forced to embrace circularity. Of course this is just what Popper said, the difference is that Musgrave wishes to bite the bullet of circularity. But are we to accept that the promised cornerstone for avoiding irrationality is no more than a declaration that there is a method M such that, by definition, M is rational? Amazingly, it seems that, according to Musgrave, we are:

> Even if it is accepted that CR withstands criticism better than rival epistemic principles (a big 'if'), another objection immediately presents itself. All this is circular! The critical rationalist is saying that it is reasonable to adopt CR by CR's own standard of when it is reasonable to adopt something! (p. 330)

Not that he is happy about it. Indeed, Musgrave concedes that an analogous circular move would countenance arguing for the reasonableness of so crazy a method as:

M_{Mus}: It is reasonable to believe anything said in a paper by Alan Musgrave

since the assertion is made in a paper by Musgrave (p. 330). (I return to this 'crazy' method in Section 5.) Nevertheless, Musgrave declares that as all epistemic principles can only be defended circularly, it is no special reason to find fault with critical rationalism! According to this, we know in advance that no epistemic principle could be faulted, thanks to the availability of its surviving a circular defence. It would follow that claims about methods are non-testable! Could this really be the long-sought for defence of Popper?

4.1 Musgrave's Remarkable Argument

It is easier to express horror at the final destination of Musgrave's reasoning than it is to show just where one is warranted in getting off his train of argument. His argument, while unsound, or so I shall argue, is subtle and interesting. His argument is this:

> Any general epistemic principle is either acceptable by its own lights (circularity), acceptable by other lights (hence irrational by its own lights and inviting an infinite regress), or not rationally acceptable at all (irrational again). So even though the rational adoption of CR involves circularity, this cannot be used to discriminate against it and in favour of some rival theory of rationality. (p. 331)

Although our interest is in CR, let us analyse this striking general argument. Any general epistemic principle is either

(A) acceptable by its own lights (circularity), or

(B) acceptable by other lights (hence irrational by its own lights and inviting an infinite regress), or

(C) not rationally acceptable at all.

So if a general epistemological principle is rationally acceptable (i.e., (C) is false), he concludes, either (A) or (B) is the case (i.e., its acceptability will be circular, or irrational and inviting a regress).

By a general epistemic principle, Musgrave has in mind an 'if-and-only-if' claim (doubtless the reason he expressed EM_{CR} as such), so as to ensure the claim about the acceptability of the principle comes under the principle itself, i.e., that it is *self-subsuming*. To make his intent clear, the form of a 'general epistemological principle' EM is this:

(EM) a claim P is acceptable iff it is classified as acceptable or beliefworthy by belief-classification method M.

EM, itself being a claim, would be subsumed under method M. By contrast, an epistemic principle concerning, say, purely mathematical claims, would not itself be a mathematical claim and so would not be self-subsuming.

To engage his argument, and see how it goes wrong, let us resist drawing any distinctions of levels and grant Musgrave's claim that a general epistemic principle itself comes under the domain of claims that M classifies as acceptable or not, let us grant that it is self-subsuming. (He equates 'self-subsuming' with 'circular' but, as we will see, the latter term has importantly different connotations.) Let us suppose, with respect to a given method M, that premise (C) is false: M *is* rationally acceptable, and allow that M is itself classified as beliefworthy by M, i.e., EM is acceptable 'by its own lights', premise (A).

But this does not yet entail that the *only* warrant for EM is EM, i.e., the only warrant for EM is that it has been classified as 'acceptable' by M! Musgrave confuses a 'self-subsuming' method with what may be called a 'self-sealing' method.

4.2 Self-Subsuming is Not Self-Sealing

Consider an example which I shall try to design so as to concede as much as possible to Musgrave. There is a principle, let us imagine, for deciding to accept or believe claims about books in print in 2004:

(EM_{BIP}) accept claims concerning which books are in print (in year 2004) if and only if they are found in the comprehensive *Handbook of Books in Print* for 2004 (BIP),

where we stipulate, for purposes of the illustration, that it really is exhaustive of the finitely many books in print in 2004.

Again, let us resist any attempt to suggest EM_{BIP} is itself a 'meta-claim', and allow that it is subsumed under itself. Suppose in fact that the assertion EM_{BIP} is found on the first page of the BIP. So EM_{BIP} is acceptable 'by its own lights', but does this entail any circularity? No. Being 'self-consistent' is not the same as being 'self-warranting', i.e., warranted only by dint of the self-subsumption: *self-subsuming is not self-sealing*. In arguing for the acceptability of EM_{BIP}, one might allude to such things as the scrupulousness with which each publisher is checked to keep the listing of books in print up to date. In other words, one would allude to the reasons that assertions find their way into the BIP handbook to begin with, the criteria which must be met before inclusion, thereby ensuring that all claims therein have certain qualities (examples such as this can easily be multiplied). According to Musgrave, appealing to these various facts about the criteria used to include assertions in the BIP handbook instantiates premise (B) rendering the warrant for BIP irrational and/or leading to an infinite regress! But this is clearly false.

Compare this with recommending *Sloppy Joe's Books in Print* which we may imagine is very sloppy, imcomplete and outdated.

($EM_{SJ'S\ BIP}$) accept claims concerning which books are in print (in year 2004) if and only if they are found in *Sloppy Joe's Books in Print.*

And again suppose this assertion is itself on page one of Sloppy Joe's volume. Following Musgrave's reasoning, even *Sloppy Joe's Books in Print* would be as acceptable as the authoritative BIP! But in fact, we would adduce many reasons for regarding its listing as unreliable, out of date, and so on.

Musgrave's argument presents us with a false dilemma: it appears to go through only by assuming that if epistemic principle EM is acceptable then the *only* warrant that may be given for the claims M classifies as beliefworthy is the fact that M classifies them as beliefworthy!

But the if and only if claim in EM does not entail this. He is confusing self-subsumption with self-sealing. In other words, the left-to-right conditional in EM is:

($EM\Rightarrow$) P is acceptable only if P is classified as beliefworthy by method M,

Musgrave conflates this with an entirely different claim, one asserting that any test of M is self-sealing:

Self-Sealing Test of M: P is acceptable *only because* P is classified as beliefworthy by method M.

The latter claim asserts that the only grounds for the acceptability of P is that P is classified as acceptable by method M. Were the 'self-sealing' test the only one available for a method (for classifying claims as acceptable), then Musgrave would be right to allege that his defence of CR is no worse off than for any other. But this is false.

4.3 Sum-up of the Confusion Between Self-Subsuming and Self-Warranting

We see that Musgrave's remarkable argument assumes and does not show that *only* a self-sealing defence is possible for method M (and thus for an epistemic principle espousing M). In so doing, Musgrave assumes the very thing he is claiming to argue for, i.e., he is guilty of question-begging. Moreover, since we have seen there are grounds to reject his claim, his question-begging adherence to it has no weight.

That a principle is not self-refuting hardly entails that there is no test or means of scrutiny (independent of the classification scheme itself) of whether the method in question is, or is not, capable of satisfying the desired aims in applying method M. Otherwise we would not have been able to mount the criticism in Section 4 of the comparativist account of severe tests, nor criticize *Sloppy Joe's Books in Print*. Nothing in Musgrave's arguments show otherwise.

It is not that Musgrave is not pained by having to assume his favored principle in order to (deductively) defend it. He is. If we permit epistemic principles whose sole support is circular, Musgrave freely admits, we can easily argue in favour of all manner of crazy procedures such as procedure M_{Mus}:

M_{Mus}: It is reasonable to believe anything said in a paper by Musgrave

since M_{Mus} occurs in Musgrave's paper. I feel his pain, and have been setting the stage for its extirpation. Before administering the anesthesia, however, let us twist the knife a bit further—to learn more about the critical-rationalist infirmity with which Musgrave saddles himself, and just how devastating the malady really is. Were self-sealing defences the most one could give for epistemic principles, then the critical rationalist should close up shop: he would have to concede there are no better grounds for CR than for any other principle, even one that ignores all evidence and counsels accepting whatever Musgrave endorses!

Thus, Musgrave's defence of critical rationalism defeats itself. For, what is wrong with a self-sealing test of a general epistemic principle EM (about method M)? What is wrong is that the epistemic principle is guaranteed to pass even if it is false. Even if method M does not satisfy its intended aim, *whatever it is*, it will nevertheless still be permissible to classify M as an acceptable method. That EM passes a self-sealing test is tantamount to its passing a test it had no risk of failing, a test that utterly lacks severity or probative power in the ordinary sense upon which CR is parasitic. What is more, his arguments are self-contradictory. My goal in showing this, I should emphasize, is a positive one: to show how to get beyond where critical rationalism thinks it can go.

In at least two places Musgrave's arguments assume the existence of reliable methods: (i) he assumes there are non-crazy methods, ones that are at least not utterly unreliable for the intended job of uncovering flaws and errors, and (ii) he also assumes there is a reliable method for distinguishing crazy methods from non-crazy methods (for classifying claims as acceptable). But since at the same time he denies these assumptions, his arguments are self-contradictory. Moreover, suppose we perform this substitution in M_{CR}: replace 'P is best tested' with the phrase 'P is endorsed in a paper by Musgrave', yielding method M_{CR*}. Now M_{CR*} is identical to M_{Mus}. Cashed out this way, Musgrave would presumably deny the corresponding principle EM_{CR*} and he would adduce non-circular reasons for this.

5. WHY BELIEVE THAT 'BELIEVING WHATEVER MUSGRAVE WRITES' IS A CRAZY RULE?

Musgrave declares that '"It is reasonable to believe anything said in a paper by Alan Musgrave"…is a crazy epistemic principle' (1999, p. 330), and I want to know why. He evidently regards its craziness as fairly obvious, and so I believe he has reasons for this judgment. I take it that he does not regard all epistemic principles for adopting beliefs as similarly crazy, else he would not be mounting efforts to argue in favor of the critical-rationalist epistemic principle (CR). For sure, it is bizarre to hold, as he seems to in his circularity concession, that CR has no better grounds than does M_{Mus}, while at the same time denying, as we may presume he does, that the two are similarly crazy.[7]

Moreover, it is implicit in Musgrave's discussion that there are some criteria for *distinguishing* (crazy from non-crazy) epistemic principles such that CR withstands this scrutiny and rules like M_{Mus} do not. The scrutiny cannot be a matter of whether they may be defended non-circularly, since we have already seen he denies that (even though we have rejected his arguments). What I want to consider is what Musgrave *could mean* in making a distinction between crazy and non-crazy belief-classification rules. Articulating the grounds behind his self-deprecating critique, ironically, takes us several steps further than he declares is possible toward an account of 'evidence transcending' or 'inductive' inference. But, in so doing, we expose a contradiction in critical rationalism, at least as he describes it, and the self-defeating nature of his defence of it.

5.1 Musgrave's Method for Classifying (belief-classification) Methods as Crazy or Not

If Musgrave does not regard all epistemic principles as crazy, if, for example, he does not regard following M_{CR} as just as crazy as following M_{Mus}, then, he must have a procedure, or criteria to apply, that discriminates crazy from non-crazy methods, or is at least capable of identifying a clearly crazy one. But then it would seem that there is at least one method that may be defended with good reason: Musgrave's method for condemning following M_{Mus} as 'crazy'. In other words, Musgrave has a method, or discriminating capacity, that pigeonholes under the rubric 'crazy' method M_{Mus} and under 'not crazy' (but rather, rational) method M_{CR}.

What makes the rule M_{Mus} crazy? Why is it *correct* to classify it under the rubric 'crazy'? Why is it *incorrect* to classify it as 'rational' (or non-crazy)?

[7] He cites two other crazy or unwarranted epistemic principles: 'Granny told me I ought to believe everything she tells me', and 'The Pope declared *ex cathedra* that everything declared *ex cathedra* by the Pope is a matter of faith' (p. 330).

Does Musgrave classify it as crazy because he thinks it an unreliable procedure to follow (i.e., that Musgrave often publishes flawed or incorrect claims)? Or because the mere fact that a paper by Musgrave claims P does not, in and of itself, provide evidence that P is correct or well-supported? But these all seem to be at odds with his insistence upon the 'wedge'. So perhaps Musgrave regards it as a crazy method because the mere fact that a paper by Musgrave claims P does not, in and of itself, make it reasonable for others to accept or believe P (where this may be construed pragmatically or otherwise).

But we need not pretend to know what classification method Musgrave is using here for the argument that I am now interested in making. Musgrave must allow there is an adequate 'metamethod' whereby he classifies M_{Mus} as crazy, in contrast to an inadequate metamethod, say, a procedure that willy-nilly classified as crazy any and all rules for adopting beliefs, or made the determination by flipping a coin. Compare two metamethods:

(MetaM_{Mus}) accept claims about whether or not a method (for belief-classification) is crazy in accordance with Musgrave's pronouncements about (crazy/non-crazy)

where this alludes to whatever classification method Musgrave is using in this paper (Musgrave 1999); and one based on, say, coin-flipping:

(MetaM_{Coin}) accept claims about whether or not a method (for belief-classification) is crazy in accordance with the outcome of a fair coin toss.

Methods like MetaM_{Coin}, presumably, would not be adequate for the task of discriminating crazy from non-crazy methods. That is because the latter procedures are poor tools for accomplishing the intended job (of correctly classifying rules as crazy). (Indeed, they are themselves crazy tools for the job!) Since Musgrave clearly thinks there are good reasons for regarding M_{Mus} as crazy, I should think he would regard as a poor metamethod one that declares M_{Mus} a non-crazy rule; or one that bases its pronouncements on irrelevancies (e.g., coin flips, or whether its adoption as a good rule would make money for Musgrave). A possible criterion, then, for evaluating metamethods might be:

(5.1) *Criterion for Evaluating Metamethods*: MetaM is a poor classification method if it erroneously classifies methods as crazy as often as not.

For example it would be poor if the test it employs to decide whether to classify a method as crazy uses a criterion with no correlation to the method's actually being crazy (however this is defined).

Now I do not know what test rule Musgrave is applying so as to declare M_{Mus} crazy, my point is only that I believe he has one and that were he to spell out the criterion behind it, we may evaluate its properties for performing the job at hand. It would follow that there are perfectly good, *non-circular*, reasons for endorsing some methods for classifying methods as crazy or not, while rejecting others as not up to the job.

5.2 Learning From the Failure of Musgrave's Defence

Our critique of Musgrave's attempted defence of critical rationalism bears positive fruits, as severe critiques should. A method for classifying methods (what I called a metamethod) is precisely on par with any method for classifying claims as acceptable or not, so from the above discussion, and the criterion in (5.1), we extract the following:

(5.2) Method M is a poor classification method (a crazy method) if it often classifies claims as acceptable when they are not, i.e., if its classification scheme is an unreliable indicator that P is acceptable (however one defines acceptability).

Even so weak an assertion as (5.2), which is itself just a start, already breaks through the critical rationalist's obstacles to progress. To begin with it gives a basis for distinguishing 'crazy' and 'non-crazy' methods, as well as grounds for criticizing arguments claiming to show why a given method is acceptable. Unless an argument gives assurance that a method avoids threats of unreliability, it fails utterly as a defence of the method. Musgrave's defence of the critical rationalist's classification method fails on these grounds: being classified as 'best-tested' by his critical rationalist makes it too easy to classify claims 'acceptable' without warrant.

Musgrave mistakenly assumes that demonstrating reliability would be tantamount to justifying enumerative induction, but enumerative induction, Musgrave declares, is 'unreliable' (p. 346), in contrast to perceptual beliefs which he claims are 'reliable'. Assuming the only kind of justification an evidence transcending claim can receive is to find it true or highly probable i.e., accepting (J), he rejects justifying ampliative inferences altogether.

> Critical rationalists deny that induction is a reliable process. Critical rationalists also deny that the process they commend is reliable—or at least, they must deny this if they [are] to avoid the widespread accusation that they smuggle into their theory either inductive reasoning or some metaphysical inductive principle. (pp. 246-247)

I shall now turn to showing how to characterise the severe testing requirement, avoiding all the shortcomings of the critical rationalist. Viewing induction in terms of severe testing, as I define it, lets us warrant induction or evidence transcending methods as reliable without smuggling in 'probabilism' or a metaphysical inductive principle!

PART II. THE SEVERE TESTING PRINCIPLE IN THE ERROR STATISTICAL PHILOSOPHY

6. HIGHLY PROBABLE VERSUS HIGHLY PROBED

The modern-day critical rationalist (Musgrave being the best among them) has not cleared away the stumbling blocks that stymied Popper; like the 'inductive logician' of old he retains the assumption that a justification or warrant for an evidence transcending inference is either to show it conclusively true (whatever that might mean) or assign it a probability. Denying the former, the inductivist looks for a probabilistic computation, most often by appealing to the statistical definition of conditional probability or *Bayes's Theorem*: $P(H|e) = P(e|H)P(H)/P(e)$[8]. Computing $P(H|e)$, the *posterior probability*, requires starting out with a probability assignment to all of the members of 'not-H,' the *prior probabilities*. Insofar as the computed degrees of confirmation are viewed as analytic and a priori—as in the 'logical probability' notions often favoured by Popperians—their relevance for predicting and learning about empirical phenomena is questionable; insofar as they measure subjective degrees of belief, they are of questionable relevance for giving objective guarantees of reliable inference. The search for an inductive logic as purely formal rules for relating statements of evidence to hypotheses has largely been abandoned, but, oddly enough the underlying conception of the nature of inductive or statistical inference appears to remain firmly entrenched in the critical rationalist's program.

The most flagrant mistake of the critical rationalists, like the inductivists and probabilists, is to suppose that an 'inconclusive' reason or warrant for an evidence-transcending claim should come in the form of a probability assignment. In fact, inductive uncertainty or inconclusiveness is not well-captured by a posterior probability assignment, in any of the senses that probability has been defined. Even if one is a frequentist about probability, as I am, the inductive job is not accomplished by attempting to assign relative frequencies to hypotheses e.g., as Reichenbach and Salmon often suggested. Even, for example, if hypothesis H has been randomly selected from an urn of hypotheses, p% of which are true, it is completely wrong-headed to suppose that the probability this particular H is true is equal to p. (I call this the fallacy of instantiating probabilities, Mayo 2003, 2004 and 2005). The severe testing intuition with which we began is at home with a very different use of probability, namely to characterize the probativeness of the testing process itself.[9]

It is time to move on. We can begin to ameliorate the current crisis by (a) rejecting the erroneous conceptions of inductive or 'evidence-transcending' inference upon which their skeptical slide is based, and (b) showing how to develop an account with the resources to define and apply 'severe or reliable error probes'. Here, I can only sketch ingredients of a full severe-testing account developed

[8] Where $P(e) = P(e|H)P(H) + P(e|\text{not-H})\,P(\text{not-H})$.

[9] Probability, in this inferential philosophy, may still be ascribed to outcomes or events, or in formal statistical modeling, to the event that a random variable takes a given value. By testing and severely passing a statistical model, one can then use it to assign probabilities to the possible outcomes.

elsewhere. To avoid confusion with probability statements, among other reasons, throughout Part II, I will replace Musgrave's P for 'proposition' with H for 'hypothesis', with qualifications to be noted.

6.1 The Common Sense Notion of Severe Testing: Isaac

Let us go back to the primitive intuition about severe testing with which we began. Ordinary considerations about testing will do. Consider a student, Isaac. If we are testing how well Isaac has mastered high school material so as to be considered sufficiently ready for work in a four-year college, then a test that covered work from 11th and 12th grade science, history, included mathematical problems (in geometry, algebra, trigonometry, and pre-calculus) required writing a critical essay, and so on, is obviously *more difficult to pass* than one which only required showing minimal proficiency in these subjects at a 6th or 7th grade level: it would be regarded as more searching, more probing, and more severe. The understanding behind this commonplace judgment is roughly this: Achieving a passing or high score is easier and more likely to have come about with the less severe test than the more severe one, *even among students who have not mastered the bulk of high-school material, and hence are not 'college-ready'*. In other words, *before regarding a passing result as genuine evidence for the correctness of a given claim or hypothesis H, it does not suffice to merely survive a test, such survival must be something that is very difficult to achieve if in fact H deviates from what is truly the case.*[10]

By the same token, if the test *is* sufficiently stringent, such that it is practically impossible for students who have not mastered at least p% of high-school material to achieve a score as high as Isaac's, then we regard his passing grade as evidence that he has mastered at least this much. The same reasoning abounds in science and statistics.

Note again the important distinction between highly probed and highly probable even in a so-called frequentist sense: Suppose Isaac had been randomly selected from a wealthy suburb in which, say, 95% of high school students are 'college ready'. Given this high (.95) 'prior' probability to H (i.e., Isaac is college-ready), even a low exam score can result in a fairly high posterior probability to H (Mayo 1997, pp. 326-29; 2004; 2005). Nevertheless, or so our severity intuitions tell us, the high posterior is not good evidence that H has withstood a severe test. In fact, we would wish to ask, What is the probability that the posterior probability would be high even if Isaac is not ready? (i.e., H false). This is an *error probability*, and if it is high, we deny we have good evidence for Isaac's readiness (H).

[10] An extremely common fallacy in other notions of severe tests is deliberately avoided in my account. At first blush, a test T that (a) regards a successful prediction as evidence for H, even though (b) a failed prediction would not have counted as disconfirming H, is typically thought 'to be about as blatant a violation of the Popperian commandment as you could commit' (Meehl 1967/1970). But in fact it might be that (a') P(test T passes H;H false) is very low, and yet (b') P(test T fails H; H true) is not low. (a) warrants (a') and (b) warrants (b'): thus there is no violation of the severity requirement. For discussion of this, see Chalmers 1999, Chapter 13 appendix.

6.2 The (Error Probabilistic) Severity Principle

We can substantiate, finally, the intuitive severity principle without baiting and switching:

SP: **x** is evidence for H iff, or just to the extent that, **x** constitutes evidence that H has survived a severe test

while demanding, quite unlike the critical rationalist, that a test method be shown to be a reliable error probe.

Test Method M: H is classified as having withstood a severe test T to the extent that H would not have survived (or survived so well), were H false (i.e., a specified flaw in H is present).

Probability may be appealed to here in characterizing the capacity of the error probe: the test that H passed would not be severe if such a passing result is fairly probable, even if H is false.

Test Method M (probabilistic): H is classified as having withstood a severe test T to the extent that H would, *very probably*, not have survived (so well), were H false (i.e., a specified discrepancy from H is present).

Except for formal statistical contexts, 'probability' here may serve merely to pay obeisance to the fact that all empirical claims are strictly fallible, even if a counterexample is never to be actually instantiated in the whole course of human history of the world. Even in technical areas, such as in engineering, it is common to work without a well-specified probability model for catastrophic events, and yet the same requirement about evidence holds. Modifying the above definition for such contexts, the engineer, Yakov Ben-Haim suggests, 'We are subjecting a proposition to a severe test if an erroneous inference concerning the truth of the proposition can result only under extraordinary circumstances.' (Ben-Haim, 2001, p. 214).[11] The kind of inference here might be H: metal buckling of more than a specified amount will not occur under conditions **x**.

7. INDUCTION AS SEVERE TESTING: THROWING OFF THE CRITICAL RATIONALIST'S SHACKLES

In the current view, data **x** provide good evidence for inferring H only if they result from a method which, *taken as a whole*, constitutes H having passed a severe test—that is, a method which would have (at least with very high probability) unearthed

[11] Ben-Haim makes this notion rigorous by means of a definition based on convex sets, which I do not understand sufficiently to explicate.

any error or flaw in the inference to H. This simple idea, once unpacked thoroughly, lets us shake off the fears and inhibitions that lead critical rationalists to ban ordinary talk of 'justification' and 'induction'. Warranted evidence-transcending inferences—i.e., justified inductive inferences—are to be regarded as cases of inferences from severe testing. A methodology for induction, accordingly, is a methodology for arriving at severe tests, and for scrutinizing inferences by considering the severity with which they have passed tests. Far from wishing to justify the familiar inductive rule from an observed correlation between A and B to an inference that all or most A's are B's (or the next A will be a B), we can see that such a rule would license inferences that had not passed severe tests: it would be a highly unreliable method. An induction following this pattern will be unwarranted, I claim, unless the inference has successfully passed a severe test. Nor, on this account, does H merit any brownie points by dint of being the least poorly tested in a crop of poorly tested hypotheses.

7.1 Taking Seriously the Need to Rule Out Errors Into Which Simple (Enumerative) Induction May Lead

Of course, critical rationalists recognize the errors into which enumerative induction may lead: that is the springboard for their skepticism about induction. Such errors, Musgrave rightly notes, are problems for 'adherents of inductive logic' (p. 346) insofar as an inductive logic is supposed to be formal and context-free. If it is a contingent matter whether given errors are sufficiently well ruled out to infer, say, from correlational data to causal claims, one cannot look to a purely formal inductive logic for evidence-transcending inferences. What he, and so many other philosophers of science, fail to see, is that the bankruptcy of the 'logicist' program for inductive logic in no way robs us from having a rich bank account of methods for reaching and warranting inductive inferences! Moreover, showing the bankruptcy of enumerative induction itself depends on being able to substantiate claims about how, in given contexts, blindly following enumerative induction readily leads one astray. In fact, 'the person of common sense', says Musgrave, is fairly savvy in avoiding such familiar foibles.

'People of sense do not argue that the more times your joke has made a person laugh the more likely it is to raise a laugh the next time you tell it.' (presumably to the same person). What warrants this assertion of Musgrave? It is not that he notices his sensible friends do not in fact make such claims, it is rather that Musgrave knows that this is one of those cases where the outcome of trials are *negatively dependent* on previous outcomes. How can Musgrave substantiate this? Is it not because there are well-known errors that would need to be ruled out before supposing the repeatability of an effect (e.g., diminishing returns)? He, like other persons of sense, are perfectly well capable of generalizing about the kinds of cases wherein the more A's that have been B's in the past, the *less likely* the next A will be a B.

Other errors that would need to be ruled out are similarly codified in good statistical reasoning as in conscientious informal critical thinking. But such 'context-dependent' tools appear as much out of reach of the critical rationalist as the inductive logician. Why else would these philosophers of science persist in overlooking the general epistemic justification for such critical tools? Of course, after our severe tester arrives at such reliable rules Musgrave can say, as I suspect he would, that he too would embrace such a method as 'best tested'. Unfortunately, this will only be a matter of *after-the-fact* reconstruction. The critical rationalist denies he is commending *forward-looking* inductive methods that are reliable. What Musgrave and other critical rationalists fail to realize, or fail to capitalize on, is that the basis of *their* criticisms of rudimentary induction rests on having general knowledge of types of situations wherein applying simple enumerative induction would readily lead to erroneous inferences.

It should come as no surprise to critical rationalists, except that they seem not to have heard the news: scientists, like people of sense, have deliberately developed models and methods for (a) checking whether this kind of temporal dependency holds in a given case, and (b) capitalising on knowledge of dependencies to develop reliable inductive rules for the kind of case at hand. This is the focus of the conglomeration of statistical methods, understood broadly as I do, to include methods of planning, collecting, modelling, and drawing evidence-transcending inferences on the basis of uncertain and limited data. An informal repertoire of day-to-day errors serves an analogous role for the 'person of sense'.

7.2 Severity in Statistical Testing

Statistical tests do not employ our notion of severity directly, but severity can be seen to provide a metastatistical concept and corresponding principles that direct the interpretation and justification of standard statistical methods (e.g., of testing and estimation). For details, see Mayo 1996, Mayo and Spanos 2006. We can encapsulate the severity requirement in statistical testing set-ups thus:

(7.2) Hypothesis H passes a severe test T with **x** if (and only if):

(i) **x** agrees with or 'fits' H (for a suitable notion of fit[12]), and

(ii) test T would (with very high probability) have produced a result that fits H less well than x does, if H were false or incorrect.

8. ERROR STATISTICS AS THE SEVERE TESTER'S THEORY OF INDUCTION

Given the importance with which Musgrave regards the 'wedge' (between reasons to accept and supplying probabilistic justification) for the critical rationalist, it is surprising that he does not take advantage of the distinct philosophical tradition that uses probability not to assign degrees of confirmation or support or belief to

[12] See note 18.

hypotheses, but rather to characterise a procedure's reliability in a series of (actual or hypothetical) experiments. Deliberately designed to reach conclusions about statistical hypotheses without invoking prior probabilities in hypotheses, indeed, explicitly denying the relevance or meaningfulness of posterior probabilities in hypotheses (as opposed to events), probability is used to quantify how *frequently* methods are capable of discriminating between alternative hypotheses and how *reliably* tests facilitate the detection of error. These probabilistic properties of statistical procedures are called *error frequencies* or *error probabilities*.[13] An account based on error probability criteria, whether formal or informal, I dub an *error statistical account* of inference.

8.1 Neyman and Popper: Finessing Induction

In Neyman-Pearson (N-P) testing methods, we see an illuminating example of the 'wedge' Musgrave lauds as the cornerstone of Popper's 'solving' the problem of induction: Neyman and Pearson ground their statistical test rules while, quite deliberately, denying 'inductive' evidence for the truth or probability of statistical hypotheses themselves. The basic rationale underlying N-P statistics was precisely to provide procedures that satisfy aims for rationally adopting an action (whether it be publishing a paper, deciding to believe, or something else) as distinct from supplying grounds for inferring the truth (or probability) of any claim or hypothesis. Neyman referred to such rules for testing as *rules of inductive behaviour* (1952; 1971).

Wishing to draw a stark contrast between this conception of tests and those of Fisher as well as Bayesians (i.e., Jeffreys), Neyman declared that the goal of tests is not to adjust our beliefs but rather to 'adjust our behavior' to limited amounts of data. Erich Lehmann (Neyman's first statistics' student at Berkeley, and eminent statistician in his own right) notes:

> It is remarkable that independently and nearly simultaneously [early 1930's] Neyman and Popper found a revolutionary way to finesse the issue [of the problem of induction] by replacing inductive reasoning with a deductive process of hypothesis testing (Lehmann 1995, p. 32).

Equally striking is that there is scant evidence of direct influences between the two.

There is, however, one exceedingly important difference between their 'finessing': The N-P tester is required to show that the statistical test procedures actually satisfy the aim of low error probabilities!

'Self-sealing' appeals will not do. Indeed, the central value of tests as rules of behavior is that 'it may often be proved that if we behave according to such a rule...we shall reject H when it is true not more, say, than once in a hundred times,

[13] Embodying a frequentist notion of probability, while denying it is useful to consider the frequency with which hypotheses like H are true (in this or other possible worlds), probability assignments are restricted to random variables (or events) associated with a probabilistic model.

and in addition we may have evidence that we shall reject H sufficiently often when it is false' (Neyman and Pearson 1933, p. 142).[14]

In simple statistical significance testing, for example, where hypothesis H might be the familiar 'null hypothesis' (e.g., of no effect or no discrepancy from a fixed parameter value, we obtain a formal exemplification of the following reasoning:

(1) If H were false (i.e., a specified flaw in H present), then (with high probability) the tests would yield evidence of a discordance of at least d (between H and **x**).

(2) There is evidence of a discordance less than *d*.

(3) Therefore, **x** is evidence that the specified flaw in H is absent.

By justifying and showing how to implement premises (1) and (2) of such arguments, one escapes the disappointing limitations of the critical rationalist's game (for a discussion of testing statistical assumptions see Mayo and Spanos 2004). Thus, these error statistical procedures have properties that would have been expected to be embraced by Popperians; and although many speak approvingly of Fisherian tests, e.g., Gillies, they have not utilized these methods to escape the most serious limitations of critical rationalism.[15]

Granted, there is still an important lacuna in the N-P test reasoning: The move from premises (1) and (2) to conclusion (3) is not deductive. Even once the work is done to accept these premises, there is a gap. What is missing is a link from the low long-run error probabilities of (1), to the specific inference in (3). It is true that following the test method is reliable in the sense that one will rarely commit errors in a long-run series of applications. But what does this say regarding the inference at hand? This is where the conception of induction as passing a severe test and the severity principle enter.

8.2 Does the Failure to Reject a Null Hypothesis Confirm It?

If we view induction as severe testing, as I propose, one has the basis for arguing from error probabilistic properties of tests to well-probed claims in the (so-called) 'single case'. In other words, it is not just low error rates in the long run that matter, it is that these may be used to attain probative tests, hence warranted inductions for the claims that withstand them. Things might have been very different if Neyman had not been so wedded to the behavioural-decision model of tests, with its low long-run error justification, once the N-P testing model got off the ground. For it

[14] Neyman regarded '"inductive Behavior" as a Basic concept of Philosophy of Science' to cite the title of a paper of his 1957a. 'Rather than speak of inductive reasoning,' Neyman remarks (1971, p. 1) 'I prefer to speak of inductive behavior'. This refers to the adjustment of our behavior to limited amounts of observation. This is an excellent example of the critical rationalist's "wedge".

[15] The differences between N-P and Fisherian tests, while important, will not concern us here: both are in the error probability tradition as I understand that term Mayo and Cox 2006.

turns out that there is ample evidence of reasoning in accord with our severity principle in little known early papers of Neyman (as well as in works of Pearson).

In one, wherein the striking title of this subsection is found[16], Neyman is addressing his remarks to none other than Carnap. 'In some sections of scientific literature the prevailing attitude is to consider that once a test, deemed to be reliable, fails to reject the hypothesis tested, then this means that the hypothesis is 'confirmed' (Neyman 1955).' Calling this 'a little rash' and 'dangerous', he claims 'a more cautious attitude would be to form one's intuitive opinion only after studying the power function of the test applied.' (p. 41).

If a non-statistically significant result occurred with a test with low power to detect discrepancies of interest, Neyman is saying, then such a non-significant result should not be taken to rule out such departures from the null. Indeed, it is a well known fallacy to go from 'no evidence against' the null hypothesis to 'evidence for' the null, and it instantiates the severity demand.

More generally, if data **x** yield a test result that is not statistically significantly different from H_0 (the null of no effect), and yet the test has low probability to reject H_0, even when discrepancy δ exists, then **x** is not good evidence for ruling out discrepancy δ.

On the other hand, in statistics as in informal reasoning, if H has managed to survive so probing, searching or *severe* a test, then this is evidence that H is true (or at least that it does not deviate from the truth by more than a given amount). Let us set it out explicitly.

Severity in the Case of Statistically Insignificant Results:

> If data **x** are not statistically significantly different from H_0, and the probability of detecting effect δ is high (low), then **x** constitutes good (poor) evidence that the actual effect is no greater than δ.

8.3 Using Severity in Scrutinizing Non-Significant Results: An Example

A common example is to collect a sample of size *n*, $\mathbf{X} = (X_1, \ldots, X_n)$, where each X_i is an independent and identically distributed Normal variable, $(N(\mu,\sigma^2))$, and run a one-sided test of the hypothesis H_0: $\mu \leq \mu_0$ versus H_1: $\mu > \mu_0$. $\bar{X}$ is the observed sample mean, and a measure of 'fit' or distance is Z: $= (\bar{X}-\mu_0)/\sigma_x$ which is distributed Normally N(0,1)), allowing us to calculate the severity associated with different outcomes and inferences. Letting $\mu_0 = 0$, we have H_0: $\mu = 0$ and H_1: $\mu > 0$. For simplicity let $\sigma_x = 1$. Suppose the test will reject H_0 iff ($\bar{X} = 2$ – a result which would be statistically significant at around the .03 level – and we observe $\bar{X}=1.5$, so H_0 is not rejected. According to the above reasoning we can interpret this result as evidence not that H_0 is exactly true, but that the discrepancy from 0 is less than δ, provided the test had sufficient high power to have detected a discrepancy this large. So, for example, consider $\delta = 1$. The power to reject H_0 given $\delta = 1$ is only .16,

[16] It is striking because it contrasts sharply with Neyman's usual disdain for talking of inductive inference, insisting, instead, on his notion of inductive behavior.

so this does not warrant inferring $\mu < 1$; by contrast, the power against $\mu = 4$ is high, around .97 and thus, following our testing method, the result is good evidence that $\mu < 4$.[17]

These formal error probabilities parallel the informal qualitative assessments that are behnd the plausibility of the idea that evidence for H is a matter of H's surviving a severe test.[18] What prevented Popper from uncovering this key, I conjecture, is his failure to take what might be called 'the error probability turn'.[19]

9. THE SEVERITY PRINCIPLE: WHAT WE LEARN FROM ERROR STATISTICS

Whether severity is understood quantitatively or qualitatively, in terms of probability or in terms of non-probabilistic notions, the overarching principle of evidence remains, and may best be expressed as:

Severity Principle: Data **x** (produced by process G) provides a good indication or evidence for hypothesis H (just) to the extent that test T severely passes H with **x**.

By expressing it this way, it is emphasised that H is regarded (or modelled) as a claim about some aspect of the process that generated the data, G. According to the severity principle, when hypothesis H has passed a highly severe test (something that may require several individual tests taken together), we can regard data **x** as evidence for inferring H because it supplies good grounds that we have ruled out the

[17] That is, the power against $\mu = 1$ is P(reject H_0; $\mu = 1$) = $P(Z > 1) = 1-\Phi(1) = .16$, where Φ is the cumulative distribution function of the standard Normal distribution, and P denotes probability. The power against $\mu = 4$ is P(reject H_0; $\mu = 4$) = $P(Z > -2) = 1-\Phi(-2) = .97$.

[18] This use of power, while reasonable when the outcome just misses rejecting the null, is too coarse, and the severity assessment gets around this. Rather than construe 'a miss as good as a mile', the severity assessment depends on the actual non-statistically significant outcome. That is, we replace the usual calculation of power against μ':

(1) $P(Z > Z_\alpha ; \mu=\mu')$,

with:

(2) $P(Z > Z_p; \mu=\mu')$, where Z_p is the observed (non-statistically significant) result Z, with corresponding p-value.

(2), quantifies the *severity* with which the test passes $\mu < \mu$.'

To illustrate, compare observing (a) $\bar{X} = 1.5$ and observing (b) $\bar{X} = .1$. Both outcomes fail to reject H_0 with our test, but intuitively we would like to reflect the fact that the latter is so much close to 0. While we saw that the power against $\mu = 1$ is low, and power does not change with the actual outcome, severity does. The severity associated with $\mu < 1$ in case (a) is $P(Z > 1.5 - 1;) = .3$ whereas in case (b) it is $P(Z > .1 - 1) = P(Z > -.9) = .8$ (all numbers are approximate here.) So in case (b), unlike case (a), the inference $\mu < 1$ is warranted with fairly high severity, .8. Note that in the case of rejecting H, high power corresponds to low severity whereas with accepting H it is the reverse. However, whatever form H takes, we can talk of the severity of a test to have uncovered that H is in error. For a detailed discussion see Mayo and Spanos 2006.

[19] In private communication Popper explained that he regretted never having had the chance to learn statistical methodology.

ways it can be a mistake to regard **x** as having been generated by the procedure described by H.

9.1 *Dangerous Misunderstandings*

Although a full understanding of how to calculate severity demands careful discussion beyond this paper, the central points I need to make require avoiding some common misunderstandings on which criticisms often rest.

9.1.1 *A Severity Assessment is always Relative to the Hypothesis that 'Passes'*

It is common to talk as if a severity assessment attaches to the test itself—as Popper does—but doing so leads to untoward results. One cannot answer the question: 'How severe is test T?' without including the particular hypothesis that is claimed to have passed, or about which one wishes to make an inference. The great advantage of relativising the assessment to the particular inference (and the particular data set) is that high severity is always what is wanted for evidence.[20] No problem occurs unless one forgets that a given test may severely pass one hypothesis and not another, even among the hypotheses under consideration. This confusion most readily takes the form of what might be called: 'The Criticism From Overly Sensitive Tests'.

Severity cannot be a sensible desiderata, so the criticism goes, because a test may be made so severe that even a trivially small departure from a hypothesis H will result in inferring H′– where H′is a rival to H, or an assertion about some anomaly or error in H. What this criticism overlooks is that the inference whose severity we would need to consider in that case is H′; but having put H to a stringent test is not to have stringently probed H′! The misunderstanding behind the criticism boils down to thinking that H′ has passed a severe test, as I am defining it, but in fact it is quite the opposite.

Consider our test for deficiencies in Isaac's college readiness and the hypothesis: H: Isaac is college-ready, as against, H′: Isaac is not college-ready. We can make the tests so hard, and the hurdle for regarding grades as evidence for H so high, that his scores are practically always going to lead to denying H and inferring H′ (he is deficient). However, H′ has passed a test with *very low* severity because it would very often lead to inferring H′, even if H′ is false and actually H is true. How to arrive at assessments of the largest discrepancy warranted by the test is formalised in a 'rule for rejection' in a severity interpretation of statistical tests (Mayo 1996).

9.1.2 *Severity Condition (ii) Differs from Saying that **x** is Very Improbable Given Not-H*

In contrast to Popper's attempted definition of severity, as well as others (e.g., likelihoodists) the second severity condition, i.e., condition (ii) is not merely to

[20] This contrasts with the use of Type I and Type II error probabilities in Neyman-Pearson tests.

assert that P(**x**; H is false) is low,[21] where 'P(**x**; H is false)' is to be read: 'the probability of **x** under the assumption that H is false'. This is called the likelihood of H given **x**. A familiar example shows why. H_1 might be that a coin is fair, and **x** the result of n flips. For any **x** one can construct a hypothesis H_2 that makes the data maximally likely, e.g., H_2 can assert that the probability of heads is 1 just on those tosses that yield heads, 0 otherwise. P(**x**; H_1) is very low and P(**x**; H_2) is high, however, H_2 has not passed a severe test because one can always construct some such maximally likely hypothesis *or other* to perfectly fit the data on coin tosses, even though it is false and the coin is perfectly fair (i.e., H_1 is true). [22] The test that H_2 passes has minimal severity. (This is a case of what I call 'gellerization'.)

In other words, what principally distinguishes the error probability account of tests is that whatever 'fit' measure is satisfied in showing H 'withstands' the test, the error statistician requires asking a question that is one level removed, as it were: How frequently would H withstand the test so well, even if H is false? (Mayo 1996, Mayo and Kruse 2001).

9.1.3 The Degree of Severity with which a Test Passes H is Not the Degree of Probability of H

Finding that a hypothesis H severely passes test T with data **x** does not license a posterior probability assignment to H, a notion which depends on having prior probability assignments to an exhaustive set of hypotheses. As already noted in Section 6, 'highly probed is not the same as highly probable'; but it bears repeating, given how flagrant is this misinterpretation. Such Bayesian calculations (from whatever school of Bayesianism one chooses) are at odds with the severity principle: high posterior probability is neither necessary nor sufficient for high severity, in any sense of probability.

9.2 What Statistical Testing Teaches Us About Severe Testing in General

9.2.1 The Need for Methods to Test the Reliability of Data Statements.

Thinking of formal error statistical testing alerts us at once that it is impossible to assess reliability or severity with just statements of data and hypotheses divorced from the experimental context in which they were generated, modeled, and selected for testing. For the critical rationalist, this recognition spells nothing but trouble. It is assumed, but not explained, how we justify the data on which the critical rationalists' claims of 'best tested H' depends, save perhaps when **x** is the most rudimentary kind of perception. That General Relativity, GTR, one of Popper's favorite examples, is

[21] The requirement of 'fit' in the severity definition, clause (i), may be defined as a requirement about likelihoods, in particular, it requires that P(x; H) be higher than P(x; H is false). It is important to see that this differs from a conditional probability; there is no assumption that a prior probability assignment to H exists or is meaningful. See following note.

[22] I am using ';' in writing P(x;H) in contrast to the notation typically used for a conditional probability P(x/H) in order to emphasize that severity does *not* use a conditional probability which, strictly speaking, requires the prior probabilities $P(H_i)$ be well-defined, for an exhaustive set of hypotheses.

best tested, for example, depends on already having an account for inductively inferring highly sophisticated hypotheses. Even assuming we knew which of many rivals 'pass' the most tests, we are offered no tools for adjudicating disagreements about what passing results actually show about the phenomenon in question. Such accounts remain irrelevant both for science and for philosophy.

9.2.2 Beyond the 'Tower Image' of Data.

Philosophers, critical rationalists included, seem stuck in what might be called the 'tower image' of empirical data: evidence claims are only as reliable as are the intermediary inferences used in arriving at or inferring them. The minute that intermediary inferences are admitted to have their own assumptions, there is a knee-jerk reaction to concede an unacceptable 'regress' without bothering to question whether this need be so. Were it so, then piling inference upon inference would leave us with increasingly less reliable results—but the opposite is true. Individual measurements, for example, may each have wide margins of error, whereas an inferred estimate (e.g., through averaging) may be highly reliable. What makes such inductive inferences about evidence work is that properly modelled and cleverly used, data can lead from less to more accurate and reliable claims: through interconnected checks of error and robustness results, garbage in need not be garbage out! This day-to-day truism of the 'person of common sense' seems overlooked by the critical rationalist!

9.2.3 Need to Partition: 'H is False' is Not the so-called Catchall Factor

The catchall factor is the disjunction of hypotheses other than H, including those not yet even thought of. 'H is false' in the definition of severity refers, instead, to a specific error that hypothesis H may be seen to be denying. Since the error probability assignments needed for formal statistical cases require hypotheses that exhaust the space of alternatives, the methods, as well as the motives, for splitting off questions and *partitioning* spaces of answers that are found in statistics are highly instructive for the broader aims of an error statistical philosophy.

This leads to an important criticism or challenge raised especially by philosophers in the Popperian tradition, whether or not they call themselves critical rationalists; namely, how does the error statistical account severely pass high level theories? (see Chalmers, Earman, Laudan, Mayo 2002a,b and forthcoming). A quick sketch must suffice.

10. SEVERE TESTING IN PROBING LARGE SCALE THEORIES

What enables this account of severity to work is that the hypothesis H under test by means of data **x** is designed to be a specific and local claim, e.g., about parameter values, about causes, about the reliability of an effect, or about experimental assumptions. 'H is false' is not a disjunction of all possible rival explanations of **x**, which would include those not yet known as just noted. This is true, even if H is part of some large scale theory *T*: the condition 'given H is false' always means 'given H is false with respect to what it says about *this particular* effect or phenomenon'. We can abbreviate this claim as $\mathcal{T}$(H) to indicate H is a piece of $\mathcal{T}$. If a hypothesis $\mathcal{T}$(H) passes a severe test we can infer something positive: that the theory $\mathcal{T}$ gets it right about the specific claim H that severely passes.

The price of this localisation is that one is not entitled to regard global or large-scale theories as having passed severe tests so long as they contain hypotheses and predictions that have not been well-probed. If scientific progress is viewed as turning on appraising high-level theories, then this type of localized account of testing will be regarded as guilty of a serious omission, unless it is supplemented with an account of theory appraisal.

> [Her] argument for scientific laws and theories boils down to the claim that they have withstood severe tests better than any available competitor. The only difference between [her] and the Popperians is that she has a superior version of what counts as a severe test. (Chalmers, 1999, p. 208)

The truth is that I never intended to provide any kind of 'argument for scientific laws and theories'; and permitting the comparativist account that he, like Popper and others champion, would conflict with the aims of severity, and preclude the very features Chalmers endorses as superior to Popper's.

Whereas we *can* give guarantees about the reliability of the piecemeal experimental test, it is highly unreliable to follow the comparativist's method: If large-scale theory $\mathcal{T}$ is 'best-tested', regard all of $\mathcal{T}$ as having withstood severe testing, and thus accept or believe $\mathcal{T}$. Still, Chalmers, like Laudan (1997), suggests that in the case of high-level theory I should define severity comparatively:

The Comparativist's Suggested Definition: A theory has been severely tested provided it has survived (severe?) tests its known rivals have failed to pass (and not vice versa).

(The question mark here is due to the fact that in Laudan's account, at least, it is not required that these lower level tests themselves be severe – or, at any rate, he does not make this point clear.) Laudan calls this the 'comparativist rescue' for my account; but no such rescue mission is required or desired. In my view, it is disingenuous to say that all of a theory has survived a good test when there are ways it can be wrong that have not been probed, that there are regions of implication not checked at all. To embrace the comparativist account is to be thrown back to the critical rationalists' problem: being unable to say what is so good about the theory that (by historical accident) happens to be best-tested so far.

10.1 Learning from Tests that Fail to be Classified as Severe

Comparativist testing accounts, eager as they are to license the entire theory, ignore what for our severe tester is the central engine for making progress, for getting ideas for fruitful things to do next, to learn more; namely, by asking, how could we be wrong in supposing all of theory $\mathcal{T}$ has severely passed? Why are we *not* allowed to say that the entire theory is severely probed as a whole? —in all the arenas in which the effects in question may occur. Even without having alternatives we can ask *how could it be a mistake to regard the existing evidence as good evidence for all of the theory?*

Although we learn a lot about phenomena from hypotheses that pass or fail severe tests, we learn at least as much from finding that our test is inadequate as a severe error probe. I know of no account of testing that recognizes this explicitly, and yet it falls out immediately from the error statistical account. One way to unearth errors in taking a passing result as evidence for a given theory $\mathcal{T}$ is to construct a suitable alternative $\mathcal{T}^*$ often using the known data **x** to ensure $\mathcal{T}^*$ accords with **x** in the respects already tested. This may serve an important role in showing why $\mathcal{T}$ fails to have passed severely as a whole. This does not mean that $\mathcal{T}^*$ has passed just as severely as $\mathcal{T}$ has (even with respect to the aspects probed). But it is instructive in finding out that aspects of $\mathcal{T}$ we might have thought were well probed by test T, were in fact not well probed. This is a crucial tool in discovering and constructing new theories. (For a discussion of how this strategy figured centrally in developing alternatives to GTR, see Mayo 2002a and forthcoming).

10.2 Reliability and Stability through Large Scale Theory Change

In addition to reliability, this account has another feature that is missing from the comparativist tester: stability and cumulativity. The severity for passing a lower level hypothesis remains the same even through changing interpretations and through changing high level theories in which it might be embedded. Suppose, for example, that a hypothesis about parameter value *m* has passed severely. This severity evaluation is not altered by the existence of another theory that agrees with this hypothesized parameter value. More generally, severely passing what theory $\mathcal{T}$ says about H, $\mathcal{T}$(H), gives us knowledge about this aspect of theory $\mathcal{T}$, and this assessment remains even though the theory undergoes repeated improvements, revisions, and even reconceptions. By contrast, as soon as an alternative theory comes to light that does as well as theory $\mathcal{T}$ does on existing tests, the comparativist would be forced to change the assessment of how well $\mathcal{T}$ has been tested (it would have to 'give back the crown' as it were.)

The error statistical tester is not precluded from talking of 'accepting' the theory, understood as accepting as severely passed some of its key hypotheses, or simply, regarding it as a fruitful basis from which to learn more and probe further the phenomenon of interest. It would be correct to regard it as a fruitful basis for learning if it allows us to say, even without having a clue about the correct large-scale theory, that any theory in the domain in question would have to include the severely tested effects.

11. CONCLUDING COMMENTS

The lack of progress in the neo-Popperian philosophy, I have argued in Part I of this paper, may be traced to its inability to characterize the severity principle (SP) underlying the epigraph of this paper:

> Observations or experiments can be accepted as supporting a theory (or a hypothesis, or a scientific assertion) only if these observations or experiments are severe tests of the theory—or in other words, only if they result from serious attempts to refute the theory". (Popper, 1994, p. 89)

Critical rationalists have failed to actually cash out what 'surviving serious criticism' demands, and why H's surviving the 'ordeal' makes it reasonable to accept H, regard H as supported, or as beliefworthy. The severity principle is at the heart of rationality in science – Popper was right about that – so long as 'H's surviving serious criticism' may be taken to mean that H has been put to a scrutiny that would have (or would very probably have) uncovered the falsity of (or errors in) H, and yet H emerged unscathed. That is to say, the epistemological force behind SP holds just to the extent that H has survived a highly reliable probe of the ways in which H might be false. However, Popper's logical account deprived him of the resources to articulate 'reliable error probes'; and surprisingly, his current day followers have yet to remove their logical empiricist blinders.

In Musgrave's "promissory notes" (p. 323), he calls for "an account of which kinds of criticism are serious criticisms and which not." Building such "a theory of criticism" is welcome but it demands empirical not purely logical assessments of the error-probing capacities of tools. The comparativist principle CR that he endorses, moreover, is unreliable —claims can easily (frequently) be 'best surviving' with **x** (at time t), even though **x** provides little or no reassurance that errors have been ruled out or even probed.

In part II of this paper, I discussed an account of testing that captures the spirit behind Popper's intuitions about severity, while enabling it to be made operational. Here, probability is used to quantify how frequently methods are capable of discriminating between alternative hypotheses and how reliably tests facilitate the detection of error. These probabilistic properties of test procedures are called error probabilities, and an account based on error probability criteria, whether formal or informal, I dub an *error statistical account of inference*. The error statistical tester agrees with Musgrave's critical rationalist in rejecting the probabilists's view of justifying claims, while being able to provide tests that are genuinely reliable error probes. On the critical rationalist's own criteria of appraisal, therefore, the error statistical approach is to be preferred to the method CR: CR fails to withstand critical scrutiny.

I have sketched how the quantitative conception of severity arising in error statistics may be carried over into more qualitative arenas and in learning about high level theories. Although much work remains in developing such qualitative severe tests, it is a research program with the properties that have much to offer the critical rationalist. If Musgrave and other critical rationalists are serious about pushing forward the stalled Popperian research program beyond the "twelve or twenty" adherents, it is to be hoped that they will at least consider the avenue for progress offered by developing the methodology of error statistics and severe testing. Not only would this help salvage the brilliant gems in Popper, it would be relevant for foundational debates in statistics as to what is really required for rational and objective scientific inquiry.

REFERENCES

Achinstein, P. (2001) *The Book of Evidence,* Oxford University Press.

Ben Haim, Y. (2001) *Information-Gap Decision Theory: Decisions Under Severe Uncertainty*, Academic Press: San Diego CA

Chalmers, A. F. (1999) *What Is This Thing Called Science?* 3rd ed., University of Queensland Press, Australia.

Earman, J. (1992) *Bayes or Bust? A Critical Examination of Bayesian Confirmation Theory,* MIT Press, Cambridge, MA.

Gillies, D. A. (1973) *An Objective Theory of Probability*, Methuem, London.

Grünbaum, A. (1978) 'Popper vs. Inductivism', in Radnitzky and Andersson (eds.), *Progress and Rationality in Science*, Boston Studies in the Philosophy of Science, vol. 58. Dordrecht, The Netherlands: Reidel, pp. 117-142.

Laudan, L. (1997) 'How about Bust? Factoring Explanatory Power Back into Theory Evaluation', *Philosophy of Science* 64(2): 306-16.

Lehmann, E. L. (1995) 'Neyman's Statistical Philosophy', *Probability and Mathematical Statistics,* 15: 29-36.

Mayo, D. G (1991) 'Novel Evidence and Severe Tests,' *Philosophy of Science* 56: 523-552.

Mayo, D. G. (1996) *Error and the Growth of Experimental Knowledge*, The University of Chicago Press, Chicago.

Mayo, D. G. (1997a) 'Duhem's Problem, The Bayesian Way, and Error Statistics, or 'What's Belief Got To Do With It'?' and 'Response to Howson and Laudan,' *Philosophy of Science*, 64(2): 222-24 and 323-33.

Mayo, D. G (1997b) 'Error Statistics and Learning from Error: Making a Virtue of Necessity,' *Philosophyof Science* 64 (Proceedings) §195-212

Mayo, D. G. (2002a) 'Theory Testing, Statistical Methodology, and the Growth of Experimental Knowledge,' *Proceedings of the International Congress for Logic, Methodology, and Philosophy of Science*, Kluwer Press, pp. 171-190.

Mayo, D. G. (2002b) 'Severe Testing as a Guide for Inductive Learning,' in Kyburg, H. E. and M. Thalos, eds., *Probability is the Very Guide of Life*, Chicago: Open Court, pp. 89-117.

Mayo, D. G. (2003) 'Could Fisher, Jeffreys and Neyman Have Agreed? Commentary on J. Berger's Fisher Address,' *Statistical Science* 18: 19-24.

Mayo, D. G. (2004) 'An Error-Statistical Philosophy of Evidence', in M. Taper and S. Lele (eds.), *The Nature of Scientific Evidence: Statistical, Philosophical, and Empirical Consideration*, Chicago: University of Chicago Press.

Mayo, D. G. (2005) 'Evidence as Passing Severe Tests: Highly Probed vs. Highly Proved' in P. Achinstein (ed.) *Scientific Evidence*, Johns Hopkins University Press: 95-127.

Mayo, D. G. (2006) 'The Philosophy of Statistics', in S. Sarkar (ed.) *Routledge Encyclopedia of the Philosophy of Science.*

Mayo, D. G. (forthcoming) 'Using Low-Level Tests to Probe High-Level Theories?: Severity Without Comparativism'

Mayo, D. G. and D. Cox (2006) 'Frequentist Statistics as a Theory of Inductive Inference,' with D.R. Cox, *Proceedings of The Second Erich L. Lehmann Symposium*, Vol xx, Institute of Mathematical Statistics (IMS) Lecture Notes-Monograph Series.

Mayo, D. G. and M. Kruse (2001) 'Principles of Inference and their Consequences,' pp. 381-403 in D. Cornfield and J. Williamson (eds.), *Foundations of Bayesianism*, Kluwer Academic Publishers, Netherlands.

Mayo, D. G. and A. Spanos (2004) 'Methodology in Practice: Statistical Misspecification Testing', *Philosophy of Science* 71: 1007-1025.

Mayo, D. G. and A. Spanos (2006) 'Severe Testing as a Basic Concept in the Neyman-Pearson Philosophy of Induction,' *British Journal of Philosophy of Science.*

Meehl, P. E. (1967/1970) 'Theory-Testing in Psychology and Physics: A Methodological Paradox,' In D. E. Morrison and R. E. Henkel (eds.), *The Significance Test Controversy* (1970), Aldine, Chicago.

Musgrave, A. (1974a) 'Logical versus historical theories of confirmation,' *British Journal for the Philosophy of Science* 25: 1-23.

Musgrave, A. (1974b) 'The Objectivism of Popper's Epistemology', in P. A. Schilpp (ed.) *The Library of Living Philosophers,* LaSalle: Open court, pp. 560-596

Musgrave, A. (1989) 'Deductive Heuristics', in K. Gavroglu, Y. Goudaroulis and P. Nicolacopoulos (eds.) *Imre Lakatos and Theories of Scientific Change,* Dordrecht' Kluwer Academic, pp. 15-32

Musgrave, A. (1999) *Essays in Realism and Rationalism*, (Chapter 16) Amersterdam: Rodopi; Atlanta, GA.

Neyman, J. (1952) *Lectures and Conferences on Mathematical Statistics and Probability*, 2nd ed. U.S. Department of Agriculture, Washington.

Neyman, J. (1955) 'The Problem of Inductive Inference,' *Communications on Pure and Applied Mathematics*, VIII, 13-46.

Neyman, J. (1957a) 'Inductive Behavior as a Basic Concept of Philosophy of Science,' *Revue Inst. Int. De Stat.,* 25: 7-22.

Neyman, J. (1957b) 'The Use of the Concept of Power in Agricultural Experimentation,' *Journal of the Indian Society of Agricultural Statistics*, IX: 9-17.

Neyman, J. (1971) 'Foundations of Behavioristic Statistics,' in V. P. Godambe and D. A. Sprott (eds.) *Foundations of Statistical Inference*, Toronto: Holt, Rinehart & Winston, pp. 1-13 (comments and reply, pp. 14-19).

Neyman, J. and E. S. Pearson (1933) 'On the problem of the most efficient tests of statistical hypotheses', *Philosophical Transactions of the Royal Society*, A, 231: 289-337. Reprinted in Neyman, J. and E. S. Pearson (1966).

Popper, K. (1959) *The Logic of Scientific Discovery,* New York: Basic Books

Popper, K. (1962) *Conjectures and Refutations: The Growth of Scientific Knowledge*, New York: Basic Books.

Popper, K. (1976) 'A Note on Verisimilitude', The British Journal for the Philosophy of Science 27: 124-159.

Popper, K. (1994) *The Myth of the Framework: In Defence of Science and Rationality* (edited by N.A. Notturno). London: Routledge.

VOLKER GADENNE

METHODOLOGICAL RULES, RATIONALITY, AND TRUTH

The methodology of critical rationalism is constituted by a set of rules, which Popper first formulated in *The Logic of Scientific Discovery* (*Logic*) and supplemented in later works. They state, for example, that scientific hypotheses should be severely tested, that they should never be regarded as finally verified, that we should not apply conventionalist stratagems, etc. In this paper, I want to discuss some problems connected with methodological rules, especially, questions concerning the status and function of such rules, and their relation to the aim of science. Many questions of this kind are still quite controversial, even among people who consider themselves as followers of Karl Popper. For instance, does science need principles about the believing or acceptance of hypotheses? Do the rules of falsificationism lead us to truth, or to progress? Can such an assumption be critically assessed and justified?

1. THE MEANS-END VIEW OF METHODOLOGICAL RULES

Methodological rules can be stated in one of the following forms: 'Apply method M'; 'Avoid procedure P'; 'Test theories critically'; 'Never regard them as finally verified'; etc. Such rules always refer to certain problem situations. Let S be a special type of problem situation in empirical science, and M a procedure or method for this type of situation. For example, an experiment is to be planned to test a theory T, or it is to be decided how to proceed when some experimental results contradict T. A methodological rule like 'Apply M' can then more exactly be stated in the form, 'If you are in situation S, apply procedure M.'

Popper realized that methodological rules couldn't be adequately conceived as analytical statements, and not as empirical statements either. He therefore declared them as *conventions* that define the game of science (1959, p. 53). As Jarvie (2001, p. 41) points out, Popper moved from the pure logic of science into aspects of scientific method that are social rather than logical. Methodological rules are social rules. Science requires cooperation of researchers; it is a 'republic of sorts' with

C. Cheyne & J. Worrall (eds.), Rationality and Reality: Conversations with Alan Musgrave, 97–107.

specialized institutions. According to Jarvie, *Logic* is actually an embryonic institutional approach, which Popper developed further in *The Open Society and Its Enemies*. The social character of science has also been emphasized by Agassi (1972), Albert (1985), and Wettersten (1992).

Let us come back to the form of methodological rules. Should they be interpreted as categorical norms or imperatives? This would not make much sense. It would be quite arbitrary and unconvincing to demand categorically, 'Be critical!' Why should we be critical? Actually, criticism causes a lot of conflict and trouble. Most people, including critical rationalists, dislike being criticised very much; some regularly feel offended. So there must be some gain associated with criticism that justifies these costs. What is it good for to be open to criticism?

Conventions cannot be true or false. As Popper argues, however, they may prove more or less fruitful (1959, p. 55). So they can be evaluated on the basis of their fruitfulness. What kind of fruitfulness had he in mind? Did he refer to the progress of empirical science? It does not seem so. He claimed that his definition of empirical science conformed to scientists 'intuitive idea of the goal of their endeavours' (p. 55). And he said his methodological rules would also prove useful for philosophers by enabling them to detect inconsistencies and inadequacies in older theories of knowledge. But is methodology only an explication of scientists' 'intuitive ideas'? And is it helpful only for the philosophy of science, not for science itself?

Provided science has a goal, it should also be possible to interpret methodological rules as hypothetical norms: Apply procedure M in order to achieve aim A (or contribute to A). Or, to take the problem situations into account: If you are in situation S, apply procedure M in order to achieve A (or contribute to A). A may be the search for truth; or, more exactly, science aims at true theories of high explanatory power. Correspondingly, Albert (1985, p. 53) regards methodology as a specific technology with respect to this aim. Methodological rules claim *means-end rationality*, or *purposive rationality*. Albert also argues that such rules should not be taken as rigorous prescriptions. Their function is similar to that of heuristic principles; they give useful recommendations and hints but never require exactly one special solution.

If we understand methodological rules this way, the proposal and acceptance of such rules obviously presupposes that certain assumptions are true. The recommendation, 'Test theories critically in order to achieve true ones', makes sense if we assume that, other things being equal, critical testing contributes to finding true theories. There may be additional aspects that influence how scientists proceed, for example, expenditure of time and money. From an epistemological viewpoint, however, the proposal and selection of methods is guided by assumptions about what method best serves a special purpose that is related to the overall aim of science. We recommend M on the basis of the assumption that M is conducive to A. When there are rival procedures M_1 and M_2, both designed to achieve the same aim A, we have to ask whether M_1 or M_2 is more effective. If we recommend M_1 instead of M_2, our proposal is based on the *epistemological hypothesis* that M_1 is more effective than M_2 with respect to A. Let us call the whole set of procedures proposed by critical rationalism the *critical method*. Critical rationalism then assumes that the critical method is conducive to the goal of achieving true, powerful theories.

Obviously, the means-end view of methodological rules is based on *epistemological hypotheses.* These hypotheses could be false. For example, too much criticism might discourage scientists. Or, even worse, God might decide to strike with blindness people who put everything into question. This seems to be quite trivial. However, some philosophers do not agree.

2. IS THE CRITICAL METHOD TRUTH-CONDUCIVE?

The means-end view of methodology is connected with the idea that following certain rules is helpful or conducive to achieving the aim of science, especially, truth. Some fallibilists vehemently object to this idea. Miller (1994, p. 46) rejects the view that if one follows Popper's methodological rules one has a better chance to get nearer to the truth than otherwise. According to Miller, such a view is not suggested by critical rationalism, not even conjecturally. He also says that Popper never claimed that, by following the principles of falsificationism, scientists would be lead to theories of greater verisimilitude. Jarvie (2001, p. 71) also emphasizes that Popper never promised that following his methodological rules leads us to progress.

Obviously, no method *guarantees* that we achieve the truth, or probably achieve it. Rules that we consider as fruitful and truth-conducive may actually be ineffective or impeding. Furthermore, it would be quite naive to specify exact probabilities with which the critical method leads to true hypotheses, or helps to avoid false ones. If this is what Miller and Jarvie want to emphasize, they are quite right. But let us assume that A is the goal of science and our task is to decide whether procedure M should be recommended with respect to A or not. In this case, a rational person will recommend M if and only if he or she *believes that M contributes to A*, or that *M gives us a greater chance to achieve A*. And this belief is reasonable if there are good arguments in favour of the hypothesis that M is the best we can do to achieve A.

Having 'good arguments' in favour of a method M does not mean that we can give a guarantee, or that we can be certain to arrive at A. Popper liked the formulation that, even if we use the rules of falsificationism, we may only *hope* to make progress. But 'hope' is not a methodological or epistemological category. We may hope something even if our rational prediction says it will not happen. 'Guarantee' is too strong for a methodology that is committed to fallibilism. One the other hand, hope is not enough. The idea that applying a method is reasonable with respect to some goal must be something less than guarantee but stronger than mere hope. Someone who denies this can hardly claim rationality, even if he or she stresses the critical method. A position that proposes the critical method, and declares truth as the aim of science, but then rejects the assumption that the critical method gives us a greater chance to arrive at true theories could appropriately be called 'critical irrationalism'.

Let us discuss this with the help of an example. In order to test a theory empirically, Popper recommended deriving from this theory a test statement that cannot be derived from other, rival theories. If possible, our test statement should even *contradict* those rival theories. Let us call this the *method of critical testing*.

'Critical' means that by tests of this kind one refutes at least one theory, or, more carefully, one gets evidence against one theory at least, either against the new theory to be tested, or against a rival theory, perhaps an established one.

We now have a choice between the following two positions. The first is:

1) We recommend the method of critical testing; but we do not form any epistemological assumption connecting critical testing and truth.

This seems to me quite unconvincing. For what reason should we recommend a method if we do not assume that it contributes to our goal? Why should critical testing then be called a reasonable or rational method? Why should we make the effort of analysing and comparing theories in order to find out test statements that discriminate between those theories? We could as well do empirical research inductively or we could recommend Feyerabend's 'anything goes'.

The second position is:

2) We recommend the method of critical testing; and we accept the assumption that critical testing contributes to achieving true theories.

In this case, we formulate an *epistemological hypothesis* (which may be false). We can argue for this hypothesis in the following way.

We want to have true theories. Since the truth of universal statements cannot be inductively proven, we try to refute false ones. We now have the sub-goal of detecting false theories. If there are several rival theories, a false one can be detected effectively by using test statements that follow from one theory and contradict another. The result will falsify at least one theory. Compare this *method of critical testing* (M_1) to some other strategies of empirical research:

(M_1) *Critical testing:* Derive from the theory under test statements that do not follow from rival theories, or, even better, that contradict them.

(M_2) *Simple testing:* Test a theory by using any observational statement that follows from it.

(M_3) *Data gathering:* Try to gather a lot of empirical data; then develop a theory that fits these data.

I think that M_1 is a more effective method than either M_2 or M_3 (or any other method yet known). We are justified in assuming that critical testing is superior to these other methods with respect to our sub-goal, that is, to identify false theories.

Similarly, some other rules of falsificationism can be justified. Such rules claim means-end rationality. Their claim can be assessed by discussing them critically and comparing them to alternative methods. Therefore, they are no mere conventions, if 'convention' means that something cannot be rationally assessed. We can argue for methodological rules, in the same way we argue for other philosophical assumptions.

Is it induction to assume that the critical method is superior to alternative methods with respect to the search for truth? If any methodological rule that cannot be deduced from principles of deductive logic is called 'inductive', then this is of course induction, and any methodology of the empirical sciences has to be inductive in this broad sense. In this case, it should however be noted that it is *not* induction as criticised by Hume and Popper. Epistemological hypotheses about the fruitfulness of procedures are neither inductive inferences nor calculi of inductive probability.

3. THE METHODOLOGICAL RULES OF CRITICAL RATIONALISM AND MUSGRAVE'S PRINCIPLE CR

What are the central methodological rules or principles of critical rationalism? Johansson (1975) listed (and criticised) more than twenty rules he found in Popper's works (most from *Logic*, some from *The Poverty of Historicism*). Jarvie (2001) extracts fifteen rules from *Logic*. They deal, among other things, with the falsifiability and falsification of theories, with conventionalist stratagems, and the acceptance of basic statements. They are governed by a *supreme rule*, saying, 'the other rules of scientific procedure must be designed in such a way that they do not protect any statement in science against falsification' (Popper 1959, p. 54).

The supreme rule (SR) expresses an idea similar to what Popper later called *openness to criticism* though the latter idea is much broader and not restricted to empirical science. We can say that SR demands openness to criticism in the realm of science. SR is also closely related to the idea of *fallibilism* since the insight that there can be no certain knowledge is the main reason for SR.

Jarvie's collection of rules does not include any principle dealing with the acceptance of hypotheses as true. This is not surprising because such a principle cannot be found in *Logic*. According to *Logic*, methodology is only concerned with the business of criticism. All rules are designed to serve falsification, and the supreme rule talks about falsification as if it was an end in itself. Does a methodology of the empirical sciences not also need a rule that allows to accept hypotheses or theories at least tentatively as true if they have stood up to critical tests? Some critical rationalists say no, for example, Agassi, Bartley, Miller, and Wettersten. They think it was Popper's ingenious idea that rationality requires the critical method but not justification of any kind. Most other philosophers have great difficulty even to understand this position and to take it seriously.

Alan Musgrave, however, in some of his publications (1993; 1999, p. 324) presents an interpretation of critical rationalism quite different from the negativism of the radical non-justificationists. According to Musgrave, the central principle of critical rationalism can be summarized as follows (1999, p. 324):

CR: It is reasonable to believe that P (at time t) if and only if P is that hypothesis that has (at time t) best withstood serious criticism.

Restricted to empirical science, the central principle becomes (1999, p. 327):

CR*: It is reasonable to adopt as true (to believe) the best-corroborated hypothesis.

If there are several hypotheses that are equally well corroborated, we have to suspend judgement and carry out further tests. We do this in order to bring about a situation in which one hypothesis is clearly superior to its rivals. In the ideal case, H should be highly corroborated while all competing hypotheses are falsified. For such a situation, CR declares it as reasonable to adopt H tentatively as true. In this case, we are also justified in believing the logical consequences of H.

To understand CR correctly, two further things should be explained. First, CR refers to evidence-transcending beliefs, not to perceptual ones. For a perceptual belief P, Musgrave holds that we are justified in believing P if and only if P has not failed to withstand criticism (1999, p. 342). Perceptual beliefs, too, can be criticised and rejected. However, they do not require critical examination to be justified. In contrast, evidence-transcending beliefs are not justified, unless they have been tested and passed these tests.

Second, the test of a proposition need not be carried out by the same person who reasonably believes it. Most things we know were discovered and tested by other people, and reported to us. We are justified in believing what other people tell us, unless we have a specific reason not to.

CR* speaks of the 'adoption' of a hypothesis as true. Does this mean a choice or decision? It seems not, since Musgrave uses the expressions 'to believe P' and 'to adopt P as true' as interchangeable. Thus CR talks about cognitive states, namely beliefs, not about deliberate action. People do not and cannot decide to believe or not to believe something. (They can of course seek new information, or think their beliefs over again, and this may lead them to new conclusions. In addition, they can try to change their belief-producing dispositions by reflection.)

In another sense, however, people can choose a hypothesis among competing ones: They can select it in order to do something with it, especially, derive a prediction, or calculate a suitable action. This raises the question whether we should accept a rule that says of actions what CR says of cognitive states. Popper (1972, p. 22) stated such a rule. When analysing the question of 'pragmatic preference', he came to the conclusion that 'we should prefer as basis for action the best-tested theory.' I suppose Musgrave agrees with this. A reasonable person will choose as basis for prediction and action that hypothesis H among competing ones that has best withstood criticism. Obviously, a person is justified in choosing H for action just in case she is justified in believing it. However, we have to bear in mind that there are some philosophers who accept such a principle for action but not a corresponding one for rational believing. Popper, too, had a problem with believing propositions, or accepting them as true (see below).

How can CR itself be justified? Musgrave argues that CR has better withstood criticism than any competing principle. And he admits that this argument involves a circle. More exactly, CR* can be justified by demonstrating that it has withstood criticism better than competing principles; arguing this way, we apply CR. And if we next argue for CR in the same way, we have to apply CR itself.

In connection with CR (or CR*), Musgrave also points out, with reference to Popper, that only the *believing* of a proposition can be justified, not the proposition itself. I found that some people have difficulty with this distinction. Some of them, when speaking of the justification of a theory, actually mean the justification of the acceptance of this theory. Couldn't we agree to call a proposition P justified just in case it is justified to believe P? Others say there might be a difference, but they doubt that it is really of epistemological relevance. Justifying a belief or a proposition, doesn't this come to the same problem? Musgrave says no. I understand his argument as follows.

Suppose a hypothesis H has been severely tested, and the result E is in accordance with H. We can then argue: E has been accepted; therefore we adopt H tentatively as true.

This argument presupposes CR* as a premise. Now compare this to the following argument that might be brought forward to justify H itself, that is, the assumption that H is true: E is true (or E has been accepted), therefore H is true.

Obviously, there is no plausible premise to support this reasoning that leads from observation statements E to a universal statement H. We would need for such an argument an acceptable calculus of inductive logic, which is not available. H could of course be deductively derived from some logically stronger hypothesis H'. But this would beg the question; we would next have to justify H'.

We started by discussing the status and function of methodological rules, then came to epistemological hypotheses, and are now considering normative principles, like CR. How are these things related? Methodological rules were reconstructed as follows: 'If you are in situation S, apply procedure M in order to achieve (or contribute to achieving) A.' Such a rule presupposes an epistemological hypothesis. The rule only makes sense if we assume that M is conducive to A. Methodological rules recommend certain procedures, and epistemological hypotheses state the rationale behind these procedures.

Now such procedures do not yet give us means to evaluate the results. Consider again the testing of competing empirical hypotheses. The critical method is designed to bring about a situation in which the rival hypotheses differ in corroboration. But the critical method, or the rules that define this method, does not tell us how such a result is to be evaluated. We therefore need, in addition, a normative principle, like CR*. Such a normative principle for the evaluation of beliefs is in turn based on epistemological hypotheses. Why should we accept CR? A possible answer is: We accept CR because critical testing best serves to detect false consequences of hypotheses; and this is the best one can do to rule out false hypotheses and retain true ones.

There is an alternative possibility: We could take normative principles like CR as conventions that do not allow for further analysis and argumentation. CR simply defines what rational believing is. You may accept this or not, but you cannot ask what lies behind CR, since this would not be a meaningful question. This view, however, would hardly be convincing, and it is not Musgrave's position, since he holds that CR is the principle that has best withstood criticism. CR can be argued for and criticised, as well as empirical hypotheses can. It is a meaningful question to ask, 'Why should we accept this normative epistemological principle?' An answer to such a question presumably involves discussing some epistemological hypotheses associated with this principle.

Now CR seems to be just the opposite of what, for example, Miller (1994) holds. It seems that the subject of belief divides critical rationalists into two groups, those who accept rational believing as part of their methodology and those who do not. What was Popper's view on this subject? This is a question not easy to answer, because Popper made a lot of remarks on corroboration, preference, and truth, some of which bear some similarity with CR, while others object to justification of any kind.

In *Logic*, Popper did not say anything about the acceptance of hypotheses as true. He was concerned with logical and methodological aspects of falsification. He also considered his theory of corroboration as an important part of his methodology, but he did not regard corroboration as an indication of truth, or as a reason to accept a hypothesis as true. In *Conjectures and Refutations*, he still kept his distance to the subject of acceptance, and said:

> Incidentally, I do not propose any 'criterion' for the choice of scientific hypotheses: every choice remains a risky guess. Moreover, the theoretician's choice is the hypothesis most worthy of *further critical discussion* (rather than *acceptance*). (1963, p. 218, note 3)

In *Objective Knowledge* (1972, p. 82), however, Popper presented a rather different view. He made a lot of statements connecting corroboration, truth, and rational preference. For example, he said the critical discussion could 'establish sufficient reasons for the following claim: 'This theory seems at present, in the light of a thorough critical discussion, and of severe and ingenious testing, by far the *best* (the strongest, the best tested); and so it seems the one nearest to truth among the competing theories.'' Since the best-tested theory not yet falsified is the best corroborated one, we can formulate Popper's statement as follows (I call it CR_v with 'v' for 'verisimilitude'):

CR_V: It is reasonable to regard the best-corroborated theory among the competing theories as the one nearest to truth.

CR_V seems to be rather similar to Musgrave's principle CR*, except that Popper speaks of the theory 'nearest to truth'. Is this difference important? From Popper's viewpoint, it is. In his *Postscript*, he emphasised:

> There simply is *no* reason to believe in the *truth* (or the probability) of any [...] theory; though there may be reasons for preferring one theory to others as a better *approximation to the truth* (which is not a probability). This makes all the difference. (1983, p. 67)

He also said (1972, p. 103) that the degree of corroboration of a theory could be taken as an *indication* of its verisimilitude, not as a measure, but as an indication.

Now philosophers have not yet been very successful in their attempts to define the concept of verisimilitude or approximation to the truth (see Kuipers 1987). All theories of verisimilitude so far developed have serious problems. As long as these are unsolved, it seems to be less problematic to talk of truth than of verisimilitude.

Why did Popper, from a certain time on, prefer to talk of approximation to truth instead of truth? He did so because he was convinced that every theory would turn out false at some time. No theory is perfectly true, but theories differ in verisimilitude, so that progress should be interpreted as approximation to truth.

But this is of course a metaphysical assumption not implied by fallibilism and not supported by science either. Fallibilism says that every theory may turn out false which is different from the claim that it will. Science textbooks present a huge number of laws that have not yet been falsified and which will perhaps never be. Fallibilism is logically compatible with the assumption that all theories not yet falsified are true. In any case, the mere possibility that all theories are false is not an argument against believing, or adopting as true, the best corroborated theories. So we can say that Popper's CR_v is hardly more convincing than CR*.

But Popper's CR_v was not his last word on corroboration, truth, and preference. In the second edition of Objective Knowledge (1979) Popper took back his CR_v, and all similar statements. He declared these statements as slightly "incautious" and tried to clarify them (see Miller 2002, p. 98). Popper wrote:

> Thus I hold that we *can* have good arguments for *preferring*—if only for the time being—T_2 to T_1 *with respect to verisimilitude*. [...] Whenever I say...that we have reasons for believing that we have made progress, I speak of course not in the factual object language of our theories (say T_1 and T_2), nor do I claim in the metalanguage that T_2 is, in fact, nearer to the truth than T_1. Rather, I give an appraisal of *the state of the discussion* of these theories, in the light of which T_2 appears to be preferable to T_1, from the point of view of aiming at the truth. (1979, p. 372)

The first part of this statement corresponds to what Popper earlier said. Then, however, he emphasises he would not claim that T_2 is, in fact, nearer to the truth than T_1. Obviously, he thinks it not possible to justify such a (fallible) claim. Thus we would not be justified in claiming (or believing) that Einstein's theory was in fact nearer to the truth than Newton's. Such a statement, either in the object language or in the metalanguage, would at best be 'incautious'. We are only justified in claiming that Einstein's theory *appears* to be nearer to the truth. But what does it then mean that we can have 'good arguments for preferring T_2 to T_1 with respect to verisimilitude'? (I must confess I have difficulty understanding the whole passage.)

It seems that this aspect of Popper's view is still unclear. This explains perhaps how it comes that critical rationalists who hold quite different views on rational acceptance refer to Popper and believe they have him right. But let us now set aside problems of interpretation and turn to the question whether CR is really necessary. Some non-justificationists argue that beliefs do not matter. The important thing to do is to apply the critical method while the scientist's beliefs are rather irrelevant. The scientist's choice is the hypothesis most worthy of *further critical discussion*.

But is acceptance for test enough? Is it the only kind of acceptance needed in science? Agassi (1985) pointed out that with respect to corroboration we have to distinguish problems in science from those in technology. Engineers or physicians have to decide which theories to rely on. They cannot put off their decisions. In basic research, however, we need not decide which of two competing, not yet falsified theories is true, or nearer to truth. We can wait until new experiments are carried out that may falsify one of them. Shortly put, corroboration is needed in application, but not in basic research.

It is true that, in basic research, scientists need not make decisions about the truth of theories. It would however be a misunderstanding to interpret CR as a rule for making such decisions. While many methodological rules, including Popper's principle of pragmatic preference, deal with procedures and decisions in certain problem situations, CR is of another kind. It is not concerned with deliberate action but with cognitive states. And epistemology has to say something about the justification of specific cognitive states, namely beliefs. Beliefs are the essential precondition of (conjectural) knowledge. Epistemology cannot restrict itself to proposing rules or procedures.

As stated above, scientists as well as other people do not *and cannot* deliberately decide to believe something nor can they deliberately avoid having certain beliefs. I cannot avoid believing that the table in front of me exists. I cannot help believing a causal law, if, in a controlled experiment, A regularly leads to B, etc. CR says that not all beliefs are equally justified. Some are more justified than others because they are in accordance with certain epistemological assumptions. It would not be justified to believe a theory that has been falsified by repeated experiments. It would be unreasonable to prefer a hypothesis just invented and not yet tested to another one that is highly corroborated. But if it happens that someone adopts H as true, and H is the hypothesis that has best stood up to severe criticism, then his or her belief is in accordance with certain *epistemological assumptions* we have accepted, saying the critical method is the most effective one for the search for truth. This is what makes that belief reasonable. Provided that we are not able to hold back any belief, it is reasonable to believe those hypotheses that have been most thoroughly tested and not yet been falsified. To require that people should believe something else would be unconvincing, and to require that they should believe nothing at all would be nonsense.

Nevertheless some advocates of non-justificationism reject CR. But what alternative is there to CR? Being fallibilists, we have given up attempts to justify knowledge as certain. However, the critics of CR not only deny certain knowledge, they reject any kind of justification, and this leads to some confusion, since, at the same time, they claim reason and rationality. The rejection of any kind of justification means that, for every proposition P, it is equally justified to believe P as to believe non-P; and this is not rationality, it is Pyrrhonian scepticism. It doesn't help to call criticism rationality as long as one does not make clear how criticism contributes to bringing about situations in which some beliefs turn out to be more acceptable than others with respect to truth.

It is however implausible that anybody really wants to maintain scepticism. Perhaps radical non-justificationists, too, have some means-end hypotheses or principles of acceptance behind their reasoning, though they do not want to state them explicitly and make them part of their methodology. They hold it as self-evident, for example, that criticism helps to detect falsities, and that hypotheses not yet falsified should be preferred to falsified ones (which is actually the same as to say that criticism contributes to the search for truth). Now isn't it possible to hold a view that consists of mere proposals and recommendations, and set aside all means-end hypotheses or principles of acceptance? Simply be open to criticism and apply the critical method; make no promise about where the critical method leads us. You may believe that it leads us to progress, but this is your private affair and not part of methodology.

This is a possible, logically consistent view. But is it satisfactory as an epistemology or methodology? Critical rationalism has always stressed that we should work out the philosophical assumptions and principles that lie behind our actions and lives, so that they can be criticised and perhaps improved. Shouldn't we do this also with those epistemological hypotheses that lie behind certain methodological rules or normative principles? An epistemology or methodology that restricts itself to proposals and rules without stating the rationale behind them appears to be essentially poor and incomplete.

Alan Musgrave is well known for his significant contributions to rationalism and realism. In most of these contributions, he sharply attacks modern versions of idealism and scepticism. His principle CR, however, was mainly addressed at philosophers of his own critical rationalist tradition. By formulating this principle, and pointing out its implications, he solved a major problem within Popperian philosophy itself. Obviously, he had to save rational science not only from scepticism but also from sceptical critical rationalists.

REFERENCES

Agassi, J. (1972) 'Sociologism in Philosophy of Science.' *Metaphilosophy*, 3: 103-122.

Agassi, J. (1985) *Technology: Philosophical and Social Aspects*. Dordrecht: Reidel.

Albert, H. (1985) *Treatise on Critical Reason*. Princeton: Princeton University Press.

Jarvie, I. J. (2001) *The Republic of Science*. Amsterdam/Atlanta: Rodopi.

Johansson, I. (1975) *A Critique of Karl Poppe's Methodology*. Stockholm: Akademiförlaget.

Kuipers, T. A. F. (ed.) (1987) *What Is Closer-to-the-Truth?* Amsterdam/Atlanta: Rodopi.

Miller, D. (1994) *Critical Rationalism: A Restatement and Defence*. Chicago: Open Court.

Miller, D. (2002) 'Induction: A Problem Solved.' In Böhm, J. M., Holweg, H. und Hoock, C. (eds.), *Karl Poppers kritischer Rationalismus heute*. Tübingen: Mohr-Siebeck, pp. 81-106.

Musgrave, A. (1993) 'Popper on Induction.' *Philosophy of the Social Sciences*, 23: 516-527.

Musgrave, A. (1999) *Essays on Realism and Rationalism*. Amsterdam/Atlanta: Rodopi.

Popper, K. R. (1959) *The Logic of Scientific Discovery*. English translation of 'Logik der Forschung' (1935). London: Hutchinson.

Popper, K. R. (1945) *The Open Society and Its Enemies*. Vol 1: *The Spell of Plato*. Vol. 2: *The High Tide of Prophecy*. London: Routledge.

Popper, K. R. (1957) *The Poverty of Historicism*. London: Routledge.

Popper, K. R. (1963) *Conjectures and Refutations*. London: Routledge.

Popper, K. R. (1972) *Objective knowledge*. 2nd rev. ed. 1979. Oxford: Clarendon Press.

Popper, K. R. (1983) *The Postscript to The Logic of Scientific Discovery*. Vol. 1: *Realism and the Aim of Science*. London: Hutchinson.

Wettersten, J. R. (1992) *The Roots of Critical Rationalism*. Amsterdam/Atlanta: Rodopi.

HOWARD SANKEY

WHY IS IT RATIONAL TO BELIEVE SCIENTIFIC THEORIES ARE TRUE?

1. INTRODUCTION

Alan Musgrave is one of the foremost contemporary defenders of scientific realism. He is also one of the leading exponents of Karl Popper's critical rationalist philosophy. In this paper, my main focus will be on Musgrave's realism. However, I will emphasize epistemological aspects of realism. This will lead me to address aspects of his critical rationalism as well.

Musgrave is both a scientific realist and a commonsense realist. 'Scientific realism,' he says, 'is a form of realism' (1999, p. 132). And realism is committed to the commonsense realist belief 'that there is a real world outside of us and largely independent of us' (1999, p. 132). 'There is,' Musgrave adds, 'a continuity between common sense and science' (1999, p. 132). But while science may lead to occasional revision and refinement of common sense, 'it does not show that it is root-and-branch mistaken' (1999, p. 133; cf. 1996, p. 23). The real world postulated by common sense is the reality that science seeks to explain. This world does not depend on human belief or experience. Nor is it relative to conceptual scheme, theoretical background or mode of description (1999, pp. 52, 173, 180 ff).

For Musgrave, though, realism is not just a thesis about reality. It is also a thesis about truth. Musgrave takes the aim of science to be truth. He 'subscribe[s] to the old-fashioned idea that scientific realism ... says that the aim of a scientific inquiry is to discover the truth about the matter inquired into' (1996, p. 19; cf. 1999, p. 52). Scientific theories are taken at face-value as genuine assertions about the world, the truth or falsity of which depends on the way the world really is (1996, p. 26). Musgrave understands truth in the classic correspondence sense that he takes to have been defined by Tarski. A theory or statement is true just in case the world is the

C. Cheyne & J. Worrall (eds.), Rationality and Reality: Conversations with Alan Musgrave, 109–132.

way it is said to be (1993, ch. 14; 1996, p. 24; 1999, p. 165). This is a 'non-epistemic conception of truth' (1996, p. 28; cf. 1999, p. 186). Given the emphasis on correspondence between theory and reality, Musgrave's realism diverges from the tendency among some scientific realists to adopt ontological rather than truth-orientated versions of the doctrine. Musgrave dismisses such 'entity-realism' as incoherent (1996, p. 20).[1]

Musgrave's realism has an epistemological dimension as well. For Musgrave, methodological considerations play a prominent role in the appraisal and acceptance of scientific theories. While a variety of methodological norms figures in Musgrave's writings, there is some tendency on his part to emphasize the testing and falsification of theories.[2] The attempt to falsify theories is the basis of the critical method in science. And criticism is the heart of rationality. A critical discussion may provide 'the best reason there is for believing (tentatively) that a hypothesis is true' (1999, p. 324). If a theory 'best withstands criticism then it is reasonable for scientists to believe that theory and to use it in practical applications' (1999, p. 325). Such belief must remain tentative, however. For Musgrave is a fallibilist who eschews the search for epistemic certainty in science and everyday affairs (cf. 1993, ch. 15; 1999, pp. 194 ff, 341-3).

But matters of method and rationality are separate matters from those of reality and truth. This is especially the case from the perspective of realism. In the first place, to believe that the world is a given way does not mean that the world is that way. Nor does it make the world that way. Reality is not subject to determination by human thought. This remains the case even if the belief that the world is a given way is a belief that is rationally justified. For one may rationally believe what is false. The point applies with equal force to scientific theories certified by the norms of scientific method. A theory that is certified by the norms of method is not thereby

[1] Entity realism is an ontological thesis about the reality of the unobservable ('theoretical') entities discovered by science. It contrasts with versions of scientific realism according to which the claims made about such unobservable entities by scientific theories are true or approximately true, or at least candidates for truth or falsity. Musgrave raises the following objection to entity realism: 'We are to believe in scientific entities ... without thinking true any theory about those entities .... This is incoherent. To believe in an entity, while believing nothing further about that entity, is to believe nothing. I tell you that I believe in hobgoblins (believe that the term 'hobgoblin' is a referring term). So, you reply, you think there are little people who creep into houses at night and do the housework. Oh no, say I, I do not believe that hobgoblins do that. Actually, I have no beliefs at all about what hobgoblins do or what they are like. I just believe *in* them' (1996, p. 20). Musgrave's point is that it is not possible to believe *in* the existence of some entity without having at least some beliefs *about* the entity. This is a crucial point to be made in relation to entity realism. But it does not entirely dispose of the doctrine. For, as Musgrave notes, entity realists may adopt a less extreme position according to which some low-level theoretical beliefs may be true of the theoretical entities.

[2] Since Musgrave often writes within the context of falsificationist philosophy of science, an emphasis on such issues as corroboration, independent testability, *ad hoc*ness and predictive novelty is perhaps understandable. However, within the context of scientific realism, Musgrave places special emphasis on the role of novel predictions, arguing that the success argument for scientific realism should be restricted to theories which correctly predict facts not employed in the construction of the theory (cf. Musgrave, 1999, pp. 55-7, 119, ch. 12). Other methodological criteria, such as simplicity or unity, also receive favourable mention (cf. 1999, 111-2, 247ff). Thus, despite the emphasis on falsification, Musgrave allows that the methodology of science consists of a plurality of methodological rules (cf. 1999, pp. 226-7, 250, fn 291).

shown to be true. A theory which satisfies methodological norms may yet be false. Nor need a theory that satisfies methodological norms be accepted as true. The methods of science are not the exclusive domain of realism. They may serve aims other than the realist aim of truth. Satisfaction of the norms of method might indicate empirical adequacy or pragmatic reliability, rather than truth.

An explanation is therefore required on the part of the realist of why certification by method provides warrant with respect to truth. I will refer to the need to provide such an explanation as *the problem of method and truth*. As a realist who holds that it may be rational to believe a theory which has been subjected to critical scrutiny in accordance with the norms of method, the problem of method and truth is one that Musgrave must address. That is, he must confront the question of why it is rational to believe theories certified by the methods of science to be true, or close to the truth.[3] In this paper, I will explore his response to the problem.

I will illustrate the problem of method and truth in section 2 by means of the examples of Lakatos's 'plea for a whiff of inductivism' and the internal realist conception of truth of Putnam and Ellis. In section 3, I will turn to Musgrave's approach to the problem of method and truth, where I will consider his treatment of inference to best explanation and critical rationalism. In section 4, I will explore a naturalistic approach to the problem which sets the issue within a broader metaphysical framework. Finally, in section 5, I shall offer some suggestions as to how Musgrave might put metaphysical aspects of his realist position to epistemological use.

2. THE PROBLEM OF METHOD AND TRUTH

Scientific realism enforces a sharp divide between method and truth. On the one hand, scientific method consists of a set of rules and procedures which govern experimental practice and inform the appraisal of scientific theories. A scientist whose acceptance of a theory or result complies with the rules and procedures of method is rationally justified in accepting the theory or result. On the other hand, truth consists in a relation of correspondence between a statement and extralinguistic reality. The relation of correspondence between statement and reality is a relation that may obtain whether or not one has methodologically warranted grounds for believing it to obtain. Indeed, it is a relation that may obtain whether or not the statement is believed to be true. Truth, in the correspondence sense, is a non-epistemic relation, which is not defined in terms of method or rational justification.

Given the separation of method and truth, the question arises of the relation between them. What bearing does method have on truth? Why should use of method lead to theories that are either true or approximately true? This is the problem of method and truth. To illustrate it, I will now turn to Lakatos's 'whiff of inductivism' and the internalist conception of truth that is due to Putnam and Ellis.

[3] The problem of method and truth is not restricted to truth-orientated forms of realism. For the entity realist must face exactly the same challenge of explaining why use of the methods of science leads to knowledge of the way the world is. The problem is the general one of explaining how a methodological procedure conduces to knowledge of an objective reality.

2.1 *Lakatos's plea for a 'whiff of inductivism'*

The problem of method and truth may be illustrated within the context of Popper's philosophy of science by means of the connection between corroboration and verisimilitude. For Popper, a theory is corroborated by successful performance in an empirical test of a prediction made by the theory. The theory receives high corroboration if it passes a range of such tests, especially ones which comprise severe tests of the theory. By contrast, the concept of verisimilitude is a measure of the truth-content relative to the falsity content of a theory, which Popper proposes as an analysis of the idea that one theory may contain more truth than another. One theory has greater verisimilitude than another if it has greater truth-content relative to falsity content than the other.

The question is whether there is any reason to believe that a theory with a higher degree of corroboration than another should also enjoy a higher degree of verisimilitude than the other. In other words, is corroboration an indication of verisimilitude?

In his contribution to *The Philosophy of Karl Popper*, edited by P.A. Schilpp, Imre Lakatos expresses the concern that Popper's 'fallibilism is nothing more than scepticism together with a eulogy of the game of science' (1974, p. 257). Lakatos's concern is precisely that, as a fallibilist and anti-inductivist, Popper is not prepared to:

> say unequivocally that the positive appraisals in his scientific game may be seen as a—conjectural—sign of the growth of conjectural knowledge; that corroboration is a synthetic—albeit conjectural—measure of verisimilitude. (1974, p. 256)

Nor may Popper assert that high corroboration provides any positive reason to believe that a theory is close to the truth.

In order to address this concern, Lakatos enters a plea for a 'whiff of inductivism' to the effect that Popper's methodology be supplemented with a 'synthetic inductive principle' (1974, pp. 254-7, 260).[4] Such a principle would connect corroboration with verisimilitude by treating the former as a 'sign' (1974, pp. 254, 256) or 'measure' (1974, p. 256) of the latter. Only in this way, Lakatos argues, can the methodological concept of corroboration and the 'logico-metaphysical' notion of verisimilitude be combined into a properly epistemological theory of the growth of scientific knowledge.

In his reply to Lakatos, Popper does not explicitly address the plea for a synthetic inductive principle. He does, however, allow that corroboration serves as an 'indication' of verisimilitude in the sense that 'we may guess that the better corroborated theory is also one that is nearer to the truth' (1974a, p. 1011). But he denies that corroboration is to be understood as in any sense a measure of verisimilitude.[5]

[4] I take Lakatos's point in describing the required inductive principle as synthetic to be that the principle is a substantive claim, the truth of which depends on facts about the way the world is. Such a principle contrasts with an analytic principle that is true in virtue of the meaning of the words 'corroboration' and 'verisimilitude'. This contrast will become clearer in section 2.2, where we will consider the internal realist conception of truth, which leads to an analytic relation between method and truth.

[5] See also Popper (1972, p. 103). For related discussion, see Newton-Smith (1981, pp. 67-70).

There is one point in the Schilpp volume, though, where Popper does seem to concede a 'whiff of inductivism'. In his reply to A. J. Ayer, Popper explains the importance of the notion of verisimilitude:

> ... there is a probabilistic though typically noninductivist argument which is invalid if it is used to establish the probability of a theory's being true, but which becomes valid (though essentially nonnumerical) if we replace truth by verisimilitude. The argument can be used only by realists who do not only assume that there is a real world but also that this world is by and large more similar to the way modern theories describe it than to the way superseded theories describe it. On this basis we can argue that it would be a highly improbable coincidence if a theory like Einstein's could correctly predict very precise measurements not predicted by its predecessors unless there is 'some truth' in it. (1974b, pp. 1192-3, fn. 165b)

Popper goes on to remark that 'there may be a "whiff of inductivism" here', which 'enters with the vague realist assumption that reality, though unknown, is in some respects similar to what science tells us' (1974b, p. 1193).

It is unclear why Popper fails to make this concession in the context of his response to Lakatos. In any event, the assumption of a real world that is 'by and large similar to the way modern theories describe it' would appear to be a metaphysical assumption of the very kind that Lakatos proposes. If there is a real world which contains the entities and laws which science tells us that it contains, then this fact is itself the explanation of why contemporary theories which say that there are such entities and laws receive high corroboration. For if the world contains things which do what a theory says they do, then that is why what the theory says about those things is true. But such an explanation may only be provided on the assumption that theories which succeed in the manner indicated by high corroboration are close to the truth.[6]

As I will attempt to show in sections 3 and 4, it is precisely such an appeal to metaphysics that is lacking from the epistemology of Musgrave's realism. In this respect, Musgrave seems to side with Popper against Lakatos in resisting the call for a metaphysical inductive principle. But, as I will attempt to show, to defend the epistemological basis of realism, the realist must put the world to good use.

2.2 Putnam on the ideal limit of inquiry

As we have seen, Lakatos proposes to bridge the gap between method and truth by means of a 'synthetic inductive principle'. An alternative approach is to close the gap in an analytic manner by defining truth in terms of method. This is the path of internal realism (e.g., Putnam, 1978, 1981; Ellis, 1980, 1990). In this section, I will

[6] This is not to say that the connection between the approximate truth or verisimilitude of a theory and its empirical success is unproblematic. In fact, it cannot be assumed that a theory with a high degree of approximate truth will be successful. For example, many of its observational claims might be false even though it contains a great deal of true theoretical claims (cf. Laudan, 1981, p. 31). But the present point is not that there is an unproblematic connection between approximate truth and success. Rather, the point is that Popper appears to make a metaphysical assumption about the nature of reality, on the basis of which some non-analytic relation between verisimilitude and corroboration might be shown to obtain.

briefly explore this path before indicating why it is not one that can be taken by the realist. Since Musgrave has forcefully argued for this conclusion, I will draw on his work in showing that realism cannot go down the internalist path.

In his (1978), Hilary Putnam notes that according to the position which he describes as 'metaphysical realism', truth is 'radically non-epistemic' (1978, p. 125).[7] For metaphysical realism, truth is a semantic relation of correspondence between linguistic items and entities in the external world. Such a concept of truth is defined independently of epistemic factors, such as evidence, confirmation or simplicity.

Putnam illustrates the non-epistemic nature of metaphysical realist truth with the example of the ideal theory which would ultimately result if science were pursued to the ideal limit of inquiry. Such a theory would maximally satisfy all methodological constraints. Putnam says the ideal theory would be:

> ... complete, consistent ... predict correctly all observation sentences ... meet whatever 'operational constraints' there are ... be 'beautiful', 'simple', 'plausible', etc... (1978, p. 125)

Given the non-epistemic nature of truth, however, it is possible that even such an ideal theory might be false. For while it might be extraordinarily unlikely for the ideal theory to be false, the fact that it maximally satisfies all methodological constraints does not entail that it is true.

Putnam rejects both metaphysical realism and the non-epistemic conception of truth.[8] He proposes instead an internal realist stance on which truth is understood in epistemic terms as an idealized form of rational justification:

> 'Truth', in an internalist view, is some sort of (idealized) rational acceptability – some sort of ideal coherence of our beliefs with each other and with our experience as those experiences are themselves represented in our belief system ... (1981, pp. 49-50)

[7] In his (1978, p. 125), Putnam describes metaphysical realism as the thesis that there is a determinate relation of reference between terms and items in a mind-independent reality. Later, in his (1981), Putnam adds that for metaphysical realism 'the world consists of some fixed totality of mind-independent objects' (1981, p. 49). While Putnam's characterization of the doctrine is perhaps intended to capture the views of many realists, it contains elements which may not be entirely acceptable to all realists. In his 'Metaphysical Realism versus Word-Magic' (2001), Musgrave argues that realists should not uncritically accept the idea of a mind-independent reality, since there is a range of mind-dependent objects (e.g., artifacts) about which one should be thoroughly realist. Musgrave also objects to the idea that there is a 'fixed totality' of mind-independent objects, since what objects there are depends on a prior specification of what sort of object is in question.

[8] Putnam presents a number of objections to metaphysical realism. One is that truth is not radically non-epistemic because the ideal theory cannot possibly be mistaken. This objection rests on his well-known model-theoretic argument against realism that since every consistent theory has at least one model, the ideal theory (which is stipulated to be consistent) must be true (1978, pp. 125-6). A second objection is that in order to describe the position of metaphysical realism it must be possible to adopt a God's eye point of view. But it is impossible to remove ourselves from our limited human perspective to adopt the external viewpoint of such an omniscient being (1981, p. 50). A third objection is that metaphysical realism opens the door to the possibility of radical scepticism, since it allows the possibility of massive illusion (e.g., evil demons, brains in vats). But such radical sceptical scenarios are not in fact possible scenarios. Hence, metaphysical realism is mistaken because it allows the possibility of such scenarios (1981, p. 15).

The internalist conception of truth differs from the metaphysical realist conception on two counts. First, it is an epistemic conception of truth which takes truth to be a form of rational acceptability. Second, because truth is idealized rational acceptability, the epistemically ideal theory produced at the ideal limit of scientific inquiry must necessarily be true.

The internal realist conception of truth provides a clear example of one way to deal with the problem of method and truth. The internalist closes the gap between method and truth by setting up an analytic or conceptual relation between method and truth. If truth just is a form of rational justification, then a theory which satisfies methodological standards of theory-acceptance is to be accepted as true, or nearly so. For that is what it is to be true. Equally, a theory which better satisfies methodological standards than a predecessor thereby displays a higher degree of truth, since increased satisfaction of such standards constitutes increase of truth.

Such an analytic resolution of the problem of method and truth is not, however, one that is open to the scientific realist. For, as Musgrave has argued, the internalist conception of truth leads to an idealist metaphysics that is unacceptable to realists. In his paper, 'The T-Scheme Plus Epistemic Truth Equals Idealism' (1999, ch. 10; cf. 1996, p. 30), Musgrave argues that epistemic theories of truth, such as internal realism, entail the dependence of reality upon belief.[9] According to Musgrave (1999, p. 188), 'the general form of an epistemic truth theory' is as follows:

(E) Necessarily, S is true if and only if S satisfies epistemic condition E.

To obtain a particular epistemic theory of truth from this general form, it suffices to replace the epistemic condition E with the preferred epistemic condition of the relevant truth theory.

Musgrave employs the example of Brian Ellis's evaluative theory of truth, which is a form of internal realism closely related to Putnam's. According to Ellis, truth is what it is epistemically right to believe. So we have:

Necessarily, S is true if and only if it is epistemically right to believe S.

Now, given the T-scheme:

(T) S is true if and only if P,

Ellis's evaluative theory of truth entails that:

Necessarily, P if and only if it is epistemically right to believe S.

Thus, to take a particular example (Musgrave 1999, p. 189)

(ET) Electrons exist if and only if it is right to believe that electrons exist.

[9] For related analysis, see Devitt and Sterelny (1987, p. 196).

But, surely, Musgrave points out, (ET) might be false. There might be no electrons even though 'our best methods optimally pursued ... lead us to think electrons exist' (1999, p. 189). The only way for (ET) to be true is for the world to depend on our methods of inquiry or our theories in idealist fashion. In this case, electrons would exist if that is what our methods of inquiry and theories lead us to believe. But that is evidently not something that a realist can accept.

2.3 *The problem restated*

Lakatos's plea for a 'whiff of inductivism' and Putnam's and Ellis's internalist conception of truth represent two different approaches to the problem of method and truth. The question is why we should suppose that the rules of method have any positive bearing on truth. The response proposed by Putnam and Ellis is to define truth in terms of method. But such a response is unavailable to the realist who takes truth to be non-epistemic, as Musgrave does. The other response which we have seen is to appeal to a synthetic metaphysical principle in the manner suggested by Lakatos with his 'plea for a whiff of inductivism'. But this response appears not to be the response favoured by Musgrave, as we shall now see.

3. MUSGRAVE ON METHOD AND TRUTH

As a scientific realist, Musgrave adheres to the view that it may be rational to believe that a scientific theory is true. A theory which passes critical scrutiny by means of the rules of scientific method may be accepted as true, where truth is understood in the non-epistemic sense of the realist. The question is why it is rational to believe that a theory which satisfies the rules of method is true. If truth is non-epistemic, then what does method have to do with it?

In this section, I will consider two answers that have been proposed by Musgrave. The first involves the idea that it is reasonable to believe the best explanation of a fact. The second is that it is rational to believe the hypothesis which best survives criticism. As we will see, neither approach succeeds in showing why it is rational to believe a theory to be true.

3.1 *'The Ultimate Argument for Scientific Realism'*

The standard argument for scientific realism is the so-called 'success argument', or, as Musgrave calls it, 'the Ultimate Argument'.[10] According to scientific realism, the entities postulated by mature scientific theories by and large exist, and the claims that theories make about those entities are by and large true, or close to the truth. Such a realist account of the relation between theories and the entities they postulate provides a compelling explanation of the empirical success of science. For if the entities postulated by a theory exist, and what the theory says about the entities is

[10] The name, 'the ultimate argument', is due to van Fraassen (1980, p. 39), who is one of the targets of Musgrave (1988).

true, then it is no surprise that the theory should meet with empirical success. By contrast, any anti-realist philosophy which rejects the realist view of the relation between theories and the entities they postulate must render the success of science an inexplicable miracle (cf. Putnam, 1975, p. 73). But to say that the success of science is a miracle is to fail to provide an adequate explanation of such success. Since realism provides a compelling explanation of success, and anti-realism fails to provide an adequate explanation, realism is evidently the best explanation of the success of science.

In his paper, 'The Ultimate Argument for Scientific Realism' (1988, pp. 232-9), Musgrave presents an analysis of the success argument.[11] It is standard practice to construe the success argument as an inference to the best explanation. In line with this practice, Musgrave also construes the argument as an inference to the best explanation. However, in a novel departure, Musgrave argues that application of the success argument is to be restricted to theories which successfully predict novel facts. He formulates the argument as an epistemic argument to the effect that it is reasonable to accept realism, rather than to the effect that realism is true. He further stipulates that in order to be acceptable, the best explanation must satisfy minimal conditions of explanatory adequacy. Otherwise, it would not be reasonable to accept the best explanation as true.

Opinion is divided over the nature of inference to the best explanation. Some take it to be a form of inductive inference. Others take it to be a *sui generis* form of inference that is more fundamental than induction (cf. Harman, 1965). Perhaps the most novel feature of Musgrave's analysis of the success argument is his suggestion that inference to the best explanation may be formulated as a deductive inference.

Musgrave proposes that inference to the best explanation be construed in deductive form as follows:

> It is reasonable to accept a *satisfactory* explanation of any fact, which is also the best available explanation of that fact, as true.
> F is a fact.
> Hypothesis H explains F.
> No available competing hypothesis explains F as well as H does.
> Therefore, it is reasonable to accept H as true. (1988, p. 239)

He then comments that 'the Ultimate Argument for scientific realism ... is an inference to the best explanation':

> The fact to be explained is the (novel) predictive success of science. And the claim is that realism ... *explains* this fact, explains it *satisfactorily*, and explains it *better* than any non-realist philosophy of science. And the conclusion is that it is reasonable to accept scientific realism ... as true. (1988, p. 239).

On such a construal, the success argument is a valid deductive argument. The fact to be explained is the novel predictive success of science. The conclusion of the

[11] I refer here to the original version of Musgrave's article in Nola (1988). The paper is reprinted in Musgrave (1999). However, the section of the article on inference to best explanation, which is of central relevance to scientific realism, has been removed. It appears, instead, in the context of a discussion of psychologism (1999, pp. 284-5).

argument is an epistemic conclusion to the effect that it is rational to believe realism to be true. For realism is the best explanation of predictive success. The conclusion depends crucially on the epistemic principle that it is reasonable to accept the best satisfactory explanation of a fact as true, which figures as the initial premise of the argument.

Musgrave's analysis of the success argument is an important advance in a number of respects. The emphasis on predictive novelty is important because it may be employed to eliminate a number of historical counterexamples which have been proposed to the success argument.[12] Musgrave's formulation of the success argument in epistemic terms makes clear that the argument must play a pivotal role in response to anti-realist critics who object to scientific realism on epistemological grounds. His emphasis on minimal conditions of explanatory adequacy is crucial, since it excludes the possibility that the best available explanation fails to be a satisfactory explanation. Finally, the explicit use of the epistemic principle in the argument makes evident the extent to which the success argument depends on the assumption of the epistemic importance of explanation.

Despite initial appearances, however, Musgrave's analysis of the success argument provides little assistance in relation to the problem of method and truth. To see this, let us further examine the notion of a *best* explanation. On what might the judgement that a theory is the best explanation be based? Musgrave does not elaborate. But it seems reasonably clear that the assessment of the explanatory merit of a scientific theory will depend upon methodological criteria of theory appraisal. Relevant criteria will include considerations of explanatory strength and unification, as well as simplicity, coherence and fit with background knowledge.[13] But since truth is understood by Musgrave in the non-epistemic, realist sense, it is unclear why theories which satisfy such methodological criteria should be accepted as true.

The question is why it is reasonable to accept the best explanation as true. Might it not be equally reasonable to accept the best explanation as empirically adequate, useful for practical purposes, or even true in some non-realist sense? Nothing Musgrave says in support of the principle that it is reasonable to accept the best explanation as true shows that the anti-realist might not accept an anti-realist

[12] It is a major weakness of earlier formulations of the success argument that the notion of success is imprecisely defined. If success is left overly vague, the success argument is vulnerable to historical counterexamples, such as those presented by Laudan of theories which attained a degree of success but were false and/or non-referential (Laudan, 1981).

[13] It is an interesting question whether evidential considerations, such as confirmation or corroboration, are of relevance to assessment of explanatory merit. Musgrave develops his analysis of inference to the best explanation as a modification of C.S. Peirce's idea of abduction. However, in his definition of abduction Peirce himself seems to exclude evidential considerations as irrelevant: 'The first stating of a hypothesis and the entertaining of it, whether as a simple interrogation or with any degree of confidence, is an inferential step which I propose to call *abduction* [or *retroduction*]. This will include a preference for any one hypothesis over others which would equally explain the facts, so long as this preference is not based upon any previous knowledge bearing upon the truth of the hypothesis, nor on any testing of any of the hypothesis, after having admitted them on probation' (Peirce, 1955, p. 151). This passage suggests that, for Peirce at least, an explanation may be evaluated *qua* explanation independently of any evidence which might be gained by empirical test of the explanation.

analogue of the principle. Nor does Musgrave provide an explanation of why it is reasonable to accept the best explanation as true.

It might, however, be thought that the issue is not whether the best explanation is to be accepted as true. Rather, the issue is whether realism is the best explanation. Musgrave addresses this issue in the pages that follow his analysis of the success argument (1988, pp. 240-4). He considers a range of anti-realist explanations of predictive success, and argues that all provide inferior explanations to the realist explanation. On the assumption that realism has been shown to be a superior explanation to anti-realism, it might therefore appear that realism is to be accepted as true.

But this only succeeds in pushing the problem back another level. Even if it is granted that realism is the best explanation of the success of science, it does not follow that it is to be accepted as true. There are other possible modes of acceptance available at this level, apart from acceptance as true. For example, one might simply agree that realism is the best explanation without proceeding to accept it as true. Alternatively, one might merely accept realism as if it were true. Or realism might be accepted as true, but truth might be understood in some non-realist sense. Nothing about best explanation, as such, clearly precludes such alternative forms of acceptance.

In sum, to show that a theory is the best explanation of a fact does not entail that the theory is to be accepted as true. Given this, Musgrave's analysis of the success argument in terms of an epistemic principle of best explanation does not succeed in showing why it is rational to accept a theory as true. It does not, in other words, provide a response to the problem of method and truth.

3.2 Critical rationalism

I turn now to a second context in which Musgrave addresses issues which relate to the problem of method and truth. In his treatment of Popper's solution of the problem of induction, Musgrave proposes a critical rationalist account of scientific theory acceptance (1999, ch. 16). I will now consider the implications of Musgrave's critical rationalism with respect to the problem of method and truth.

Popper's philosophy of science is sometimes described as 'negativist' (cf. Lakatos, 1974, p. 258). In an attempt to solve Hume's problem of induction, Popper dismisses induction as a myth. Instead of offering a positive justification of induction, Popper argues that the attempted falsification of a theory may provide rational grounds for tentative acceptance of the theory. It is possible neither to prove that a theory is true nor to provide inductive support for the theory. However, if a theory has survived rigorous empirical tests, then it may be rational to tentatively accept the theory.

Since Popper denies that there may be any grounds which provide positive support for a theory, the question arises of how his claim that it may be rational to accept a theory is to be understood. To address this question, it is necessary to introduce a distinction between Popper's critical rationalist account of rationality and the traditional justificationist conception of rationality to which Popper's account is opposed. Perhaps what most fundamentally characterizes Popper's account of rationality is his outright dismissal of the justificationist conception of rationality.

The justificationist conception of rationality is the conception of rationality that underlies most traditional and contemporary thinking about rational belief. According to justificationism, in order to have a rational belief the belief itself must be rationally justified. There must be reasons which provide support for the belief.

Musgrave characterizes justificationism by means of the following principle:

(J) A's believing that P is reasonable if and only if A can justify P, that is, give a conclusive or inconclusive reason for P, that is, establish that P is true or probable. (1999, p. 321)

As this formulation of justificationism makes clear, reasons may either be conclusive or inconclusive. Conclusive reasons are reasons which show that a belief is true. Inconclusive ones merely show it to be likely or probable. In either case, rational belief requires there to be reasons which support the belief itself.

By contrast with justificationism, critical rationalists deny that there may be reasons for a belief or theory. But this does not mean that there is no rationality. On the contrary, as Popper remarked, 'there is nothing more "rational" than the method of critical discussion, which is the method of science' (1972, p. 27). Criticism, rather than justification, is the key to rationality.

Accordingly, Musgrave offers the following principle as formulation of critical rationalism:

(CR) It is reasonable to believe that P (at time t) if and only if P is that hypothesis which has (at time t) best withstood serious criticism. (1999, p. 324)

In other words, if a hypothesis is subjected to serious criticism and survives, while alternative hypotheses do not, there is good reason to accept the hypothesis which stands up to criticism in favour of those which succumb to it. By contrast with justificationism, such a conception of rationality does not involve good reasons for a hypothesis. It is belief in the hypothesis, rather than the hypothesis itself, for which there may be good reason. Critical rationalism alters the locus of rationality. 'It is', Musgrave explains, 'acts of belief (actions of believing?) that are reasonable or rational, not the things we believe, belief-contents, propositions, theories, or whatever' (1999, p. 322).

On Musgrave's analysis of critical rationalism, it is rational to believe 'the theory which best survives critical scrutiny' (1999, p. 330). To believe a theory is to believe that it is true (cf. 1999, pp. 321, 326). And the method of criticism is the method of science. The critical rationalist account of theory acceptance is therefore of clear relevance to the problem of method and truth. For the critical rationalist asserts that survival of critical scrutiny provides the basis for rational belief in the truth of scientific theories.

But what is it for the method of criticism to be the method of science? As earlier noted, within the context of a Popperian falsificationist theory of method, the primary means of criticism is the attempt to falsify a theory by rigorous empirical test. Within a strictly falsificationist framework, it is possible to criticize a theory in a variety of ways. A theory may entail a false prediction or it may be unfalsifiable. It might predict no novel facts, be poorly corroborated, or be *ad hoc*. But there is no need for the method of criticism to be restricted to strictly falsificationist resources. A theory might also be criticised on grounds which have no immediate connection with empirical falsification as such. For example, a theory might lack coherence, be overly complex, have limited explanatory scope, or be inelegant.

A variety of methodological considerations may therefore play a role in the critical method. But it remains to be asked how the critical method warrants belief in the truth of a theory. By itself, the rejection of justificationism does not suffice to resolve the problem of method and truth. If truth is non-epistemic, and the critical method is the basis of theory acceptance, the connection between method and belief in the truth is left entirely unexplained.

It would be misguided to suppose that survival of criticism provides positive support for a theory. For the critical rationalist, survival of rigorous test or other attempts to criticize a theory does not lend positive support to a theory. To assume that criticism yields positive support is to assume a justificationist conception of rationality. But, for the critical rationalist, survival of criticism does not prove that a theory is true, nor does it render the theory more likely to be true. It does not provide any positive justification for the theory at all. Rather, survival of criticism provides one with a basis to tentatively believe in the truth of a theory, as opposed to alternative theories which have been exposed to criticism and failed to survive.

The trouble is that nothing has been done to secure belief in truth as the unique mode of theory acceptance. It is possible to agree with the critical rationalist conception of scientific inquiry, but to deny that theories are to be accepted as true. To take but one example, it would be perfectly consistent for an anti-realist to endorse the critical method while at the same time embracing a constructive empiricist view of theory acceptance along the lines of Bas van Fraassen (1980).[14] On such an account, it would be rational to accept a theory which best withstands critical scrutiny. But the theory is to be accepted as empirically adequate, rather than as true. That is, it is to be accepted as true at the observable level, without commitment to the truth of its non-observational content.

Nothing about the critical method entails that a theory which survives criticism is to be accepted as true. Critical rationalists are fallibilists. As such, critical rationalists themselves insist that a theory which survives rigorous empirical test may fail to be true. But, if it does not follow from survival of criticism that a theory is true, then neither does it follow that the theory is to be accepted as true. There is

[14] Indeed, van Fraassen comes close to such a position when he remarks that 'the success of current scientific theories is no miracle. It is not even surprising to the scientific (Darwinist) mind. For any scientific theory is born into a life of fierce competition, a jungle red in tooth and claw' (1980, p. 40). Of course, this remark is made in the context of van Fraassen's discussion of the realist's success argument. But the talk of fierce competition suggests that van Fraassen approaches the question of theory acceptance with a decidedly Popperian cast of mind.

nothing about the notion of criticism as such which requires one to believe that a theory which survives criticism is true.

Musgrave introduces a modification of critical rationalism which may seem to go some way toward disarming this objection. The modification relates to the 'epistemic primacy' of perception (1999, p. 342). Perception is the source of the empirical evidence which is employed to test our theories. But on what basis are perceptual reports accepted? In ordinary circumstances, perceptual reports are not accepted as the result of test. Rather, they are accepted at face value. Perception is only subjected to test when something goes wrong. As Musgrave notes, "only when we have some specific reason to suspect perceptual error do we 'check out' a perceptual belief"(1999, p. 342). But if it may be rational to accept a perceptual report which has not been subjected to test, then survival of criticism cannot be necessary for rational belief.

This point requires that critical rationalism be amended. For if it may be rational to accept a perceptual belief without submitting it to test, then it may be rational to accept such a belief without it having survived criticism. Musgrave, therefore, introduces a distinction between perceptual and non-perceptual beliefs:

> A non-perceptual belief is reasonable if it has best withstood criticism—a perceptual belief is reasonable if it has not failed to withstand criticism. The latter is just the commonsense view 'Trust your senses unless you have a specific reason not to'. (1999, p. 342)

On the modified version of critical rationalism to which this distinction gives rise, rational theory acceptance requires survival of criticism. But perceptual belief is rational provided only that no problem has so far arisen with respect to the perception on which it is based.

But even if the primacy of perception is granted, this does not affect the objection. It may simply be conceded that perception provides a *prima facie* rationale for the acceptance of a perceptual report. No such rationale is thereby provided for theory acceptance. This is particularly apparent in light of Musgrave's epistemic distinction between perceptual and non-perceptual belief. The primacy of perception specifically relates to perceptual belief. Nothing follows from the primacy of perception with respect to the rationality of non-perceptual belief. If the primacy of perception is to be of any relevance to theory acceptance, then an additional assumption is required which extends the primacy of perception to the non-perceptual realm.

The point may be illustrated by means of the earlier example of the constructive empiricist version of critical rationalism. Such a constructive empiricist accepts the critical rationalist account of theory acceptance with the qualification that theories which survive criticism are to be accepted as empirically adequate. It is entirely consistent with such a position to grant the epistemic primacy of perception, and to agree that perception provides a *prima facie* rationale for perceptual belief. But the primacy of perception only entails that perceptual beliefs be accepted as true. It does not extend to the level of theory. Hence, the constructive empiricist may restrict theory acceptance to empirical adequacy.

Thus, even if the primacy of perception is granted, it does not follow that theories which pass critical scrutiny need be accepted as true. Given this, and the earlier point that survival of critical scrutiny does not entail belief in the truth of a theory, I conclude that the critical rationalist position presented by Musgrave does not resolve the problem of method and truth. It remains to be shown why use of the critical method provides any reason to believe that a theory is true.

3.3 Epistemic versus metaphysical principles

We have now considered two approaches proposed by Musgrave which are of relevance to the problem of method and truth. Both of the approaches are based on epistemic principles of rational belief. As such, both of the approaches proposed by Musgrave contrast with the approaches to the problem of method and truth canvassed in sections 2.1 and 2.2.

In section 2.1, we considered Lakatos's 'plea for a whiff of inductivism' that Popper's methodology be supplemented by a metaphysical principle which connects corroboration with verisimilitude. Such a principle would consist of a substantive synthetic claim about the world in the light of which corroboration is revealed to be an indication of verisimilitude. By contrast, Musgrave's epistemic principles say nothing about the world. Instead, they specify conditions under which it may be rational to believe a proposition or hypothesis to be true.

In section 2.2, we considered the analytic approach to the problem of method and truth that is due to internal realism. The internalist identifies truth with satisfaction of methodological criteria. Given such an identification, it may be rational to believe that a theory which satisfies methodological criteria is true. For that is what it is to be true.

By contrast with internal realism, Musgrave is a realist for whom truth is a non-epistemic correspondence relation. As such, Musgrave must reject the analytic approach on two counts. As a realist, he must reject the internalist conception of truth because of the idealism to which it leads. And as an advocate of a non-epistemic conception of truth, he must reject the internalist identification of truth with satisfaction of epistemic criteria.

But while it is clear that Musgrave must reject the analytic approach, it is not entirely clear why he rejects metaphysical principles in favour of epistemic principles of rational belief. It may be that Musgrave rejects metaphysical principles because he takes them to be inductive principles of the uniformity of nature of a kind that Hume showed to be unjustified (cf. Musgrave, 1993, pp. 157ff). It may be that he takes the rejection of justificationism to entail the rejection of metaphysical principles (cf. 1999, p. 327). It may be that he takes there to be no need for metaphysical principles over and above scientific theories which may be accepted on critical rationalist grounds (1999, pp. 328-9). It may be that he takes such principles to rest on an anthropocentric metaphysics (1999, pp. 283, 285). Or perhaps the point is simply that realism should avoid excess metaphysical commitment (1999, p.131).

Whatever Musgrave's exact reason for rejecting metaphysical principles may be, I shall now attempt to show that such principles are necessary in order to solve the problem of method and truth. The truth of an empirical claim about the world depends upon the way that the world in fact is. In order to show that use of an epistemic method leads to such truth about the world, it is necessary to say something about the world. Otherwise, no connection is made between method and truth. In short, the problem of method and truth is at least partly one of metaphysics.[15]

4. METAPHYSICS AND NATURALISM

In my own recent work, I have sought to develop a naturalistic response to the problem of method and truth. I understand the rules of method in instrumental fashion as means for the pursuit of the aims of inquiry. The relation between epistemic means and ends is a synthetic relation, rather than an analytic one. Hence, the reliability of rules of method may be subject to empirical appraisal. For it is an empirical matter whether use of a particular method reliably conduces to a given cognitive goal.[16] Empirical evidence cannot directly reveal use of a method to lead to truth at the theoretical level. However, I argue that the best explanation of the role played by method in the success of theoretical science is that the rules of method are reliable means of promoting the realist aim of truth (Sankey, 2000, 2002).

I shall say nothing further about my approach to this issue, other than to locate it within the broader perspective of which it forms part. This perspective reflects a non-anthropocentric conception of human inquirers and their place in our environing reality. We humans are organisms who inhabit a pre-existing natural world. We interact with this world. But we did not create it. Its basic structure and composition are independent of us. Yet our survival requires that we act in the world. To promote survival, our actions must be informed by reliable knowledge of our environment. But it cannot be known *a priori* how best to acquire such knowledge. This is a contingent matter which depends on our epistemic capacities and their relation to the world. We can only learn such things by empirical investigation of ourselves and our surroundings.

This perspective is a blend of epistemological and metaphysical ingredients. It combines claims about reality with claims about our knowledge of reality. Within such a perspective, epistemological claims may derive support from metaphysical claims. For example, general considerations about the nature of reality may be

[15] Musgrave is not completely dismissive of metaphysical principles. Against those who treat laws and theories as inference licenses, Musgrave claims that they may be under the influence of a positivistic bias against metaphysics (1999, p. 283). Moreover, he notes against positivism that metaphysical principles of theory construction may play a significant role in science and may even be subject to rational appraisal (1999, p. 309).

[16] I follow Laudan (e.g., 1996, ch. 7) in endorsing a form of normative naturalism. According to normative naturalism, the epistemic warrant for a rule of method derives from empirical evidence of reliable promotion of the cognitive aims served by the rule. In contrast with Laudan, however, I set normative naturalism within a realist framework on which the methods of science are seen as reliable means of advancing toward the realist aim of truth (cf. Sankey, 2000). As will become apparent in section 4.1, my approach also has certain affinities with the methodological pragmatism of Rescher (e.g., 1977), who treats methods as cognitive instruments subject to empirical appraisal and pragmatic justification.

employed to explain why certain methods of inquiry constitute a reliable means of inquiry into that reality.

To illustrate the relevance of metaphysical considerations to the problem of method and truth, I will now examine two examples of the epistemological application of metaphysical considerations. The first case is that of Nicholas Rescher's methodological pragmatism. The second is Hilary Kornblith's grounding of inductive inference in natural kinds.

4.1 Rescher's methodological pragmatism

For the classical pragmatist, a true proposition is one the acceptance of which leads to practical success. Rescher refers to such pragmatism as thesis pragmatism, since it relates to specific propositions or theses. He rejects the pragmatist view of truth in favour of a correspondence conception. Instead of thesis pragmatism, he proposes a methodological pragmatism, which applies the criterion of practical success at the level of the methods of inquiry (Rescher, 1977). The rules of method are to be evaluated in the manner of instruments in terms of their success in practical application. If a rule reliably performs the function for which it is designed, it thereby receives pragmatic justification (1977, pp. 3-4). By contrast, individual claims are not practically justified, but receive indirect support from the methods by which they are certified (1977, pp. 71-2).

For Rescher, pragmatically warranted methods of inquiry are to be regarded as 'truth-indicative' (1977, p. 83).[17] A proposition which satisfies a rule of method is therefore to be accepted as true. Thus, while truth and utility are distinct at the level of propositions, Rescher takes pragmatic success to have a bearing on truth at the level of method. Because Rescher takes certification by rules of method to warrant acceptance as true, his methodological pragmatism is therefore of relevance to the problem of method and truth. The question is why practical justification of method should be taken to be truth-indicative. The answer, as we shall now see, turns on metaphysical considerations.

In order to explain how practical success relates to truth, Rescher places the use of method within a broader metaphysical setting. This is characterized by the following principles which relate to human agency, the community of inquirers and the nature of reality (1977, pp. 84-9). *Activism*: our survival and welfare require action on our part; since we act on the basis of beliefs, our beliefs are of practical relevance. *Reasonableness*: belief guides action in a way that coordinates action with beliefs and needs. *Interactionism*: our active intervention in the world produces outcomes which may either satisfy or frustrate our intentions. *Purposive constancy*: to establish the reliability of a method, inquirers must employ the same method for

[17] Rescher's expression 'truth-indicative' may seem to suggest that a proposition that satisfies a methodological rule is thereby definitively shown to be true. Indeed, Rescher sometimes uses the expression 'truth-criterion' (e.g., 1977, p. 81), which may suggest that satisfaction of a rule suffices to establish the truth of a proposition. But I do not think that Rescher takes satisfaction of a rule to be criterial for truth in the sense that it either constitutes or demonstrates truth. Rather, satisfaction of a rule provides a warrant or justification for acceptance of the proposition as true (cf. 1977, pp. 79-80).

the same purpose. *Uniformity of nature*: continued use of a method depends upon the underlying constancy of nature and the conditions of application of the method. *Nonconspiratorial nature of reality*: nature is indifferent to our beliefs and needs, neither conspiring for nor against belief-based actions.

Against this metaphysical backdrop, Rescher argues that a method of inquiry whose use systematically meets with success is to be seen as truth-indicative. False belief may sometimes lead to success, but it could hardly be supposed to do so on a routine basis:

> Isolated successes can be gratuitous and probatively impotent, but the situation will be otherwise when what is at issue is not isolated actions based on particular beliefs, but a general policy of acting, based on a generic and methodologically universalised standard of belief-validation. When one views man as a vulnerable creature in close interaction with a hostile (or at best neutral) environment, it is—to be sure—conceivable that action on a false belief or even set of beliefs might be successful, but it surpasses the bounds of credibility to suppose that this might occur systematically, on a wholesale rather than retail basis. Given a suitable framework of metaphysical assumptions, it is effectively impossible that success should crown the products of systematically error-producing cognitive procedures. (1977, pp. 89-90)

Here, in a manner that recalls the rejection of miracles in the success argument (cf. section 3.1), Rescher dismisses the idea of a pragmatically successful but systematically erroneous method as incredible. The crucial factor is the rational implementation of belief in what Rescher describes as a 'highly reactive environment' (1977, p. 84), 'a duly responsive nature' that is 'complex and volatile' (1977, p. 91). In such a world, a method of belief-formation that regularly gives rise to successful practical action cannot, in Rescher's words, be 'systematically error-producing'. Quite the contrary, it must surely be 'truth-indicative'.

I shall delve no further into the intricacies of Rescher's methodological pragmatism, though pertinent questions might usefully be raised regarding the line of reasoning that underlies the proposed metaphysical rationale for the truth-indicativeness of method.[18] The purpose of my discussion of Rescher is simply to illustrate how metaphysical considerations may be brought to bear on the problem of method and truth. To further illustrate this, I will now turn to Kornblith's account of the ground of inductive inference.

[18] It is, however, important to note two issues to which Rescher's approach immediately gives rise. The first is the apparent circularity involved in drawing upon substantive principles about the world in arguing that methods of inquiry yield truths about the world. Rescher admits the circularity. Instead of being vicious, however, he seeks to show that the justification of method by practice is cyclical and self-supporting (1977, ch. 7). The second is the nature of the reasoning from metaphysical principles to the truth-indicativeness of inquiry procedures. In our (2000a, p. 51), Robert Nola and I assimilate the reasoning involved to inference to the best explanation. However, Rescher resists this interpretation (private communication). He argues that it is instead an inference to best systematization. (See Rescher, 2001, ch. 10 for comparison of inference to the best explanation with inference to the best systematization.)

4.2 Kornblith's natural ground of induction

In his book, *Inductive Inference and its Natural Ground*, Hilary Kornblith proposes a naturalistic account of the reliability of induction. The account combines psychologically informed epistemology with a realist metaphysics of natural kinds. Kornblith takes epistemology to be directed to two questions: '(1) What is the world that we may know it?; and (2) What are we that we may know the world?' (1993, p. 2). His reply is that mind and world fit together. On the one hand, properties which occur together in natural kinds make reliable induction possible. On the other hand, our minds are naturally equipped with a conceptual and inferential apparatus tuned to the natural kind structure of the world.

Kornblith adopts Richard Boyd's account of natural kinds as homeostatic property clusters (Boyd, 1991). According to this account, natural kinds comprise complexes of properties which form relationships of homeostatic equilibrium (Kornblith, 1993, pp. 35-6). Such cohesive properties work together to maintain the stability of a substance or organism. However, not all sets of properties may enter homeostatic equilibrium, since 'only certain arrangements will form stable configurations in a homeostatic relationship' (1993, p. 36). It is precisely because the formation of homeostatic relationships is subject to constraints that natural kinds may ground induction. Given that properties may only be conjoined in limited ways, it is possible to'reliably infer the presence of some of these properties from the presence of others' (1993, p. 36).

Kornblith takes the success of science to show that natural kinds are the ground of induction (1993, pp. 41-2). Such success is due to the development of theories about the unobservable structures that underlie the observable properties of things. The classifications devised on the basis of such theories reflect real divisions between natural kinds of things, rather than merely nominal or interest-relative kinds.

> Inductive inferences can only work, short of divine intervention, if there is something in nature binding together the properties which we use to identify kinds. Our inductive inferences in science have worked remarkably well, and, moreover, we have succeeded in identifying the ways in which the observable properties which draw kinds to our attention are bound together in nature. In light of these successes, we can hardly go on to doubt the existence of the very kinds which serve to explain how such successes were even possible (1993, p. 42).

Thus, Kornblith argues that the reliable use of induction in science can only be explained by means of real natural kinds which support induction. It is only if the properties of a member of a kind form a union on the basis of which they must co-occur that induction which projects such properties to unobserved members of the kind could possibly succeed on a reliable basis.

To complete the fit between mind and reality, Kornblith argues that the human mind is disposed to form concepts and draw inferences in ways that reflect real natural kinds. However, I shall not discuss this issue here, since my principal aim in discussing Kornblith is to draw attention to the role of metaphysics in dealing with the problem of method and truth. Kornblith explains the reliability of induction on the basis of real kinds in nature. It is because members of a natural kind share properties in common with other members of the kind that our inductions about the properties of members of the kind prove to be reliable. Thus, Kornblith employs facts about the nature of reality to explain why induction is reliable. He therefore employs metaphysical considerations to explain why use of a method of inquiry leads to truth.

4.3 The moral of the metaphysical story

The approaches of Rescher and Kornblith represent two contrasting approaches to the problem of method and truth. Rescher argues that success in practical application reveals the truth-indicative character of rules of method. Kornblith takes successful use of induction to require the existence of real kinds in nature which make reliable induction possible. Rescher emphasizes the practical implementation of method, while Kornblith draws on empirical research. Rescher's approach forms part of a general theory of the nature and justification of method, whereas Kornblith's account is restricted to the reliable use of induction.

But, despite the contrasts, the approaches of Rescher and Kornblith are united by a deeper commonality. For both approaches exemplify a synthetic solution to the problem of method and truth, which employs metaphysical considerations to establish a connection between method and truth. Both Rescher and Kornblith appeal to the success of science and action in order to argue that our methods provide epistemic warrant with respect to the truth of our beliefs and theories. Both approaches locate the success of method within a broader metaphysical framework which involves assumptions about the nature of the world we inhabit as well as about ourselves as actors and inquirers. Moreover, the metaphysical assumptions employed by both approaches are all broadly consonant with realism.[19]

The latter point deserves emphasis. In their attempt to connect method with truth, both Rescher and Kornblith deploy metaphysical assumptions that are realist in spirit. Such assumptions cannot therefore be rejected by the realist on metaphysical grounds. The question is whether such metaphysical assumptions should be allowed to play the epistemological role which Rescher and Kornblith ascribe to them. Yet it is entirely unclear how to solve the problem of method and truth in the absence of metaphysical assumptions. I therefore see no alternative but to put the realist's metaphysical assumptions to epistemological use in a manner such as that illustrated by Rescher and Kornblith.

[19] That the metaphysical considerations to which Rescher and Kornblith appeal are broadly consonant with realism is perhaps most tellingly illustrated by noting that both of their approaches are compatible with a metaphysical realist commitment to an objective, mind-independent reality. Rescher adopts a general principle of uniformity of nature, while Kornblith opts for a somewhat more substantive metaphysics of natural kinds. But both the commitment to the uniformity of nature and to the reality of natural kinds are entirely consonant with a metaphysical realist commitment to mind-independence.

5. CONCLUSION

In this paper I have sought to raise the problem of method and truth as a challenge to epistemological aspects of Alan Musgrave's scientific realism. The paper has been largely an exercise in comparative epistemology, which examines alternative solutions to the problem. In line with Musgrave's analysis of the inherent idealism of internal realism, I have argued that the internal realist solution to the problem is not available to the scientific realist. I have also sought to show that Musgrave's own appeal to strictly epistemic principles fails to provide a satisfactory solution to the problem, since such principles do not preclude anti-realist forms of theory-acceptance. By contrast, I have attempted to show that metaphysical considerations are necessary in order to explain why satisfaction of methodological norms warrants acceptance of a theory as true. In this final section, I seek to extract relevant lessons from my analysis with respect to the epistemology of Musgrave's scientific realism.

In the first place, as a realist, Musgrave should have no particular cause to baulk at metaphysical assumptions of the sort described in the previous section. For example, the metaphysical principles introduced by Rescher in relation to human agency, causal interaction and the nonconspiratorial nature of reality, are in full accord with realism.[20] The principles are compatible with a realist commitment to an objective, mind-independent reality. They are no more, at base, than an articulation of a commonsense view of ourselves, our surroundings and our relationship to those surroundings. And, in Musgrave's view at least, the scientific realist is not just a realist about science but a realist about common sense as well.

But Musgrave might baulk at appeal to the uniformity of nature. The reason would not be his realism, though, but his anti-inductivism (see Musgrave, 1993, ch. 9). Here Musgrave's realism must simply be played off against his anti-inductivism. For what is it to be a scientific realist, if it is not to say that there is a real world in which observed phenomena are brought about by the action of unobservable entities? Of course, we might wrongly identify the causal processes and laws of nature which govern the phenomena. Or the world might be radically transformed overnight. But these are merely sceptical points. The world that we inhabit is a world of objectively existing things, real causal relations and law-governed phenomena. Such a world is characterized by underlying natural uniformities which it is the business of science to discover. A realism that denies this is realism in name only. Indeed, it is realism without the real world.

In section 3.1, I objected to Musgrave's epistemic principle of best explanation that nothing prevents the adoption of an anti-realist analogue of the principle. Yet, as we saw in section 4, metaphysical resolution of the problem of method and truth

[20] As for the natural kinds to which Kornblith appeals in his account of induction, here the realist might have cause to object either to the particular account of natural kinds that Kornblith employs or to the existence of natural kinds, as such. But the idea that there is a real world, in which there are real, non-conventional differences between different sorts of things, is not something to which any realist should seriously wish to raise objections.

proceeds by way of inference to the best explanation of success.[21] It might appear inconsistent to object to inference to the best explanation in one context while embracing it in another. My point, however, is not that the realist may do without an explanatory pattern of inference altogether. Given the gap between method and truth, some form of explanatory reasoning must play a role in the epistemology of scientific realism. My point, rather, is that inference to the best explanation as such is not the exclusive domain of the realist. The anti-realist may take it to be justified to accept the best explanation but decline to accept it as true in the realist sense.

However, a realist outcome may be secured once explanatory inference is set within an appropriate metaphysical framework. In the spirit of the approaches considered in section 4, I suggest that the problem of method and truth is to be dealt with along the following lines. Realism at the level of common sense may be taken as our point of departure. The world of common sense is an independently existing reality of causally interacting objects. These objects may or may not be observable by us. We employ a variety of methods to inquire into the ways of this world. Some methods are purely observational, while others are rules of theory appraisal. On the whole, our sense experience provides us with true beliefs about the observable world. In addition, our theoretical reasoning about unobservable states of affairs is frequently rewarded with success at the level of observation and practical action. Given the sort of world we inhabit, the best explanation of the systematically successful implementation of a method of inquiry is that the method provides a reliable means of discovery of truth about the world. Like us, our methods are fallible. But in a world such as ours the use of such methods could not consistently meet with success, if they were not for the most part a reliable guide to the truth.

In section 3.2, I objected that critical rationalism does not explain why survival of criticism warrants truth as the unique mode of theory acceptance. Yet I do not oppose the method of criticism as such. Indeed, I take the method of criticism to be largely constitutive of the methodology of science. For, as pointed out previously, both falsificationist norms of empirical test and non-falsificationist criteria of theory appraisal may serve as the basis of the critical method in science. The question is simply one of why a theory which survives criticism need be accepted as true.

As with the previous point, this question becomes manageable if the critical method is placed within a broader metaphysical context. If a theory is subjected to a battery of demanding tests, consistently yielding accurate predictions in a range of different circumstances, such performance under test is to be accorded evidential weight with regard to the truth of the theory. It is true, of course, that occasional predictive success may occur as the result of good fortune or accident. But in the sort of world that we inhabit pervasive error is not rewarded by systematic success. A theory which survives a range of rigorous tests may ultimately fail as a result of deeper and more detailed investigation. But in order to sustain systematic success across a great variety of tests, it must either contain a considerable portion of truth or

[21] To be more precise, metaphysical resolution of the problem of method and truth proceeds by way of inference to best explanation *or similar form of inference*. For, as we saw in note 18, Rescher prefers to characterize his pragmatic account of method in terms of inference to best systematization.

approximate the truth sufficiently closely for it to be empirically indistinguishable from the truth.

It might, finally, be objected that appeal to metaphysical considerations in an epistemological context must proceed in a circle. In order for a claim about reality to justify a method of inquiry there must be reason to accept the claim about reality. But there can be no reason to accept a claim about reality until some method of inquiry is justified.

Such circularity is surely to be avoided. But to insist that epistemology proceed without metaphysics is to fail to appreciate the task with which the realist is confronted. It is not just that the methods of inquiry must be shown to be rationally justified. Since the purpose of inquiry is to discover truth, the methods must be shown to promote the search for truth. But since truth is a matter of how the world is, it must be shown that the methods lead to truth about a mind-independent world. But this requires that something substantive be said about the nature of the world in virtue of which the world is accessible to our methods of inquiry.

The ultimate aim of such an account is a coherent structure in which claims about methods and claims about reality fit together in relations of mutual support. To suppose that such relations of mutual support must result in circular justification is to mistake the nature of epistemology. For human knowledge is a natural phenomenon like any other. To explain how humans know the world requires that we explain how human inquirers may be related to reality in such a way that they may know it. Thus metaphysics and epistemology go hand in hand. For the realist, at least, facts about reality must be brought to bear on facts about inquiry if we are to explain how inquiry yields truth about reality.[22]

REFERENCES

Boyd, Richard (1991) 'Realism, Anti-Foundationalism and the Enthusiasm for Natural Kinds.' *Philosophical Studies* 61: 127-148.
Devitt, Michael and Sterelny, Kim (1987), *Language and Reality*. Oxford: Blackwell.
Ellis, Brian (1980) 'Truth as a Mode of Evaluation.' *Pacific Philosophical Quarterly* 61: 85-99.
Ellis, Brian (1990) *Truth and Objectivity*. Oxford: Blackwell.
Harman, Gilbert (1965) 'Inference to the Best Explanation.' *Philosophical Review* 74: 88-95.
Kornblith, Hilary (1993), *Inductive Inference and its Natural Ground*, Cambridge, Mass: MIT Press.
Lakatos, Imre (1974) 'Popper on Demarcation and Induction.' in Schilpp (1974): 241-273.
Laudan, Larry (1981) 'A Confutation of Convergent Realism.' *Philosophy of Science* 48: 19-49.
Laudan, Larry (1996) *Beyond Positivism and Relativism*. Boulder: Westview Press.
Musgrave, Alan (1988) 'The Ultimate Argument for Scientific Realism.' in Nola (1988): 229-252.
Musgrave, Alan (1993) *Common Sense, Science and Scepticism*. Cambridge: Cambridge University Press.
Musgrave, Alan (1996) 'Realism, Truth and Objectivity.' in R.S. Cohen, R. Hilpinen and Qiu Renzong (eds.) *Realism and Anti-Realism in the Philosophy of Science*. Dordrecht: Kluwer Academic Publishers, pp.19-44.
Musgrave, Alan (1999) *Essays on Realism and Rationalism*. Amsterdam and Atlanta: Rodopi.
Musgrave, Alan (2001) 'Metaphysical Realism versus Word-Magic.' in D. Aleksandrowicz and H. G. Russ (eds.) *Realismus Disziplin Interdisziplinaritat*. Amsterdam and Atlanta: Rodopi, pp. 29-54.

[22] *Acknowledgments:* I am grateful to Harold Brown, Paul Hoyningen-Huene, Robert Nola and Marcel Weber for comments on earlier drafts of this paper.

Newton-Smith, W.H. (1981) *The Rationality of Science.* London: Routledge & Kegan Paul.
Nola, Robert (ed.) (1988) *Relativism and Realism in Science.* Dordrecht: Kluwer Academic Publishers.
Nola, Robert and Sankey, Howard (eds.) (2000) *After Popper, Kuhn and Feyerabend: Recent Issues in Theories of Scientific Method.* Dordrecht: Kluwer Academic Publishers.
Nola, Robert and Sankey, Howard (2000a) 'A Selective Survey of Theories of Scientific Method.' in Nola and Sankey (2000): 1-65.
Peirce, Charles S. (1955) 'Abduction and Induction' in J. Buchler (ed.) *Philosophical Writings of Peirce.* New York: Dover Publications.
Popper, Karl (1972) *Objective Knowledge.* Oxford: Oxford University Press.
Popper, Karl (1974a) 'Lakatos on the Equal Status of Newton's and Freud's Theories.' in Schilpp (1974): 999-1013.
Popper, Karl (1974b) 'Ayer on Empiricism and Against Verisimilitude.' in Schilpp (1974): 1100-1114.
Putnam, Hilary (1975) 'What is Mathematical Truth?' in *Mathematics Matter and Method: Philosophical Papers Volume 1.* Cambridge: Cambridge University Press, pp. 60-78.
Putnam, Hilary (1978) *Meaning and the Moral Sciences.* London: Routledge & Kegan Paul.
Putnam, Hilary (1981) *Reason, Truth and History.* Cambridge: Cambridge University Press.
Rescher, Nicholas (1977) *Methodological Pragmatism.* Oxford: Blackwell Publishers.
Rescher, Nicholas (2001) *Philosophical Reasoning: A Study in the Methodology of Philosophizing.* Oxford: Blackwell Publishers.
Sankey, Howard (2000) 'Methodological Pluralism, Normative Naturalism and the Realist Aim of Science.' in Nola and Sankey (2000): 211-229.
Sankey, Howard (2002) 'Realism, Method and Truth.' in M. Marsonet (ed.) *The Problem of Realism* Aldershot: Ashgate, pp. 64-81.
Schilpp, Paul A. (ed.) (1974) *The Philosophy of Karl Popper: Library of Living Philosophers Volume 14.* La Salle, Illinois: Open Court.
van Fraassen, Bas (1980) *The Scientific Image.* Oxford: Oxford University Press.

STATHIS PSILLOS

THINKING ABOUT THE ULTIMATE ARGUMENT FOR REALISM

1. INTRODUCTION

Alan Musgrave has been one of the most passionate defenders of scientific realism.[1] Most of his papers in this area are, by now, classics. The title of my paper alludes to Musgrave's piece 'The Ultimate Argument for Realism', though the expression is Bas van Fraassen's (1980, p. 39), and the argument is Hilary Putnam's (1975, p. 73): realism 'is the only philosophy of science that does not make the success of science a miracle'. Hence, the code-name 'no-miracles' argument (henceforth, NMA). In fact, NMA has quite a history and a variety of formulations. I have documented all this in my (1999, chapter 4). But, no matter how exactly the argument is formulated, its thrust is that the success of scientific theories lends credence to the following two theses: a) that scientific theories should be interpreted realistically and b) that, so interpreted, these theories are approximately true. The original authors of the argument, however, did not put an extra stress on *novel* predictions, which, as Musgrave (1988) makes plain, is the litmus test for the ability of any approach to science to explain the success of science.

Here is why reference to novel predictions is crucial. Realistically understood, theories entail too many novel claims, most of them about unobservables (e.g., that there are electrons, that light bends near massive bodies, etc.). It is no surprise that some of the novel theoretical facts a theory predicts may give rise to novel *observable* phenomena, or may reveal hitherto unforeseen connections between known phenomena. Indeed, it would be surprising if the causal powers of the entities

[1] I want to dedicate this paper to Alan Musgrave. His exceptional combination of clear-headed and profound philosophical thinking has been a model for me. His commitment to, and defence of, realism have inspired and guided my own work in this area. I hope that our residual disagreements will not obscure our deep agreement. Sections 5 to 8 were inspired by a paper by P. D. Magnus and Craig Callender, titled 'Retail Realism and Base Rate Neglect'. I want to thank Magnus and Callender for many useful comments.

C. Cheyne & J. Worrall (eds.), Rationality and Reality: Conversations with Alan Musgrave, 133–156.

posited by scientific theories were exhausted in the generation of the already *known* empirical phenomena that led to the introduction of the theory. So, on a realist understanding of theories, novel predictions and genuine empirical success is to be expected (given of course that the world co-operates).

The aim of this paper is to rebut two major criticisms of NMA. The first comes from Musgrave (1988). The second comes from Colin Howson (2000). Interestingly enough, these criticisms are the mirror image of each other. Yet, they both point to the conclusion that NMA is *fallacious*. Musgrave's misgiving against NMA is that if it is seen as an *inference to the best explanation*, it is deductively fallacious. Being a deductivist, he tries to correct it by turning it into a *valid deductive argument*. Howson's misgiving against NMA is that if it is seen as an *inference to the best explanation*, it is *inductively* fallacious. Being a subjective Bayesian, he tries to correct it by turning it into a *sound subjective Bayesian argument*. I will argue that both criticisms are unwarranted.

Actually, I would have no problem with Musgrave's version of NMA if deductivism were correct. But, as I will try to argue, the deductivist stance is both descriptively and normatively wrong. To avoid a possible misunderstanding, let me note that I have no problem with deductive logic (how could I?). My problem is with deductivism, that is the view that, as Musgrave (1999a, p. 395) puts it, 'the only valid arguments are deductively valid arguments, and that deductive logic is the only logic that we have or need'. One could cite Bayesianism as a live example of why deductivism is wrong. But, I think, there are important problems with Bayesianism too.[2] Put in a nutshell, the Bayesian critique of NMA is that it commits the base-rate fallacy. Howson tries to rectify this by arguing that a 'sounder' version of NMA should rely explicitly on *subjective* prior probabilities. Against the Bayesian critique of NMA I will primarily argue that we should resist the temptation to cast the no-miracles argument in a subjective Bayesian form. However, I will also explore the possibility of accepting a more objective account of prior probabilities, if one is bent on casting NMA in a Bayesian form.

Here is a brief summary of the menu. Section 2 defines scientific realism and investigates Musgrave's own understanding of it. Section 3 explains, rather briefly, what I take the form and the aim of the no-miracles argument to be. Section 4 criticises Musgrave's deductivism and his attempt to show that NMA is best understood as a deductive enthymeme. Section 5 explains how NMA (as an inductive argument) is supposed to commit the base-rate fallacy. Section 6 argues that there are ways to give a more objective account of the prior probabilities that are supposed to be necessary for NMA to be inductively sound. Section 7 explores some features of the base-rate fallacy and explains why it is reasonable to ignore the base-rates (let's say the prior probabilities, though they are not the same) on certain occasions. Section 8 argues that if we look at case histories we can have strong reasons to be realists about several theories. Section 9 explores two ways to think of NMA that do not involve prior probabilities.

[2] I have tried to explore some of these problems in my (2006).

2. WHAT IS SCIENTIFIC REALISM?

I take the following three theses as constitutive of scientific realism (cf. my 1999, xix-xxi; 2000).

The Metaphysical Thesis: The world has a definite and mind-independent structure.

The Semantic Thesis: Scientific theories are truth-conditioned descriptions of their intended domain. Hence, they are capable of being true or false. The theoretical terms featuring in theories have putative factual reference. So if scientific theories are true, the unobservable entities they posit populate the world.

The Epistemic Thesis: Mature and predictively successful scientific theories are well-confirmed and approximately true. So entities posited by them, or, at any rate entities very similar to those posited, inhabit the world.

Musgrave (1996, p. 23) agrees that realism involves the *Semantic Thesis*. He is not very explicit about the *Metaphysical Thesis*. Actually, he is quite critical of the realist view which 'erects current science into a metaphysic and ties scientific realism too closely to that metaphysic' (1996, p. 21). As I understand it, the *Metaphysical Thesis* means to make scientific realism distinct from all those anti-realist accounts of science, be they traditional idealist and phenomenalist or the more modern verificationist accounts which, based on epistemic accounts of truth, allow no divergence between what there is in the world and what is issued as existing by a suitable set of epistemic practices and conditions. It implies that if the unobservable natural kinds posited by theories exist at all, they exist independently of our ability to be in a position to know, verify, recognise etc. that they do. Musgrave does accept all this. Throughout his work on realism, he has defended a non-epistemic conception of truth and has argued very persuasively against epistemic conceptions of truth. He has also defended the mind-independent existence of the world (see, for instance his 1989; 1996). So he does, after all, accept a version of the *Metaphysical Thesis* above.

When it comes to the *Epistemic Thesis*, Musgrave seems to distinguish between two versions of it: a weak and a strong one. He does accept the weak version. For, he thinks 'that some scientific entities do exist and that some of what science tells us about them is true' (1996, p. 21). He calls 'ludicrous' the view that 'all scientific theories are false' (1996, p. 22). But he (1996, pp. 19-21) seems to take the strong version of the *Epistemic Thesis*, which he associates with what he calls 'mad-dog realism', to imply commitment to *all* entities posited by current theories and belief in *everything* they say about them. He is quite clear that he denies this strong version. He protests that this view is overly optimistic and unwarranted. I think he is quite right when he says: 'We should be more confident about atoms and molecules than we are about electrons, and more confident about electrons than we are about quarks and gluons' (1996, p. 22). He is equally right when he adds: 'Realism about the entities and theories of current science should rather be guarded' (1996, p. 22).

Guarded realism is still realism! Guarded realists need not take current science uncritically. They need not commit themselves to everything that current science asserts. They can have a differentiated attitude towards the theoretical constituents of modern science: some of them are better supported by the evidence than others; some of them play an indispensable explanatory role, while others do not; some contribute to the successes of theories, while others do not. But, I think, we should not lose sight of the general philosophical issue at stake. I take it to be this: is there any strong reason to believe that science cannot achieve theoretical truth? That is, is there any reason to believe that after we have understood the theoretical statements of scientific theories as expressing genuine propositions, we can never be in a warranted position to claim that they are true (or at least, more likely to be true than false)? What the *Epistemic Thesis* means to assert is that theoretical truth is achievable (and knowable) no less than is observational truth. So, the *Epistemic Thesis* is meant to be optimistic: science has succeeded in tracking truth. To be sure, this requires a certain *epistemic luck*: it's not a priori true that science has been, or has to be, successful in truth-tracking. If science does succeed in truth-tracking, this is a *radically contingent fact* about the way the world is and the way scientific method and theories have managed to 'latch onto' it.

The debate about the *Epistemic Thesis* has brought to focus one central issue: are the ampliative-abductive methods of science reliable and can they confer justification on theoretical assertions? The defence of the *Epistemic Thesis* requires a positive answer to this question. For, it is part of the realist thesis that the ampliative-abductive methods employed by scientists to arrive at their theoretical beliefs are reliable: they tend to generate approximately true beliefs and theories. The *no-miracles argument* (NMA) has played a pivotal role in this defence.

3. THE NO-MIRACLES ARGUMENT

How does NMA support the *Epistemic Thesis?* As I have argued elsewhere (cf. Psillos 1999, chapter 4), the structure and role of NMA in the realism debate is quite complex. To a good approximation, it should be seen as a grand Inference to the Best Explanation (IBE). The way I read it, NMA is a philosophical argument which aims to defend the reliability of scientific methodology in producing approximately true theories and hypotheses. I don't want to repeat here the exact formulation of the argument (see Psillos 1999, pp. 78-81). However, I want to emphasise that its conclusion has two parts. The *first* part is that we should accept as (approximately) true the theories that are implicated in the (best) explanation of the *instrumental* reliability of first-order scientific methodology. The *second* part is that since, typically, these theories have been arrived at by means of IBE, IBE is reliable. Both parts are necessary for my version of NMA.

The main strength of NMA rests on the first part of the conclusion. Following more concrete types of explanatory reasoning which occur all the time in science, it suggests that it is reasonable to accept certain theories as approximately true, at least in the respects relevant to their theory-led predictions. So, it is successful instances of explanatory reasoning in science which provide the *basis* for the grand abductive argument. However, NMA is not *just* a generalisation over the scientists' abductive inferences. Although itself an *instance* of the method that scientists employ, it aims at a much broader target: to defend the thesis that Inference to the Best Explanation, (that is, a *type* of inferential method) is reliable. This relates to the second part of its conclusion. What, I think, makes NMA distinctive as an argument for realism is that it defends the achievability of theoretical truth. The second part of the conclusion is supposed to secure this. The background scientific theories, which are deemed approximately true by the first part of the conclusion, have themselves been arrived at by abductive reasoning. Hence, it is reasonable to believe that abductive reasoning is reliable: it tends to generate approximately true theories. This conclusion is not meant to state an a priori truth. The reliability of abductive reasoning is an empirical claim, and if true, it is contingently so.

It should be noted that, as I conceive of it, NMA needs a qualification. Although most realists would acknowledge that there is an explanatory connection between a theory's being empirically successful and its being, in some respects, right about the unobservable world, it is far too optimistic—if defensible at all—to claim that *everything* that the theory asserts about the world is thereby vindicated. So, realists should *refine* the explanatory connection between empirical and predictive success, on the one hand, and truthlikeness, on the other. They should assert that these successes are best explained by the fact that the theories which enjoyed them have had *truthlike theoretical constituents* (i.e., truthlike descriptions of causal mechanisms, entities and laws). The theoretical constituents whose truthlikeness can best explain empirical successes are precisely those that are essentially and ineliminably involved in the generation of predictions and the design of the methodology which brought these predictions about. From the fact that not every theoretical constituent of a successful theory does and should get credit from the successes of the theory, it certainly does *not* follow that none do (or should) get some credit.

There are a number of objections to this explanationist version of NMA. One of them has also been pressed by Musgrave (1988, p. 249; 1999, pp. 289-90), and this is particularly hurtful. The objection is that NMA is *viciously* circular: it employs a second-order IBE in defence of the reliability of first-order IBEs. As is explained in detail in my (1999, chapter 4), the abductive defence of realism proceeds within a broad naturalistic framework. Within this framework, the charge of circularity loses most of its bite because what is sought is not justification of inferential methods and practices (at least in the neo-Cartesian internalist sense) but their explanation and defence (in the epistemological externalist sense). In any case, I (1999, pp. 81-90) argued that (a) there is a difference between premise-circularity and rule-circularity (a premise-circular argument employs its conclusion as one of its premises; a rule-circular argument conforms to the *rule* which is vindicated in its conclusion); (b) rule-circularity is *not* vicious; and (c) the circularity involved in the defence of basic rules of inference is rule-circularity. Though these points had already been made with regard to basic deductive and inductive rules, I showed how the above defence of IBE is rule-circular. So, the employment of IBE in an abductive defence of the reliability of IBE is *not* viciously circular. As a support of all this consider the following case. Many (if not all) use modus ponens unreflectively as an inferential rule and yet the establishment of the *soundness* of modus ponens proceeds with an argument which effectively uses modus ponens. This procedure can still explain to modus ponens-users why and in virtue of what features deductive reasoning is sound.

Being a deductivist, Musgrave thinks that the *only* kind of validity is deductive validity. He denies that there are such things as non-deductive cogent arguments (cf. 1999a). He takes it that rule-circular arguments in favour of inferential rules may have only some *psychological* force (cf. 1999, pp. 289-90). But he (1999, p. 295) is aware of the point that the proof of the soundness of modus ponens requires the use of modus ponens. How does he react to this? It seems that he has wavered between two thoughts. The *first* is that 'there is little future in the project of 'justifying deduction"' (1999, p. 296). As he acknowledges, 'Any "justification" which is non-psychologistic will itself be a deductive argument of some kind, whose premises will be more problematic than the conclusion they are meant to justify' (*ibid.*) To be sure, he immediately adds that there is a difference between deductive rules and non-deductive (ampliative) ones in that, even if neither of them can be 'justified', non-deductive rules can be *criticised*. But how much pause should this give us? Let us grant, as we should, that none of our basic inferential rules (both deductive and non-deductive) can be 'justified' without rule-circular arguments. The fact that the non-deductive rules can be criticised more severely than the deductive ones may make us be much more cautious when we employ the former. That's all there is to it. The *second* thought that Musgrave has (cf. 1980, pp. 93-5; 1999, pp. 96-7) is that there is a sense in which deduction *can* be 'justified', but this requires an appeal to 'deductive intuitions'. As he (1980, p. 95) graphically puts it: 'In learning logic we pull ourselves up by our bootstraps, exploit the intuitive logical knowledge we already possess. Somebody who lacks bootstraps ('deductive intuition') cannot get off the ground'. This is, I think, exactly right. But, as I have argued in some detail in my (1999, pp. 87-9), exactly the same response can be given to calls for 'justifying' non-deductive rules. When it comes to issues concerning the vindication of inference to the best explanation, if one lacks 'abductive' intuitions, one lacks the necessary bootstraps to pull oneself up.

4. DEDUCTIVISM

To realists, it might come as a surprise that Musgrave (1996, p. 19) takes realism to be, 'first and foremost a thesis about the aim of science. It says that the aim of a scientific inquiry is to discover the truth about the matter inquired into'. So he takes realism to be an 'axiological thesis': 'science aims for true theories'.[3] There is clear motivation for this view: even if *all* theories we ever came up with were false, realism wouldn't be threatened (cf. 1996, p. 21). As we have seen, Musgrave does

[3] This axiological thesis has been a constant pillar of his realism. For some early formulation of it, see his (1977).

not think that all our theories have been, or will be, outright false. But he does take this issue (whatever its outcome may be) to have no bearing on whether realism is a correct attitude to science. There are, however, inevitable philosophical worries about the axiological characterisation of realism. *First*, it seems rather vacuous. Realism is rendered immune to any serious criticism which stems from the empirical claim that the science we all love has a poor record in truth-tracking (cf. Laudan 1984). *Second*, aiming at a goal (truth) whose achievability by the scientific method is left unspecified makes its supposed regulative role totally mysterious. Finally, all the excitement of the realist claim that science engages in a cognitive activity which pushes back the frontiers of ignorance and error is lost.

Though Musgrave does not address these worries explicitly, he does so implicitly. For, he does try to defend the prime realist argument for epistemic optimism, viz., the no-miracles argument. He (1988, p. 237; 1999, p. 60) takes NMA to be an inference to the best explanation. Besides, he (1988, p. 232; 1999, p. 119) has been one of the first to stress that what needs to be explained is *novel* success (that is, the ability of theories to yield successful novel predictions). And he has been one of the first to note that NMA should focus on the novel success of *particular* theories (cf. 1988, p. 249). He has also produced some powerful arguments to the effect that non-realists' explanations of the success of science are less satisfactory than the realist one. Most of them appear in his (1988). In fact, he (1988, p. 249) concludes that the realist explanation is the *best*. The issue then is this: Does Musgrave endorse NMA? The answer to this question is *not* straightforward.

Precisely because Musgrave takes NMA to be an inference to the best explanation, he takes it to be deductively invalid, and hence fallacious. Being a deductivist, he takes it that the only arguments worth their salt are deductive arguments. So he cannot endorse NMA, at least as it stands. Musgrave takes all *prima facie* non-deductive arguments to be *enthymemes*. An enthymematic argument is an argument with a missing or suppressed premise. After the premise is supplied (or made explicit), the argument becomes deductively valid. But it may or may not be sound (cf. his 1999, pp. 87 & 281ff). According to Musgrave, non-deductive arguments are really deductive enthymemes, with 'inductive principles' as their missing premises.

As it is typically presented, IBE has the following form (cf. Musgrave 1988, p. 239; 1999, p. 285):

(*IBE*)

1. F is the fact to be explained.
2. Hypothesis H explains F.
3. Hypothesis H satisfactorily explains F.
4. No available competing hypothesis explains F as well as H does.
5. Therefore, it is reasonable to accept H as true.

Given that this argument-pattern is invalid, Musgrave proposes that it should be taken to be enthymematic. The missing premise is the following epistemic principle (cf. *ibid.*):

(*missing premise*) 'It is reasonable to accept a satisfactory explanation of any fact, which is also the best explanation of that fact, as true'.

Add to (*IBE*) the missing premise, and you get a valid argument. Briefly put, the deductive version of IBE is this:

(*D-IBE*)

1. If hypothesis *H* is the best explanation of the fact to be explained[4], then it is reasonable to accept *H* as true.
2. *H* is the best explanation of the evidence.
3. Therefore, it is reasonable to accept *H* as true.

This is a valid argument. Besides, Musgrave (1999, p. 285) thinks that 'instances of the scheme might be sound as well'. In any case, he thinks that the missing premise 'is an epistemic principle which is not obviously absurd' (1999, p. 285). In light of this, it's no surprise that Musgrave reconstructs NMA as an enthymeme. That's how he (1988, p. 239) puts it:

> The fact to be explained is the (novel) predictive success of science. And the claim is that realism (more precisely, the conjecture that the realist aim for science has actually been achieved) *explains* this fact, explains it *satisfactorily*, and explains it *better* than any nor-realist philosophy of science. And the conclusion is that it is reasonable to accept scientific realism (more precisely, the conjecture that the realist aim for science has actually been achieved) as true.

This is a deductive enthymeme, whose suppressed premise is the aforementioned epistemic principle (*missing premise*). What is worth stressing is that Musgrave takes NMA to aim to tell in favour of the *Epistemic Thesis* (see section 2). Though he formulates the argument in terms of his own axiological thesis, he takes it that, if successful, NMA makes it reasonable to accept that truth has been achieved.

I would have no problem with (*D-IBE*) if deductivism were correct. But, I think, the deductivist stance is so radically at odds with the practice of science (as well as of everyday life) that it would have to give even the most dedicated deductivist pause. Human reasoning is much broader than deductivists allow. It is *defeasible*, while deductive reasoning is not. That is, it is sensitive to new information, evidence and reasons in a way that is not captured by deductive arguments. The latter are monotonic: when further premises are added to a valid deductive argument, the original conclusion still follows. But human reasoning is non-monotonic: when new information, evidence and reasons are added as premises to a non-deductive argument, the warrant there was for the original conclusion may be removed (or enhanced). Human reasoning is also *ampliative*, while deductive reasoning is not.

[4] This, in effect, sums up premises (2) to (4) of (*IBE*).

That is, the conclusions we adopt, given certain premises, have excess content over the premises. Deductive reasoning is not content-increasing. In a (logical) sense, the conclusion of a valid deductive argument is already 'contained' in its premises.[5] This is not to belittle deductive reasoning. It's the only kind of reasoning that is truth-preserving. The importance of truth-preservation can hardly be exaggerated. But we should not forget that, though deductive reasoning preserves truth, it cannot establish truth. In particular, it cannot establish the truth of the premises. If we are not talking about logical (and mathematical and analytical—if there are such things—truths), the premises of deductive arguments will be synthetic propositions, whose own truth can be asserted, if at all, on the basis of ampliative and non-deductive reasoning. So, though deductive reasoning is indispensable, it can hardly exhaust the content and scope of human (and scientific) reasoning.[6] As a *descriptive* thesis, deductivism is simply false.

Is then deductivism to be construed as a *normative* thesis? I am aware of no argument to the effect that deductivism is normatively correct. This is not to imply that deductive logic has no normative force. It does. But recall that deductivism is the thesis that *all* arguments worth their salt *should* be construed as deductive enthymemes. Whence could this thesis derive its supposed normative force? I don't see a straightforward answer to this question. Musgrave suggests that reconstructing supposed non-deductive arguments as deductive enthymemes 'conduces to clarity' (1999, pp. 284-5). That is, it makes their premises explicit. Hence, it also makes explicit what is required for the premises to be true, and for the argument to be sound. I think, however, that this point is problematic. Non-deductive arguments (e.g., simple enumerative induction, or inference to the best explanation) are *not* unclear. If anything, the problem with them is how to justify them. But a similar problem occurs with deduction, as we saw at the end of the previous section. Suppose, however, that we leave this problem to one side. Suppose that we grant that turning a non-deductive argument into a deductively valid one conduces to clarity since it makes its premises explicit. Deductivists still face a problem: what, if anything, justifies the missing premise? To fix our ideas, consider the major premise of (*D-IBE*) above. What justifies the principle 'If hypothesis *H* is the best explanation of the fact to be explained, then it is reasonable to accept *H* as true'? The sceptic can always object to *this* principle that it is question-begging. How can a deductivist reply to this charge?

Musgrave (1999a, p. 408) does consider this problem. He takes the sceptic to rely on the following idea, which Musgrave calls 'justificationism': 'a reason for

[5] For more on non-deductive reasoning and on the way IBE should be understood as a *genus* of ampliative reasoning, see my (2002).

[6] Musgrave might reply to this by saying that scientists employ 'demonstrative inductions', which are really deductions, though not deductions *from* the phenomena, as Newton thought (cf. his 1999, pp. 303 & 306). I don't want to discuss this issue here, though it certainly needs attention. Briefly put, the thrust of demonstrative induction is that premises of greater generality and premises of lesser generality will yield a conclusion of intermediate generality. But this must be noted: it is wrong to think that demonstrative induction frees us from the need to engage in ampliative inference. As Norton (1994, p. 12) notes: 'Typically, ampliative inference will be needed to justify "the premises of greater generality"'.

believing P must justify P, show that P is true or at least probably true'. Not surprisingly, he rejects justificationism. So, if justificationism is abandoned, the fact that the reasons which support the major premise of (*D-IBE*) are not conclusive is *not* a reason not to believe in the major premise. I think this is exactly right. But it has a repercussion which Musgrave does not seem to appreciate. Justificationism has also been assumed by the sceptics in their critique of inductive (or non-deductive) reasoning. One way to put their point is that the premises of a non-deductive argument do not establish the truth of its conclusion. If justificationism is to be abandoned, as it should be, it should be abandoned in all contexts. That is, it should be abandoned for deductivism as well as inductivism. It seems, then, that Musgrave himself offers us a strong reason to hold onto inductivism.

Perhaps, deductivism is a fall-back position. It says that arguments can be *reconstructed* as deductively valid arguments. But this thesis is trivial. *Any* argument can be turned into a deductively valid one by adding suitable premises. In particular, any *invalid* argument can be rendered valid by adding suitable premises. Consider the fallacy of affirming the consequent. The argument:

1. If (if a and b) and b, then a
2. If a then b
3. b
4. Therefore, a

is perfectly valid. If all logically invalid arguments were considered enthymemes, there would be no such thing as invalidity. Musgrave is aware of this objection, too. His reply is this: '[Y]ou cannot allow anything whatever to count as a 'missing premise'; what the 'missing premise' is must be clear from the context of the production of the argument in question' (1999a, p. 399; 1999, p. 87, n. 106). But, surely, the context underdetermines the possible 'missing premises'. More importantly, for any 'missing premise', there will be *some* contexts in which it is appropriate.

To sum up, Musgrave's misgivings against NMA were motivated by the thought that if it is seen as an inference to the best explanation, it is deductively fallacious. He tried to correct it, as we have seen, by turning it into a *valid deductive argument*. We found his attempt wanting because we found deductivism wrong. What is interesting is that others, most notably Colin Howson, think that if it is seen as an inference to the best explanation, NMA is *inductively* fallacious. He tries to correct it, by turning it into a sound subjective Bayesian argument. All this will leave Musgrave totally unmoved, since he thinks there is no such think as inductive logic (cf. 1999a). Still, for those of us who a) think that there is more to reasoning than deduction, b) are critical of subjective Bayesianism, and c) want to defend some form of NMA, it will be important to examine whether the Bayesian criticism of NMA succeeds or fails.

5. SUBJECTIVE BAYESIANISM TO THE RESCUE?

Howson (2000, 36) formulates the 'no-miracles' argument (NMA) as follows:

(A)

1. If a theory T is not substantially true then its predictive success can only be accidental, a chance occurrence.
2. A chance agreement with the facts predicted by T is very improbable—of the order of a miracle.
3. Since this small chance is so extraordinarily unlikely, the hypothesis that the predictive success of T is accidental should be rejected (especially in light of the fact that there is an alternative explanation—viz., that T is true—which accounts better for the predictive success).
4. Therefore, T is substantially true.[7]

He then argues in some detail that (A) is inductively fallacious. He contests the soundness of all if its premises (cf. 2000, 43). However, the novelty of Howson's view relates to his criticism of premise (3) and of the inferential move to (4). His prime point is that (A) is wrong because it commits the base-rate fallacy.

Let me introduce the base-rate fallacy with a standard example in the literature, which is known as the Harvard Medical School test.

(*Harvard Medical School test*)

A test for the presence of a disease has two outcomes, 'positive' and 'negative' (call them + and –). Let a subject (Joan) take the test and let H be the hypothesis that Joan has the disease and –H the hypothesis that Joan doesn't have the disease. The test is highly reliable: it has zero false negative rate. That is, the likelihood that the subject tested negative given that she does have the disease is zero (i.e., prob(–/H)=0). Consequently, the true positive rate, i.e., the likelihood of being tested positive given that she has the disease is unity, (prob (+/H)=1). The test also has a very small false positive rate: the likelihood that Joan is tested positive though she doesn't have the disease is, say, 5% (prob(+/–H) =.05). Joan tests positive. What is the probability that Joan has the disease given that she tested positive? That is, what is the posterior probability prob(H/+)?

When this problem was posed to experimental subjects, they tended, with overwhelming majority, to answer that the probability that Joan has the disease given that she tested positive was very high—very close to 95%.

This answer is wrong. Given only information about the likelihoods prob (+/H) and prob (+/–H), the question above—what is the posterior probability prob (H/+)?—is indeterminate. This is so because there is some crucial information *missing*: we are not given the incidence rate (base-rate) of the disease in the population. If this

[7] This formulation does not exactly match the way Howson puts the argument, but it closely resembles it.

incidence rate is very low, e.g., if only 1 person in 1,000 has the disease, then it is very *unlikely* that Joan has the disease even though she tested positive: prob(H/+) would be less than .02.[8] For prob(H/+) to be high, it must be the case that prob(H) be not too small. But if prob(H) is low, then it can dominate over a high likelihood of true positives and lead to a very low posterior probability prob(H/+). The lesson that many have drawn from cases such as this is that it is a fallacy to ignore the base-rates because it yields wrong results in probabilistic reasoning. The so-called base-rate fallacy is that experimental subjects who are given problems such as the above tend to neglect base-rate information (that is, the prior probabilities), *even when* they are given this information explicitly.[9]

With this in mind, let us take a look at NMA. To simplify matters, let S stands for predictive success and T for a theory. According to (A) above, the thrust of NMA is the comparison of *two* likelihoods, viz., prob(S/-T) and prob(S/T). The following argument captures the essence of Howson's formulation of NMA (see (A) above).

(B)

1. prob(S/T) is high.
2. prob(S/-T) is very low.
3. S is the case.
4. Therefore, prob(T/S) is high.[10]

What's explicit in (B) is that alternative theories (or the falsity of T) fail(s) to support the evidence. Let us say that the false-positive rate is low and the false-negative rate is zero. That is, the probability of T being successful given that it is false is very small (say, prob(S/-T)=.05)) and the probability of T being unsuccessful given that it is true is zero (i.e., prob(-S/T)=0). Hence, the true-positive rate (prob(S/T)) is 1. Does it follow that prob(T/S) is high? NMA is portrayed to answer affirmatively. But if so, it is fallacious: it has neglected the base-rate of truth (that is, prob(T)). Without this information, it is impossible to estimate correctly the required posterior probability. If the base-rate of true theories is low, then prob(T/S) will be very low too. Assuming that base-rate of true theories is 1 in 100 (i.e., prob(T)=.01), prob(T/S)=.17. (The calculation mimics the one offered in note 7). The conclusion seems irresistible: as it stands, (B) commits the base-rate fallacy—it has neglected prob(T), or as the jargon goes, the base-rate.

Every cloud has a silver lining, however. So, Howson (2000, pp. 55-9) urges us to think how NMA could become 'sounder' within a Bayesian framework. We are

[8] By Bayes's theorem, prob(H/+) = prob(+/H)prob(H)/prob(+), where prob(+) = prob(+/H)prob(H)+ prob(+/-H)prob(-H). Plug in the following values: prob(+/H) =.95, prob(H)=.001, prob(-H) =.999, prob(+/-H) =.05. Then, prob(H/+) is roughly equal to .02.

[9] This problem was first investigated by Tversky and Kahneman (1982). It was dubbed 'the base-rate fallacy' by Bar-Hillel (1980).

[10] To be more precise, we need to state the conclusion thus: Therefore, $prob_{new}(T)$ is high, where $prob_{new}(T) = prob_{old}(T/S)$.

invited to accept that NMA can succeed only if information about base-rates (or prior probabilities) is taken into account. In effect, the idea is this:

(B^1)

1. prob(S/T) is high.
2. prob(S/-T) is very low.
3. S is the case.
4. prob(T) is not very low.
5. Therefore, prob(T/S) is high.

What has been added is an explicit premise that refers to the *prior probability* of true theories. For (B^1) to be sound, this probability should not be low. How low prob(T) can be will vary with the values of prob(S/T) and prob(S/-T). But it is noteworthy that, with the values of the likelihoods as above, if prob(T) is only 5%, then prob(T/S) is over 50%. To be sure, (B^1) is not valid. But, as Howson (2000, p. 57) notes, it is 'a sound probabilistic argument'. Of course, (B^1) rests also on the assumption that prob(S/-T) is very low. This can be contested. But, Howson notes, there may be occasions on which this low probability can be justified, e.g., when, for instance, we think of -T as a disjunction of *n* theories T_i (i=1,...,n) whose own prior probabilities prob(T_i) are negligible. In any case, his point is that NMA can be a sound argument only when we see that it is based on some substantive assumptions about *prior probabilities*. Being a subjective Bayesian, he takes these prior probabilities to be 'necessarily subjective and a priori' (2000, p. 55).

6. A WHIFF OF OBJECTIVISM

I will start my criticism of Howson's argument by resisting the view that one needs to rely on *subjective* prior probabilities in formulating NMA. So for the time being at least, I will assume the foregoing Bayesian reformulation of NMA. Actually, let us reformulate (B^1), based on what has been called the *Bayes factor*. This is the ratio:

(Bayes factor): $f = prob(S/\text{-}T)/prob(S/T).$

Recall Bayes's theorem:

$$prob(T/S) = prob(S/T)prob(T)/prob(S) \quad (1)$$

where:

$$prob(S) = prob(S/T)prob(T)+prob(S/\text{-}T)prob(\text{-}T).$$

Using this factor, (1) becomes this:

$$prob(T/S) = prob(T)/ prob(T) + f prob(-T). \qquad (2)$$

(B^1) can then be written thus:

(B^2)

1. f is very small.
2. S is the case.
3. prob(T) is not very low.
4. Therefore, prob(T/S) is high.

The Bayes factor is small if prob(S/-T) << prob(S/T). Now, whether the conclusion follows from the premises depends on the prior probability prob(T). So, the Bayes factor, on its own, tells us little. But it *does* tell us something of interest. Actually, it tells us something that can take out *some* to the sting of subjectivism in Bayesianism. Two things are relevant here. The *first* is that there is a case in which the prior probability of a theory does not matter. This is when the Bayes factor is *zero*. Then, no matter what the prior prob(T) is, the posterior probability prob(T/S) is unity. The Bayes factor is zero if prob(S/-T) is zero. This happens when just one theory can explain the evidence. Then, we can dispense with the priors. This situation may be unlikely. But it is not a priori impossible. After all, the claim that evidence underdetermines the theory is not a logical truth! Put in a different way, one quick problem that Howson's reconstructions of NMA faces is that it equates, at least implicitly, explanation with deduction. Given this equation, it is trivially true that there cannot be just one theory that explains the evidence, since there will be many (an infinite number of?) theories that entail it. In many places (cf., for instance 2000, pp. 40-1), Howson does make this equation. But this is a Phyrric victory over NMA. There is more to explanation than the deduction of (descriptions of) the phenomena from the theory (and deduction is not even necessary for explanation). So, it may well be the case that many theories entail (descriptions of) the relevant phenomena, while only one of them *explains* them. I won't argue for this claim now. Suffice it for the present purposes to note that equating explanation with deduction is question-begging.[11]

Be that as it may, let us grant, for the sake of the argument, that the case in which the Bayes factor is zero is exceptional. There is a *second* thing in relation to Bayes factor that needs to be noted. Assume some kind of indifference (or a flat probability distribution) between prob(T) and prob(-T); that is, assume that prob(T) = prob(-T) = 1/2. Then (2) above becomes:

$$prob(T/S) = 1/ 1+f \qquad (3)$$

Assuming indifference, the Bayes factor shows that likelihood considerations (especially the fact, if it is fact, that f is close to zero) can make T much more likely

[11] For more on the realist reply to the argument from the underdetermination of theories by evidence, see my (1999, chapter 8).

to be true. The point here is not that we can altogether dispense with the priors. Rather, the point is that we are not compelled to take a *subjective* view of the prior probabilities. So, there is a version of NMA which, though close to (B^2) above, does *not* assume anything other than indifference as to the prior probability of T being true.

(B^3)

1. *f* is close to zero.
2. S is the case.
3. prob(T) = prob(-T) =1/2.
4. Therefore, prob(T/S) is high.

(B^3) strikes me as fine. If one wanted to capture the *thrust* of NMA within a Bayesian framework, one could hold onto (B^3). This does not commit the base-rate fallacy. Besides, it avoids the excesses of subjective Bayesianism.

So far, I have assumed that prior probabilities and base-rates are one and the same thing. In fact, Howson does assume this too. He (2000, p. 57, n.5) calls the prior probabilities 'the epistemic analogue of the base-rate'. Normally, base-rates are given by reliable statistics. Hence, they are quite objective. When a subject is asked how probable it is that Jim (a young adult male) suffers from hypothyroidism, given that he has the symptoms, she doesn't commit a fallacy if she ignores her own prior degree of belief that Jim has hypothyroidism. After all, she might not have any prior degree of belief in this matter. The fallacy consists in her claiming that the probability is high while ignoring some relevant factual information about hypothyroidism, viz., that it is quite rare, even among people who have the relevant symptoms. This is some objective statistical information, e.g., that only 1 in 1,000 young adult male suffers from hypothyroidism. Base-rates of this form can (and should) be the input of a prior probability distribution. But they are *not* the prior subjective degrees of belief that Bayesians are fond of. In incorporating them, Bayesians move away from a purely subjective account of prior probabilities. But what about the converse? If prior probabilities are purely (and necessarily, as Howson says) subjective, then why should an agent rely on base-rates to fix her prior probabilities? That is, why should an agent's subjective prior probability of an event to occur be equated with the rate of the occurrence of this event in a certain population? Purely subjective priors might be assigned in many ways (and, presumably, there is no fact of the matter as to which way is the correct, or rational, one). An agent might know a relevant-base rate but, being a purely subjective Bayesian, she might decide to disregard it. She won't be probabilistically incoherent, if she makes suitable adjustments elsewhere in her belief corpus. Or, though the base-rate of hypothyroidism in the population is very low, her subjective prior probability that Jim suffers from hypothyroidism may be quite high, given that she believes that Jim has a family history of hypothyroidism. The point here is that *if* prior probabilities are purely subjective, it seems within the rights of a Bayesian agent to fix her prior probabilities in a way different from the relevant base-rates. So, prior probabilities are not, necessarily, base-rates. Or, more provocatively, ba(y)se rates are *not* base-rates.

In light of this, something stronger can be maintained. Subjective Bayesians had better have a more objective account of prior probabilities, if they are to reason correctly (according to their own standards) and to avoid falling victims of the base-rate fallacy. For if prior probabilities are totally up to the agent to specify, then the agent seems entitled to *neglect* the base-rate information, or to adopt a prior probability which is significantly lower or higher than the base-rate. If anything, base-rates should act as an external constraint on Bayesian reasoning, by way of fixing the *right* prior probabilities. The need to take account of base-rates seems to make Bayesianism more prescriptive than it intends to be. The call to rely on the base-rates is a substantive piece of advice, which goes beyond the mere call for synchronic and diachronic coherence.

7. IGNORING BASE-RATES

As we have seen, the Bayesian critique of NMA (see argument (B) above) consists in the claim that it *ignores* the base-rates of truth and falsity. But there is a sense in which this is not quite correct. The Bayesian criticism presupposes that *there are* base-rates for truth and falsity. However, it is hard, if not outright impossible, to get the relevant base-rates. The issue is not *really* statistical. That is, it's not really that we don't have a list of true and false theories at our disposal. Nor, of course, is the issue that the advocates of NMA fail to take account of such a list. The issue is philosophical. The very idea of a base-rate of truth and falsity depends on how the *relevant* population of theories is fixed. This is where many philosophical problems loom large. For one, we don't know how exactly we should individuate and count theories. For another, we don't even have, strictly speaking, outright true and false theories. But suppose that we leave all this to one side. A more intractable problem concerns the concept of success. What is it for a theory to be successful? There is no reason here to repeat well-known points (see my 1999, pp. 104-8). But the general idea is clear. By choosing a loose notion of success, the size of the relevant population might increase and a lot of false theories might creep in. True theories won't be left out, but they may be vastly outnumbered by false ones. There will be many more false positives than otherwise. In this population, the probability of a randomly selected theory being true will be low. By choosing a stricter notion of success, e.g., by focusing on *novel* predictions, fewer theories will be admitted into the relevant population. The number of true theories will exceed the number of false theories. The number of false positives will be low, too. In that population, the probability of a randomly selected theory being true will be high. In sum, base-rates are unavailable not because we don't have enough statistics, but because we don't have clear and unambiguous reference classes. And we don't have the latter because our central individuating concepts (theory, success, etc.) are not precise enough.[12]

[12] In connection with the base-rate fallacy, L. J. Cohen (1981) has made the general point that there is no such thing as *the* relevant base-rate.

I want to add one more reason why I think that Howson's reformulation of NMA as a probabilistic argument is *deeply* problematic: it fails to capture the rich structure of theory-change in science. Recall the Pessimistic Induction. Laudan (1984) has invited us to see that if the history of science is the waste-land of aborted 'best theoretical explanations' of the evidence, it might well be that current best explanatory theories might take the route to this waste-land in due course.[13] In response to this argument, realists (cf. Kitcher 1993; Psillos 1999) have argued that theory-change is not as radical and discontinuous as the opponents of scientific realism have suggested. They have aimed to show that there are ways to identify the theoretical constituents of abandoned scientific theories which essentially contributed to their successes, separate them from others that were 'idle'—or as Kitcher has put it, merely 'presuppositional posits'—and demonstrate that those components which made essential contributions to the theory's empirical success were those that were retained in subsequent theories of the same domain. What follows from the relevant realist arguments is this: the fact that our current best theories may well be replaced by others does not, necessarily, undermine scientific realism. All it shows is that a) we cannot get at the truth all at once; and b) our judgements from empirical support to approximate truth should be more refined and cautious in that they should only commit us to the theoretical constituents that do enjoy evidential support and contribute to the empirical successes of the theory. Realists ground their epistemic optimism on the fact that newer theories incorporate many theoretical constituents of their superseded predecessors, especially those constituents that have led to empirical successes. The substantive continuity in theory-change suggests that a rather stable network of theoretical principles and explanatory hypotheses has emerged, which has survived revolutionary changes, and has become part and parcel of our evolving scientific image of the world. I think it is obvious that this rich structure cannot be captured by Howson's reformulations of NMA. In fact, it is not clear at all in what sense we can talk about base-rates of truth and falsity any more. The static picture of some percentages of true and false theories is replaced by a *dynamic* one, according to which theories improve on their predecessors, explain their successes, incorporate their well-supported constituents and lead to a truer description of the deep structure of the world.

These considerations make me very sceptical about the prospects of even starting to formulate the no-miracles argument as a probabilistic argument in the first place. It makes me even more sceptical about the cogency of the Bayesian charge that realists *ignore* base-rate information. But suppose that there are base-rates available. Is it always a bad idea to ignore them?

To address this question, let us go back to the original setting of the base-rate fallacy and take a look at another standard case in which this fallacy is to be committed. This is the Blue Cab/Green Cab case.

[13] It might be ironic that Lewis (2001) argues that the pessimistic induction is fallacious because it commits the base-rate fallacy.

> (*Blue cab/Green cab*)
>
> There is a city in which there are two cab companies, the Green cabs and the Blue cabs. Of the total number of cabs in the city, 85% are green and 15% are blue. There was a late-night hit-and-run car accident and the sole eyewitness said that it was a blue cab involved. The eye-witness is very reliable: in test situations involving blue and green objects at night, he made the correct identifications in 80% of the cases and he was mistaken in 20% of cases. What is the probability that the culprit was a blue cab?

When asked the foregoing question, subjects involved in psychological experiments, tended to trust the eyewitness and said, in an overwhelming percentage, that the probability that the culprit was a *blue* cab was very high. This is supposed to be a standard case of the base-rate fallacy, since, given the base-rates for blue and green cabs, the probability that the culprit was a blue cab is low (.41). It's more likely that the culprit was a green cab, since there are many more of those around.

There are two points that need to be noted. *First*, it is one thing to reason correctly probabilistically (the subjects, obviously, didn't). It is quite another thing to get at the truth. For, it may well be that the eyewitness really *saw* a blue cab and that a blue cab was involved in the accident. Unlikely things do happen, and we should be able to identify them no less than we are able to form a belief about what it is likely to happen and what it is not. What is important here is that the base-rate information might have to be ignored, if what we want to get at is the *truth*. There is not, of course, any definite answer to the question: when are the base-rates to be ignored and when are not? But there is an interesting observation to be made. In the case at hand, there is some crucial information to be taken into account, viz., that the situation is *ambiguous*. After all, it was dark and, in the dark, our observations are not very reliable. Actually, as Birnbaum (1983) has noted, if a witness is aware that there are many more green cabs than blue cabs in the city, he is predisposed to see green cabs in ambiguous situations. This, it should be noted, is a piece of information (or background knowledge) that the subjects of the experiment also have. So, the very fact that, despite the prevailing disposition, the witness is reported to have seen a *blue* cab carries *more* weight than the relevant base-rates. So, there is a sense in which the subjects commit a fallacy (since they are asked to reason probabilistically but fail to take account of the base-rates), but there is another sense in which they reason correctly because the salient features of the *case history* can get them closer to the truth.

Transpose all this to the problem of truth and success. If we take the base-rates into account, we may get at the correct probability of a theory's (chosen at *random*) being approximately true, given that it is successful. And this probability may be quite low, if the base-rate of truth is very low. Suppose we conclude from this that this theory is *not* approximately true (because it is very unlikely that it is). But it may well *be* approximately true. The fact that it appears unlikely to be approximately true is not due to the fact that the theory fails to approximately fit with its domain, but rather due to the fact that the very few approximately true theories are swamped by the very many plainly false, but successful. If the theory *is* approximately true, but—due to the correct probabilistic reasoning—we don't believe so, our beliefs will have been led away from the truth. In fact, we may reason as above. Suppose we grant the prevalence of false theories among the successful ones. Then, one might well be predisposed to say that a theory T is false, given its success. When, then, the eyewitnesses (the scientists, in this case) say that a specific theory T is approximately true (despite that this is unlikely, given the base-rates), they should be trusted—at the expense of the base-rates.

The *second* point can be motivated by a certain modification of the Green cab/Blue cab example. The situation is as above, with the following difference: the subjects are told that 85% of the car accidents are *caused* by blue cabs and 15% by green cabs. In these circumstances, the subjects did use the base-rates in their reasoning concerning the probability that the culprit was a blue cab (see Koehler 1996, p. 10). It is easy to see why they did: they thought that the base-rate information, viz., that blue cabs cause accidents much more often than green cabs, was *causally relevant* to the issue at hand. What needs to be emphasised is that in cases such as these there is an *explanation* as to why the base-rate information is relied upon. It's not just because the subjects want to get the probabilities right. It is also because this causally relevant information has a better chance to lead them to true beliefs.

Transpose this case to the problem of truth and success. Suppose that there is indeed a high base-rate for false theories. This would be relevant information if it were indicative (or explanatory) of success. If falsity did explain success, then, clearly, the small base-rate for truth would undermine belief in a connection between success and approximate truth. But falsity does *not* explain success. What is more, among the false theories some will be successful and some will be unsuccessful. In fact, it is expected that from a population of false theories (shall we say of all possible false theories?), most of them will be *unsuccessful*, while some will be successful. In terms of percentages, it might well be a bit of a fluke that some false theories are successful. The likelihood prob(S/-T) will be low. In fact, it can be so low as to dominate over the high base-rate of false theories. So, suppose that prob(S/-T) =.05, prob(-T) =.9 and prob(S) =.99. Then, prob(-T/S) is .045. A false theory would get no credit at all from success. Conversely, even if the base-rate of truth is low, there is an explanation as to why true theories are successful.[14] This might well be enough to show why, despite the low base-rate, a certain successful theory may well be deemed approximately true. Its posterior probability may be low,

[14] There is a worry here, voiced by Levin (1984), viz., that the truth of the theory does not explain its success. He asks: '[w]hat kind of *mechanism* is truth? How does the truth of a theory bring about, cause or create, its issuance of successful predictions? Here, I think, we are stumped. Truth (...) has nothing to do with it' (1984, p. 126). Musgrave (1999, pp. 68-9) has answered this worry very effectively. What does the explaining is the theory. But, Musgrave adds: 'Semantic *ascent* being what it is, we do not have *rival* explanations here, but rather equivalent formulations of the *same* explanation. "*H* believed that *G* and *G*" is equivalent to "*H* believed *truly* that *G*" (given the theory of truth that Levin and the realists both accept' (1999, p. 69). He then goes on to claim, correctly I think, that the explanation of the success of an action in terms of the *truth* of the agent's relevant beliefs is a mechanical or causal explanation.

but this will be attributed to the rareness of truth and not to any fault of the individual theory.

Here is another reason why it is, at least occasionally, right to ignore the base-rates. To motivate it, consider again the original Green cab/Blue cab case. As above, 85% of the cabs belong to the Green cab company and 15% to the Blue cab one. Imagine that people involved in car accidents are set on taking the cab companies to court. Suppose that on each occasion of the lawsuit, the court takes account of the base-rates and concludes that the cab was green, despite the fact that the eye-witness testified otherwise. Let's say that the court judges that it is always more likely (given the base-rates) that the cab was green (recall that the probability of the cab being blue is .41) and hence it decides to press charges against the Green cab company.[15] If courts acted like that, then the Green company would pay in 100% of such cases, whereas its cabs were responsible for only 59% of such accidents. Fairness and justice seem to give us *some* reason to ignore the base-rates![16]

If we transpose this to the problem of truth and success, the moral should be quite clear. If scientists acted as the imagined judges above, they would be unfair and unjust to their own theories. If, as it happened, the base-rate of false theories were much higher than the base-rate of true ones, they would deem false theories that were true. Conversely, if the base-rate of true theories were much higher than the base-rate of false ones, they would deem true theories that were false.[17]

8. TAKING ACCOUNT OF CASE HISTORIES

If we leave base-rates behind us, what is left? There are always the case histories to look into. Though, as we saw in section 3, it does make sense to raise the grand question 'why is science successful (as an enterprise) as opposed to paradigmatically unsuccessful?', what really matters is the *particular* successes of individual theories, e.g., the discovering of the structure of the DNA molecules, or the explanation of the anomalous perihelion of Mercury. Now, if we think of it, it does *not* matter for the truth of the double helix model that truth is hard to get. The base-rate of truth (or of falsity)—even if we can make sense of it—is outweighed by the case history. We have lots of detail information about the DNA-molecule case to convince us that the double helix model is approximately true, even if, were we to factor in the base-rate of true theories, the probability of this model being approximately true would be very low. We are right in this case to ignore the base-rate, precisely because we know that this model's being approximately true does *not* depend on how many other true or false theories are around.

This last observation seems to me quite critical. The approximate truth of each and every theory will *not* be affected by the number (or the presence) of other

[15] If probability .59 is too low to capture the court's call that the case should be proven 'beyond reasonable doubt', then we can alter the numbers a bit so that the probability that the cab was green is high enough.

[16] A similar point is made by Windschitl and Wells (1996, 41).

[17] The base-rate fallacy has been subjected to very detailed and informative scrutiny by Jonathan Koehler (1996).

theories (even more so if these are independent of the given theory). Approximate truth, after all, is a relation between the theory and its domain (a relation of approximate fit). This relation is independent of what other (true or false) theories are available. In fact, we can see that there is an ambiguity in the probabilistic formulations of NMA. Though I have hinted at this above, it is now time to make it explicit.

There are *two* ways to think of arguments such as (A) and (B). The first is to apply the argument to a *specific* theory T (say, the electron theory, or Newtonian mechanics or the special theory of relativity). Then we ask the question: how likely is this *specific* theory T to be true, given that it has been successful? The second way is to apply the argument to an *arbitrary* theory T. Then we ask the question: how likely is an *arbitrary* (randomly selected) theory T to be true, given that it has been successful? If the issue is posed according to this second way, then it does follow from Bayes's theorem that the probability of a theory's being approximately true will depend on (and vary with) the base-rate of true theories. But if the issue is raised for a specific theory, then base-rates have no bite at all. Even if we had the base-rates, there are good reasons to neglect them—and scientists do neglect them—when the case history offers abundant information about the approximate truth of a given theory.[18]

9. LIKELIHOODISM

We are not done yet. The subjective Bayesian might now come back with a vengeance. He might say: ditch the base-rates, and go for purely subjective estimates of how likely it is that a theory is true. Consider what Howson (2000, p. 58) says: '[F]ar from showing that we can ignore even possibly highly subjective estimates of prior probabilities, the consideration of these quantities is *indispensable* if we are to avoid fallacious reasoning.' So, can we do away with priors altogether? Let us recall the Bayes factor from section 6. As Kevin Korb (forthcoming, p. 4) has argued, this factor reports the 'normative *impact* of the evidence on the posterior probability, rather than the posterior probability itself'. To get the posterior probability, we also need the prior. If the Bayes factor $f = \text{prob}(S/\text{-}T)/\text{prob}(S/T) = 1$, then $\text{prob}(S/\text{-}T) = \text{prob}(S/T)$, that is, the success of a theory makes no difference to its truth or falsity. But, the further from unity f is, the greater is the *impact* of the evidence. If $f = 0$, as we saw in section 6, then $\text{prob}(T/S) = 1$. And if f tends to infinity, then, given that $\text{prob}(T) > 0$, $\text{prob}(T/S)$ tends to 0. Given all this, it seems that we can reformulate Howson's NMA ((B^1) in section 5) in such a way that it avoids base-rates (prior probabilities). The idea is that NMA need not tell us how probable a

[18] I don't want to deny that high probability is sufficient for warranted belief. But is it necessary? I don't think so. One of the prime messages of the statistical relevance model of explanation is that *increase* in probability does count for warranted belief. Now, empirical success does increase the probability of a theory's being approximately true, even with a low base-rate for truth. This can be easily seen by looking again at the example which preceded argument (B^1) in section 5. There, the prior probability prob(T) of T was 1% but the posterior probability prob(T/S) rose to 17%. So, success does make a difference to the probability of theory's being true.

theory is, given the evidence (or its success). Rather, it tells us what the impact of the evidence (or the success) is on the posterior probability of the theory (without assuming that there is need to *specify* this posterior probability, and hence need to rely on a prior probability).

(B^4)

1. *f* is close to zero (i.e., prob(S/-T) is close to zero and prob(S/T) is close to 1).
2. S is the case.
3. Therefore, the impact of S on prob(T/S) is greater than its impact on prob(-T/S).

(B^4) can be supplemented with some specification of prior probabilities and hence it can yield a concrete posterior probability. Thus, it can then become either (B^2) or (B^3) above. But, even as it stands, it is suitable for *modest* Bayesians, who just want to capture the comparative impact of the evidence on competing hypotheses.

But we should also take a look at what has been called 'likelihoodism' (Sober 2002, p. 24). As Sober (2002) understands it, likelihoodism is a modest philosophical view. It does not aim to capture all epistemic concepts. It uses the *likelihood ratio* to capture the strength by which the evidence supports a hypothesis over another, but it does not issue in judgements as to what the probability of a hypothesis in light of the evidence is. In particular, likelihoodism does not require the determination of prior probabilities. So, it does not tell us what to believe or which hypothesis is probably true. Given two hypotheses H_1 and H_2, and evidence *e*, likelihoodism tells us that *e* supports H_1 more than H_2 if $prob(e/H_1)>prob(e/H_2)$. The likelihood ratio $prob(e/H_1)/prob(e/H_2)$ is said to capture the strength of the evidence.

Note that the likelihood ratio $f^*= prob(e/H_1)/prob(e/H_2)$ is he converse of the Bayes factor *f*, as defined above. So likelihoodists can adopt a variant of (B^4):

(B^5)

1. f^* is greater than one (i.e., prob(S/T) is close to 1 and prob(S/-T) is close to zero).
2. S is the case.
3. Therefore, S supports T over -T.

It is not my aim here to defend either (B^4) or (B^5). But it should be stressed that if we have in mind a more modest version of NMA, that is, that success tells more strongly in favour of truth than of falsity, then we can take (B^4) as a version of NMA suitable for modest Bayesians and (B^5) as a version of NMA suitable for non-Bayesians.[19]

[19] For a critique of likelihoodism, see Achinstein (2001, pp. 125-131).

10. CONCLUDING THOUGHTS

The moral of sections 3 and 4 is that there is no reason to think of the Ultimate Argument for realism as a deductive argument, contrary to what Musgrave suggests. The moral of sections 5 to 8 is that we should also resist the temptation to cast the no-miracles argument in a(n) (immodest) subjective Bayesian form. Once we free ourselves from both deductivism and subjective Bayesianism, there is no reason to think that NMA is either deductively or inductively fallacious. Many will remain unpersuaded. Both deductivism and Bayesianism are all-encompassing (shall I say imperialistic?) approaches to reasoning and they have many attractions (and a number of well-known successes). In fact, they share a common central theme: reasoning has a certain formal structure (given by deductive rules and Bayes's theorem—or better, Bayesian conditionalisation). So the substantive assumptions that are employed in reasoning have to do either with the truth of the premises (in deductivism) or with the prior probabilities (in Bayesianism). But perhaps, the simplicity of both schemes of reasoning is their major weakness. Reasoning is much more complex than either of them admits.

So, what sort of argument is the Ultimate Argument for realism? I know of no more informative answer than this: it is an inference to the best explanation (IBE). And what kind of inference is IBE? I know of no more informative answer than this: it is the kind of inference which authorises the acceptance of a hypothesis H as true, on the basis that it is the best explanation of the evidence. The rationale for IBE is that explanatory considerations should inform (perhaps, determine) what is reasonable to believe. I know all this is too crude to count as an explication. Further explication can be given, as I tried to show in my (2002). In any case, even if the Ultimate Argument for realism were to be found wanting as an *explanatory* argument, it would still be the case that the realist explanation of the success of science remains the best. Musgrave's 'The Ultimate Argument for Realism' is to be credited for making a very compelling—perhaps unparalleled—case for this.

ACKNOWLEDGEMENT

Earlier versions of this paper were presented in seminars in the Universities of Stockholm, Oslo, Gothenburg and Rotterdam and in a meeting of the Belgian Society for the Philosophy of Science. I want to thank all participants in these events for very useful comments and criticism, and in particular Janneke van Lith (who acted as a commentator in the Rotterdam meeting), Igor Douven, Michel Ghins, Olav Gjelsvik, Leon Horsten, Uskali Maki, Oystein Linnebo, Nils Roll-Hansen, Dag Westerstahl and Asa Wikforss. Research for this paper was funded by the framework EPEAEK II in the programme Pythagoras II.

REFERENCES

Achinstein, P. (2001) *The Book of Evidence*. New York: Oxford University Press.
Bar-Hillel, M. (1980) 'The Base-Rate Fallacy in Probability Judgements.' *Acta Psychologica* 44: 211-33.

Birnbaum, M. H. (1983) 'Base Rates in Bayesian Inference: Signal Detection Analysis of the Cab Problem.' *American Journal of Psychology* 96: 85-94.
Cohen, L. J. (1981) 'Can Human Irrationality be Experimentally Demonstrated?' *Behavioural and Brain Sciences* 4: 317-31.
Howson, C. (2000) *Hume's Problem*. New York: Oxford University Press.
Koehler, J. J. (1996) 'The Base Rate Fallacy Reconsidered: Descriptive, Normative and Methodological Challenges.' *Behavioural and Brain Sciences* 19: 1-17.
Korb, K. (forthcoming) 'Bayesian Informal Logic and Fallacy.' Melbourne: School of Computer Science and Software Engineering, Monash University, Technical Report 2002/120, 22pp.
Laudan, L. (1984) *Science and Values*. Berkeley: University of California Press.
Levin, M. (1984) 'What Kind of Explanation is Truth?' in Leplin, J. (ed.) *Scientific Realism*. Berkeley and Los Angeles: University of California Press.
Lewis, P. (2001) 'Why the Pessimistic Induction is a Fallacy.' *Synthese* 129: 371-80.
Musgrave, A. (1977) 'Explanation, Description and Scientific Realism.' *Scientia* 112: 99-127.
Musgrave, A. (1980) 'Wittgensteinian Instrumentalism.' *Theoria* 46: 65-105.
Musgrave, A. (1988) 'The Ultimate Argument for Scientific Realism.' in Nola, R. (ed.) *Relativism and Realism in Science*. Dordrecht/Boston: Kluwer.
Musgrave, A. (1989) 'NOA's Ark—Fine for Realism.' *The Philosophical Quarterly* 39: 383-398.
Musgrave, A. (1996) 'Realism, Truth and Objectivity.' in R. S. Cohen *et al.* (eds) *Realism and Anti-Realism in the Philosophy of Science*. Dordrecht: Kluwer.
Musgrave, A. (1999) *Essays on Realism and Rationalism*. Amsterdam: Rodopi.
Musgrave, A. (1999a) 'How to do without Inductive Logic.' *Science & Education* 8: 395-412.
Norton, J. (1994) 'Science and Certainty.' *Synthese* 99: 3-22.
Psillos, S. (1999) *Scientific Realism: How Science Tracks Truth*. London: Routledge.
Psillos, S. (2000) 'The Present State of the Scientific Realism Debate.' *The British Journal for the Philosophy of Science* 51: 705-28—reprinted in Clark, P. & Hawley, K. (eds) (2003) *Philosophy of Science Today*. Oxford: Clarendon Press.
Psillos, S. (2002) 'Simply the Best: a case for Abduction.' in Sadri, F. & Kakas, A. (eds), *Computational Logic: From Logic Programming into the Future*. LNAI 2408, Berlin-Heidelberg: Springer-Verlag.
Psillos, S. (2006) 'Putting a Bridle on Irrationality: an Appraisal of van Fraassen's new epistemology.' in B. Monton (ed.) *The Many Faces of Empiricism*, Oxford: Oxford University Press.
Putnam, H. (1975) *Mathematics, Matter and Method*. Cambridge: Cambridge University Press.
Sober, E. (2002) 'Bayesianism—its Scope and Limits.' *Proceedings of the British Academy* 113: 21-38.
Tversky, A. & Kahneman, D. (1982) 'Judgment under Uncertainty: Heuristics and Biases' in D. Kahneman, P. Slovic & A. Tversky (eds) *Judgment Under Uncertainty: Heuristics and Biases*. Cambridge: Cambridge University Press.
van Fraassen, B. C. (1980) *The Scientific Image*. Oxford: Clarendon Press.
Windschitl, P. D, & Wells, G. L. (1996) 'Base Rates do not Constrain Nonprobability Judgments.' *Behavioural and Brain Sciences* 19: 40-1.

MICHAEL REDHEAD

THE UNSEEN WORLD

Science deals with many things we cannot *directly* observe.[1] By directly I mean with the unaided senses. For example there are the elementary particles such as electrons and quarks which are supposed to provide the microscopic building blocks of matter, but also the mysterious photons and gluons etc. which mediate interactions between the microscopic building blocks. And then of course in molecular biology there are the proteins and genes and so on which explain the processes underlying living organisms. But also there are more abstract entities such as energy and entropy which are not part of our immediate sensory experience, and still more abstract entities, like numbers and mathematical points, not just indeed in physical space, but in still more abstract mathematical spaces, such as Hilbert space in quantum mechanics.

So much of modern science seems concerned with what I will call the Unseen World (using sight as a generic term covering all the senses). Indeed the Unseen World effectively constitutes what we may call the scientific world-view. This was famously illustrated by Eddington with his talk of 'the two tables', the table of everyday experience, firm and solid in front of him, and the scientific table, mostly empty space permeated by the force-fields of elementary particles. Which is the 'real' table? And which is the true story, the scientific story or the everyday story?

In this paper I want to explore the cognitive credentials of the Unseen World from both an historical and a modern perspective. Hume famously warned that 'the ultimate springs and principles are totally shut up from human curiosity and enquiry'. But science seems not to have heeded Hume's warning, and let me begin by reminding you of a famous medieval woodcut, in which a curious person peers beyond the vault of the heavens to learn of the hidden mechanisms and contraptions that lie beyond.

But my first question is: what *can* we directly observe with the unaided senses? Microscopic objects in our immediate vicinity perhaps, such as the table in front of me or the chair next to me. But is it the table we see, or the light reflected off the table, or is it the electrical stimulation in the retina caused by the light, or in the optic nerve, or what is it exactly that we see?

[1] This paper is dedicated to Alan Musgrave, whose robust sense of realism about the unseen world has been a source of inspiration to us all. The paper was originally published in the Discussion Paper Series DP 61/02, of the Centre for Philosophy of Natural and Social Science, LSE.

C. Cheyne & J. Worrall (eds.), Rationality and Reality: Conversations with Alan Musgrave, 157–164.

Naively we can think of a sort of homunculus inside our brains (our conscious selves) reading out and interpreting the input signals, but if our brains (and minds?) are just part of nature, then the whole idea of a homunculus, or the ghost in the machine, as the philosopher Gilbert Ryle called it, seems patently absurd. This is the problem of consciousness, but it is not the problem I am going to consider today, interesting and important thought it is.

Let us start with the assumption that we do, in some sense, SEE tables and chairs in a good light possessing normal eyesight and so on. Even if we don't actually see them, i.e. they are not actually *being* observed, nevertheless they are *observable* in the sense that it is *possible* to see them.

Some philosophers of science, and indeed historically many scientists, have thought that science is concerned with discovering regularities in the behaviour of observable entities. Such people are generally called positivists. Scientific knowledge can be checked out in a positive fashion by direct observation. Labels such as 'positivist', and more particularly its cognate 'empiricist', are used with many shades of meaning in philosophy. I shall use such terms with a broad brush, just to give you the general idea.

At first blush the positivist position sounds attractive. The scientific attitude has progressed by getting rid first of supernatural spirits and gods controlling the world, then of theoretical metaphysical concepts like dormative virtues and other mysterious substantial forms beloved of the Aristotelians, and finally arriving at the culmination of what the nineteenth century philosopher Auguste Compte called positive (i.e. non-speculative) knowledge.

But has science really followed the positivist programme? There are all kinds of difficulties. If we are restricted to direct observation then what is the point of scientific instruments like telescopes and microscopes? Surely these are supposed to enable us to see things that we can't directly observe?

There is a significant difference here between the telescope and the microscope. The optical telescope enables us to see things that we could see directly if we were differently located, i.e. moved closer to the distant tower or close up to the moons of Jupiter or whatever. But for the microscope it is not a matter of relocating ourselves. For the virus or the cell to become directly visible to us we would have to change our normal sensory apparatus or adopt the perspective of the Incredible Shrinking Man. So to count the virus or the cell as observable needs rather more science fiction than the case of the telescope.

Historically the first practical versions of the telescope and the compound microscope were employed by Galileo at the beginning of the seventeenth century. The telescope revealed all sorts of oddities in the heavens, from mountains on the moon to the satellites of Jupiter, announced by Galileo in his famous book *The Starry Messenger* (1610). What was the reaction of Galileo's Jesuit opponents? Some refused even to look through the telescope, averring that if God had intended us to inspect the heavens so closely he would have equipped us with telescopic eyes! Others claimed Galileo's observations were artefacts of the instrument.

With the microscope, amazing detail was exposed; for example, the famous drawing dated 1625 of a bee, made by Francesco Stelluti looking through an early microscope. But sometimes people saw what they wanted or expected to see.

Preformationists, like Nicholaas Hartsoeker, in embryology at the end of the seventeenth century claimed to see the homunculus sitting perfectly preformed in the head of the spermatozoon!

What we see is largely determined by the overall theoretical background of our thinking. The slogan here is the theory-ladenness of observation.

We have already had occasion to question whether the table or chair is *directly* observable. Is not observation always a case of probing or interacting with the physical world, and don't we always observe things by the *effects* they produce ultimately in our conscious minds? We often talk loosely of observing fundamental particle reactions, for example, with a bubble chamber or suchlike, but it's only when we *look* at the photographic plate recording the tracks that the observation is translated into positive knowledge *for us*.

From this perspective electrons, quarks, genes and viruses are after all observable. So do they really belong to the Unseen World, and on *that* account should they be eschewed by the scientist? This debate was carried on particularly vigorously at the end of the nineteenth century in respect of the reality of atoms. For Mach, Ostwald and others, the atoms of the physicist and the chemist were just fictional entities introduced as speculative mechanisms for explaining empirical regularities about chemical combination or the properties of gases. They were not to be thought of as 'real' in any robust philosophical sense.

To the modern scientist it is usually assumed that these debates have long been settled in favour of a realist conception of so-called theoretical entities rather than their positivist dismissal. But again things are less simple than they seem.

If we look at the history of science we can see it as a series of U-turns about the explanatory theoretical structures that lie behind or beneath the world of macroscopic experience. Entities like phlogiston or the luminiferous aether or caloric have simply disappeared from the scientific vocabulary and the nature of atoms and molecules is quite different from the modern perspective of quantum mechanics than from the billiard ball conception of the nineteenth century. This leads to the famous pessimistic induction. If we have been so often wrong in the past, is it not pure hubris to believe that our present scientific theories won't look equally ridiculous a hundred years from now?

To defuse the pessimistic induction philosophers have tried to read the history of science in a more continuous and progressive fashion. It has been argued by John Worrall (1989), for example, that although the ontology of physical theory changes abruptly, nevertheless there may be what might be called structural continuity in the sense that in many cases the mathematical equations survive. Only the *interpretation* of the quantities entering into the equations changes. There are two versions of this structuralist philosophy. In an extreme, even bizarre, ontological version, it is only structure which really exists. Everything else is just imaginative fiction. In a more prosaic epistemic version, structure is all that we can claim reliably to *know*. We don't deny that atoms or quarks exist, just that we never know what their true natures are, only the *mathematical* description of how they are constructed, related to one another, behave in various experimental contexts and so on. The basic argument here is that the continuity of mathematical structure defeats the argument of the pessimistic induction. There are various comments I would like to make. Does it make sense to talk about things we can *never* come to know? This line of thought would drive us towards ontological structuralism. This of course is linked to the verificationist theory of meaning espoused by the old logical positivists. Statements that cannot be verified are simply meaningless. Of course any strict interpretation of such a principle would arguably render *every* statement in science, just as much as, for example, in theology, meaningless. We never know anything for certain except *perhaps* in logic or mathematics. I say 'perhaps' because even these claims are not entirely clear but that is another story. So, if there are so many things I am not certain about, by the same token I personally am quite happy to accept that there are things I am ineluctably ignorant about.

But is it true that mathematical structure really survives in tact? In the most revolutionary episodes in modern physics, relativity theory and quantum mechanics, that is just not right. The new mathematics involves parameters like the velocity of light c in the case of relativity, or Planck's constant h in the case of quantum mechanics. It is only by letting c tend to infinity or h to zero that we recover something like the old mathematics of classical physics. But these limits are in general highly singular. A world in which h is actually zero is qualitatively quite different from a world in which h is different from zero, however small in magnitude it might be. To illustrate what I have in mind consider squeezing a circle so as to try and turn it into a line. But a line just is *not* a very elongated circle—it has no inside and whether a curve is open or closed is an all or nothing matter. This is what mathematicians mean when they talk about singularities.

As another example, which is relevant to quantum mechanics, let us consider the limit of the classical wave equation of an elastic string for example, as the velocity of the waves tends to infinity. The character of the equation changes dramatically from what mathematicians call a hyperbolic equation to what they call a parabolic equation.

Suppose the two ends of the string, of length L, are fixed, then the solution for the displacement y of the 'limit equation' is just $y = 0$. But for any finite velocity c, the solution of the original wave equation at an antinodal point is $y = \sin 2\pi\nu t$, where $\nu = c/2L$ for the fundamental mode of the string. Consider the time average:

$$\ddot{y} = 1/T \int_0^T \sin 2\pi\nu t.dt \qquad \text{over a time interval T.}$$

Then:

$$\ddot{y} = 1/2\pi\nu T\ (1\text{-}\cos 2\pi\nu T)$$

$$= L/\pi cT\ (1\text{-}\cos \pi cT/L)$$

For fixed T, however small, $\ddot{y} \rightarrow 0$ as $t \rightarrow \infty$. But for fixed c, however large, we can always chose a T small enough to keep $\ddot{y}$ *unequal* to zero. So the oscillatory behaviour of the string can always be revealed by averaging the motion over sufficiently short resolution times.

So in structural terms, relativity and quantum mechanics genuinely involve new structure, not just the preservation of old structure. So is this not another example of a U-turn, like the abandonment of caloric or phlogiston? The best I can do here is to say that the way mathematical structures 'develop' in physical theory has a certain natural, although not of course inevitable, aspect to it—natural, that is to say, to a mathematician.

There is of course a long tradition in natural philosophy that the physical world is constructed according to mathematical principles. This has a certain mystical appeal about it. For Plato, in the *Timaeus*, everything is constructed out of two sorts of triangle, a kind of mathematical atomism, and Galileo famously remarked that 'the book of nature is written in the language of mathematics'. For the cosmologist James Jeans, God was a mathematician. So in this vein, in discovering the new mathematical structures are we learning to read the mind of God, as Stephen Hawking claimed in his famous best-seller *A Brief History of Time*?

Let us pursue this question of the role of mathematics in physics for a moment. There are two quite distinct cases to consider. In the first case mathematics provides a language to *represent* physical reality or at any rate some emasculated, idealised version of physical reality. We translate a physical problem into a mathematical problem and then, when we get the mathematical answer, just translate back into physics again. But in other cases we embed the physics in a wider mathematical framework, involving what I call surplus structure, which *controls* the bit of mathematics actually used to represent the physical world itself. What do we mean by one bit of mathematics controlling another bit? In pure mathematics this is a familiar idea. Let me give two simple examples.

To prove Desargue's theorem in plane projective geometry, the usual method is to introduce a point which does not lie on the plane, i.e. move to a three-dimensional geometry. In this setting we need only to assume the axioms of incidence to prove the theorem in the plane. If we restrict ourselves entirely to the plane we have to invoke a more powerful principle such as Pappus's theorem concerning properties of hexagons in the plane to get the proof. In a sense the third dimension is controlling, i.e. explaining, what is going on in the plane.

Or again consider the binomial expansion of the function $1/1\text{-}x^2$:

$$1 + x^2 + x^4 + \ldots$$

This only converges for $|x|<1$, and the reason is clearly related to the singular behaviour of the function at $x = \pm 1$. But what about the binomal expansion of $1/1+x^2$:

$$1 - x^2 + x^4 - \ldots ?$$

This function is perfectly well behaved for $x = \pm 1$, but the convergence properties of the series are now controlled (explained) by the singularity at $x = \pm \sqrt{-1}$, i.e. by the extension of the real line to the complex plane.

All this is familiar in pure mathematics. The surprising thing is that this sort of thing is also going on in modern theoretical physics. In particular in modern gauge theories of elementary particle interactions, the explanatory principles all operate in the realm of surplus structure! Let me quote from a well-known monograph by Henneaux and Teitelboim (1992, p. xxiii):

> Physical theories of fundamental significance tend to be gauge theories. These are theories in which the physical system being dealt with is described by more variables than there are physically independent degrees of freedom. The physically significant degrees of freedom then re-emerge as being those invariant under a transformation connecting the variables (gauge transformation). Thus one introduces extra variables to make the description more transparent, and brings in at the same time a gauge symmetry to extract the physically relevant content.
>
> It is a remarkable occurrence that the road to progress has invariably been toward enlarging the number of variables and introducing a more powerful symmetry rather than conversely aiming at reducing the number of variables and eliminating the symmetry.

Gauge theories are complicated by so-called ghost particles associated with these unphysical degrees of freedom. This is how the famous physicist Steven Weinberg (1996, p. 27) explains the role of ghost particles:

> [E]ach ghost field…represents something like a negative degree of freedom. These negative degrees of freedom are necessary because…we are really over-counting; the physical degrees of freedom are the components of [the gauge field] *less* the parameters needed to describe a gauge transformation.

So ghosts (and indeed antighosts!) play a vital role in modern non-Abelian gauge theories. But these ghosts are not intended to have a real physical existence. They belong to the Unseen World in a more extreme sense than electrons or photons. One cannot but be reminded here of the famous Tibetan ghost traps that were supposed to ensnare the, to us non-existent, ghosts!

But what sort of world is the Unseen World? There is an ongoing theme in writing about science that behind and beyond the complex, variegated, diverse world of sensory experience there lies a simple, unified, integrated world that science is gradually revealing, that the Unseen World knits together the patchwork structure of the world of appearances, and provides the true account of the reality referred to in Plato's famous simile of the cave. As T. H. Huxley put it: 'The aim of science is to reduce the fundamental incomprehensibilities to the smallest possible number.' This theme of unification has generally been expressed by a scheme of reduction in which the sciences are arranged in a hierarchy, with sociology and psychology somewhere at the top, below that biology and then chemistry, the whole tower resting on the bedrock of physics. And physics itself is reduced to a unitary theory of everything, a TOE.

Such is the rhetoric particularly espoused by Nobel prize winners in physics applying for huge government grants to work on problems in fundamental physics. You might be forgiven for believing that the ultimate aim of science is to achieve a sort of one-off Humperdinck's Law from which everything else would be accounted for and explained.

But a strong reaction against this sort of wild talk has set in recently in philosophy of science. The pendulum has swung strongly in the opposite direction, promoting the disunity of science and the virtues of the Dappled World, the title of Nancy Cartwright's recent book. The arguments here look at detailed case studies of what science is *really* like, and not just, in moments of wishful thinking, how we would *like* it to be. The description of real science provided by this work is much closer to the experience of the research worker at the cutting edge of the sciences than the sanitised account given in much of the popular science literature.

To be sure, warnings about the tendency of human beings to jump to conclusions about unification go back at least to the seventeenth century when Francis Bacon (1620/1960, p. 51) wrote:

> The human understanding is of its own nature prone to suppose the existence of more order and regularity in the world than it finds. And though there may be many things in nature which are singular and unmatched, yet devises for them parallels and conjugates and relations which do not exist.

But has the pendulum swung too far? I would like to explain my own point of view on this question. The idea of unification is essentially a *regulative* ideal. We may even want to *define* a concept of *scientific* rationality as one which invokes the simplest, most unified theory, to explain empirical phenomena. On this account creationism, for example, is to be rejected, not because science shows it to be false, but 'because its acceptance would violate the canons of scientific rationality'. This argument in defence of the scientific account is by itself clearly viciously circular. Its justification can, however, be provided in terms of the past record of scientific theories based on the pragmatic explanatory virtues of simplicity and unification, in producing successful novel predictions, the usual gold standard of scientific progress. So is it not rational to expect the same criteria to produce more successful science in the future? But such meta-inductions are always liable to fallibility. Perhaps at some deep level of explanation physics will get more complicated rather than increasingly simple. But that is why I talk of a regulative ideal. It does not have to be indefinitely achievable, but its past successes provide justification for pursuing the ideal as a leading principle of scientific investigation.

The difference between myself and Cartwright is essentially that she *likes* the Dappled World *à la* Gerard Manley Hopkins, whereas I want to get out my needle and thread and try to stitch the whole thing together.

So, let me try to summarise the status of the Unseen World. In philosophy there have always been two attitudes to the senses. The first is that the senses are linked not to reality, but to mere appearances. In the words of Parmenides' poem they access the Way of Seeming, not the Way of Truth. The senses are in effect a *barrier* interposed between us and reality. Reality can only be known, if at all, by reason or rational insight. The other view, a liberal and relaxed form of empiricism, is that the senses *link* us in an admittedly tenuous and fallible way with reality, and that science, in pursuing that link has at any rate in part revealed to us the Unseen World that lies behind and beyond the world of everyday experience.

REFERENCES

Bacon, F. (1620/1960) 'Novum Organum', in *The New Organum and Related Writing*, edited by F.H. Anderson. New York: Liberal Arts Press.

Cartwright, N. (1999) *The Dappled World: A Study of the Boundaries of Science*. Cambridge: Cambridge University Press.

Henneaux, M. & Teitelboim, C. (1992) *Quantization of Gauge Systems*. Princeton: Princeton University Press.

Weinberg, S. (1996) *The Quantum Theory of Fields: Vol II Modern Applications*. Cambridge: Cambridge University Press.

Worrall, J. (1989) 'Structural Realism: The Best of Both Worlds?' *Dialectica* 43: 99-124.

ALAN CHALMERS

WHY ALAN MUSGRAVE SHOULD BECOME AN ESSENTIALIST

1. INTRODUCTION.

It was at a conference at La Trobe University in Melbourne that I heard John Fox characterise Alan Musgrave as a lap-dog realist. This was in response to a paper by Musgrave in which he insisted that the realism he defended was of a quite modest kind, and not warranting the label of 'mad dog realist' which was pinned on him at Virginia State Polytechnic in 1986 and which has since stuck. There are unsolved problems in Musgrave's version of realism which a realist should be able to solve and which I believe can be solved by adding to it an element of essentialism. A version of the latter position has been clearly articulated and defended recently by Brian Ellis (2001). However, I believe Ellis's position is too strong and can be weakened in various ways without destroying its main thrust. I aim to persuade Alan Musgrave to become a lap dog, rather than mad dog, essentialist and thereby become twice as modest as he already is.

There are two problems for which Musgrave has no adequate solution, as he himself acknowledges. One concerns an adequate grounding for the intuitive distinction between scientific deductions that are explanatory and those that are not. (The deduction of the range of a projectile from Galileo's laws of motion plus initial conditions explains why the projectile has that range, but the deduction of the height of a cliff from those laws plus the time of fall of a stone from top to bottom does not explain why the cliff has that height.) The other is a need for an account of physical necessity that will serve to ground the distinction between scientific laws and true, but only accidentally true, generalisations. Essentialism, the idea that things necessarily behave in the ways that they do on account of the kinds of thing that they are, can solve both problems straightforwardly. Musgrave raises the possibility of an essentialist solution to problems with realism only to dismiss it. However, the account of essentialism that he dismisses is unduly unsympathetic. There is a version which is immune to his criticism and which, I argue, is just what he needs to strengthen the weak points in his realism.

C. Cheyne & J. Worrall (eds.), Rationality and Reality: Conversations with Alan Musgrave, 165–181.

2. PHYSICAL NECESSITY

A widely held intuition has it that laws of nature are necessarily true, as opposed to accidentally true, universal generalisations. 'All the coins in my pocket are silver' may be true, but if it is, it is so only accidentally, and would cease to be true if I were to add a copper one. By contrast, 'all metals expand when heated' is a genuine law and is true, according to a common intuition at least, because there is something about the nature of metals that makes it physically necessary that they expand when heated. Appeal to the law governing the expansion of metals can help explain why the bottle top is loosened when held under the hot water tap, whereas 'all the coins in my pocket are silver' cannot help to explain why any one of them is silver.

Some accidental regularities can be ruled out as genuine laws by the demand that laws must involve strict universals whose range is not restricted to limited temporal or spatial regions and which may be replaced by a conjunction of singular statements. This is sufficient to rule out the example referring to coins in my pocket, where the 'law' can be replaced by a finite conjunction of statements of the form 'coin c_n is silver', where the n's range over the n coins in my pocket. However, there are true accidental generalisations that are strictly universal, such as in the following example due to Popper (1972, p. 427). 'All moas die before they are 50' is a strict universal statement if we understand a moa as a particular biological structure exemplified by the large birds that once inhabited New Zealand but which are now extinct. Popper invites us to make the plausible assumptions that all moas that have existed and ever will exist are those that for a period inhabited New Zealand, that none of them lived beyond fifty years and that, had the environmental conditions been more favourable, it is possible that some moas would have lived beyond fifty. Given all this, 'all moas die before they are 50' is a strict universal which is true, but only accidentally so since it might well have been falsified had the environmental conditions been different. It does not qualify as a genuine law.

The foregoing example suggests how the distinction between lawlike and accidental regularities can be explored by appeal to counterfactuals. Genuine laws, it is said, support counterfactuals in a way that regularities that are only accidentally true do not. 'If this coin were made of metal it would expand when heated' is supported by 'metals expand when heated' in a way that 'this copper coin would be silver were it to be placed in my pocket' is not supported by 'all the coins in my pocket are silver'. Genuine laws rule out certain happenings, such as the contraction of metals on heating, as physically impossible in a way that accidentally true generalisations do not.

Possible worlds are frequently invoked to illustrate physical necessity. Genuine laws are said to differ from accidental regularities insofar as the former, but not the latter, are true in all possible worlds. It is assumed that the 'possible' here relates to some kind of physical, as opposed to logical, possibility. A world in which metals contract when heated is logically possible, it is typically claimed, but is not physically possible. A genuine law is true in all *physically* possible worlds. In what way is a physically possible world to be characterised? A natural answer is that a world is physically possible if it does not violate the laws of nature. But there is clearly a circularity here that diminishes the informativeness of spelling out the meaning of physical necessity by appeal to possible worlds. Popper (1972, pp. 433-37) used the distinction between laws and initial conditions in his attempt to utilise possible worlds to explicate physical necessity. A possible world is one that differs from ours, if at all, only in the initial conditions. A world like ours but with only three planets in the solar system is a possible world in this sense. But the planets in this possible world will move in orbits that are close to ellipses just as in our world, for the laws of inertia and gravity must apply for the world to be a physically possible one. Talk of initial conditions does help to clarify what is implied by physical necessity. But, as Popper for one acknowledged, it does not escape circularity, because the separation of those variations of initial conditions that are allowed from those that are not requires appeal to the laws of nature. We can postulate a change in the number of planets without violating the laws of nature but we cannot vary the shape of a planetary orbit whilst keeping its mass and speed unchanged without violating those laws.

Physical necessity is to be identified as conformity with the laws of nature. But, certainly since Hume, it has been assumed that there is some strong sense in which the laws of nature are contingent rather than necessary, where contingency is contrasted with logical necessity. There is no logical contradiction in the assertion that a planet orbit the sun in an elliptical orbit with an unchanging speed, and so the law of conservation of angular momentum is not logically necessary. So what kind of necessity is physical necessity? What exactly is involved in the claim that laws of nature are necessary in a way that accidental regularities are not?

The status of scientific laws has a bearing on how explanation in science is to be understood. Occurrences can be explained by showing them to be consequences of laws of nature as they apply in the circumstances giving rise to the occurrence. We can explain why a stone dropped from the top of a tower has a certain velocity when it reaches the foot of the tower by showing that that velocity follows from the law of fall given the height of the tower. But there are counter examples that strongly suggest that this account of explanation is insufficient as it stands. 'This coin is silver' follows from 'this coin is in my pocket' and 'all the coins in my pocket are silver', but the derivation does not explain why the coin is silver. The height of the tower in our earlier example can be derived from the law of fall and the time of decent but that does not explain why the tower has the height that it does.

While the first of the above examples seems connected with the distinction between law-like and accidental generalisations, the second seems connected with the notion of causality. Our intuitions tell us that gravity causes the falling stone to accelerate but does not cause the tower to have a particular height. What is the relation between laws and causality? Bertrand Russell (1912-13) argued that causality is a vague and confused notion that should be eliminated from philosophy of science and all the explanatory work done by laws. Nancy Cartwright in *How the Laws of Physics Lie* (1983) took the opposite view, giving priority to causes at the expense of laws. We need the notion of cause, she argued, to understand the effectiveness of our interventions in nature. There may well be a lawlike connection between smoking and lung cancer, but to understand why we can hope to decrease incidences of lung cancer by eliminating smoking, but cannot expect a cure for lung cancer to reduce the prevalence of smoking, we need to appreciate that smoking causes lung cancer and not the reverse.

3. MUSGRAVE'S RESPONSE

Musgrave's writings indicate that he is well aware of the problems I have sketched in the previous paragraph. My sketch relied very much on intuitions about the distinction between lawlike and accidental regularities and about what causes and explains what. But Musgrave (1999, p. 6) rightly warns us that in such matters 'intuition unsupported by argument is never very satisfactory'. However, he is unable to go much beyond intuition himself to argue for a decisive position on physical necessity, explanation and causation and candidly acknowledges this.

In a context where he is investigating the status of ideal laws in science, Musgrave offers an interesting discussion of the relationship between laws and counterfactuals. An implication is that the mere ability to support counterfactuals is not sufficient to distinguish laws from accidental regularities. If it is true that no film stars live in New York, then 'If Lindsey were a film star then Lindsey would not live in New York' is a true counterfactual supported by 'No film stars live in New York'. A counterfactual 'If it were the case that A then it would be the case that B' is supported by the generalisation L just in case A and L implies B. This example about film stars shows that accidental generalisations can support counterfactuals in this sense. The example employs a generalisation that refers to specific locations and so can be dismissed as a definite law on that ground. But Musgrave's point can be sustained by utilising Popper's example referring to moas. Confronted by the skeleton of a bird in a museum I can affirm the true counterfactual 'If this is a moa then it died before it was 50' where the generalisation supporting the counterfactual is the true, but not law-like, generalisation 'All moas die before they are fifty'.

What seems to be called for is a distinction between counterfactuals that are supported by genuine laws and those supported by generalisations that or only accidentally true. As Musgrave (1999, p. 152) observes 'one could hardly say that for a generalisation to support a counterfactual it must express a genuine law, and then go on to characterise laws by saying that they alone support counterfactuals.' Then comes the frank admittance 'I have no independent account of law-hood to offer here.'

The next two paragraphs involve a digression. As I mentioned above, Musgrave's discussion of counterfactuals occurs in a context where he is analysing the status of ideal laws in science. He proposes that they are counterfactuals that are supported by underlying theories. The ideal gas laws, for instance, are counterfactuals that are supported by deriving them from the kinetic theory of gases together with assumptions such as perfectly elastic collisions and the absence of interaction between molecules except during collision. My quibble is that this is an unsatisfactory analysis of the ideal gas laws and related idealisations as they function in physics. I do not deny that the ideal gas laws can be explained by the kinetic theory in some strong sense. What I do deny is that the status of the ideal gas laws as laws depends on that explanation. One set of reasons for my position is historical. The gas laws were formulated, experimentally supported and proved their worth in thermodynamics prior to, and independently of, their explanation by the kinetic theory. From Carnot onwards, ideal gases figured in the derivation of inequalities associated with the second law of thermodynamics. The point is not merely historical. It is regarded as important in contemporary physics to formulate phenomenological thermodynamics in a way that is independent of underlying matter theory. When ideal gases are introduced in such a context they are regarded as the limit that real gases tend to as the pressure is reduced.

The key problem that Musgrave addresses with respect to ideal gases laws concerns their epistemological status. How can claims about ideal states of affairs be tested in the non-ideal setting of our experiments? Musgrave's answer is that the kinetic theory is testable, and so support for it is transmitted to the ideal gas laws that follow from it given appropriate idealising assumptions. But this is not the only way in which the ideal gas laws can, and did, gain empirical support, and Musgrave's own remarks in the final paragraph of the relevant chapter (1999, p. 153) shows how this can be so. If ideal gases are understood to be the limit that real gases tend to as the pressure decreases, then empirical support for the ideal laws is obtained by investigating experimentally the limit towards which real gases tend as their pressure is reduced. In fact the deviation of the behaviour of real gases from that of ideal ones is below the limits of experimental error for pressures very readily produced in the laboratory. How else can we account for the fact that the ideal gas laws were *experimental discoveries*?

Let us return to our main theme. So far we have observed that Musgrave has pointed to the limitations of attempts to single out laws through their ability to support counterfactuals, and admits to having no other account of law-hood to offer.

Equally indecisive is his discussion of the distinction between derivations that are explanatory and those that are not. 'Realists need some account of when a deduction is explanatory. And I have to confess I have no such account' (1999, p. 6). He shares the intuition that the issue involves causality (gravity causes the stone to accelerate but does not cause the tower to have the height that it has) but declares himself to be reluctant to add to the ink that has been spilled attempting to address this issue by fashioning a distinction between causal laws and accidental generalisations. He conjectures that causal explanations might be marked by specific features of the initial conditions rather than by a special kind of law. We shall have occasion to pursue the matter of initial conditions below.

Soon I will be ready to investigate essentialism as a metaphysical position capable of responding to our range of problems associated with physical necessity. Musgrave himself raises this prospect only to dismiss it. He offers two main reasons for doing so. One stems from the fact that, like Popper, he identifies essential explanations with ultimate ones. Once the identification is dropped, objections to ultimate explanations do not necessarily constitute arguments against essentialism. Musgrave's second objection is based on the idea that knowledge of essences must be a priori knowledge. This objection has more purchase on those positions more generally referred to as essentialist and needs to be explored.

Essentialists hold that items in the world possess properties that they must *essentially* possess if they are to be the kinds of thing that they are. A silver coin must possess various properties by virtue of being silver. Expanding when heated and combining with hydrochloric acid to yield silver chloride and hydrogen are two of them. By contrast, the physical location of a silver coin will be an accidental rather than an essential property, so a silver coin in my pocket is no more nor less a silver coin than one in the bank. Genuine laws describe how things necessarily behave because of the properties they essentially have. So 'silver expands when heated' is a law whereas 'all the coins in my pocket are silver', even if true, is not. I will be exploring the extent to which this can be articulated into a defendable position in later sections. Here I wish only to say enough to make sense of Musgrave's objection to it. His objection is that essentialism renders scientific knowledge a priori because it involves a fundamental reliance on definitions. If the definition of the essential properties of silver contains or implies the fact that it will expand when heated then the law of expansion is true by definition. If anything lacks that property we know in advance that it is not silver. So, if 'statements of essences are definitions then they cannot do the job expected of them' (1999, p. 13). That is, essentialism is incompatible with the empirical character of science. In the next section I outline a recently articulated version of essentialism that is not open to Musgrave's objections, although I will later argue that it has other undesirable features.

4. MAD DOG ESSENTIALISM

In his articulation of the mechanical philosophy Robert Boyle found laws of nature to be problematic and felt it necessary to invoke God's powers to sustain them.

> And it is intelligible to me that God should at the beginning impress determinate motions upon the parts of matter, and guide them as he thought requisite for the primordial constitution of things, and that, ever since, he should by his ordinary concourse maintain those powers which he gave the parts of matter to transmit their motion thus and thus to one another. But I cannot conceive how a body devoid of sense, truly so called, can moderate and determine its own motion, especially so as to make them conformable to laws that is has no knowledge or apprehension of. (Stewart,1979, pp. 181-2)

Assumptions implicit in this passage have become orthodox since that time, and are well established in Humean and neo-Humean philosophy of science. It is assumed that laws are contingent, imposed on an otherwise passive world. The basic ontology of the world at any one instant is the spatial distribution of objects, occurrences or events that comprise it. The relationship between that distribution at one time and at another is a separate matter. All kinds of relationships are logically possible, and the regular relationships that happen to obtain from amongst the logical possibilities are the laws of nature.

Brian Ellis, in his recent book *Scientific Essentialism,* rejects such a view as fundamentally mistaken. He sees the world as inherently active. Objects in the world are what they are by virtue of how they are disposed to act and interact with other objects, so this aspect of their being is ontologically fundamental. A charged body will attract or repel other charged bodies, give rise to a magnetic field when moving and radiate when accelerating because it is in the nature of charged bodies to do such things. Precise statements of these modes of acting, such as Coulomb's law or the Lorentz force law, describe the laws of nature. They are not something imposed on charged bodies because they are already implied in what it is to be a charged body. So charged bodies necessarily obey the laws that they do.

From this point if view, the basic ontology of the world includes such things as dispositional properties, powers and capacities, and it is such entities that play the role of truth-makers for the laws of nature. This contrasts with the more conventional view, common since Hume, according to which it is necessary to reduce dispositional properties and the like to non-dispositional (categorical) properties plus contingent laws of nature imposed on systems described in such terms. For Ellis, laws are immanent in an inherently active world.

A second key element of Ellis's essentialism is its use of natural kinds. The membership of natural kinds must be unambiguously established by nature, rather than by any human classificatory scheme. Any scepticism about the existence of natural kinds is best countered, as Ellis does, by invoking fundamental particles such as electrons and protons as constituting such kinds. Electrons are identical in the strong sense that they obey Fermi-Dirac rather than Boltzmann statistics (there are only three ways of distributing two electrons over two containers, not four) and the Pauli exclusion principle. Their identity is involved in the explanation of why metals conduct electricity in the way that they do and why atoms have the spectra and chemical properties that they do. Anyone that accepts these explanations and agrees that metals and chemical elements have their properties independent of human knowledge of them must accept that electrons are identical and form a natural kind in the strong sense characterised by Ellis.

With electrons, protons and neutrons recognised as natural kinds it would seem to be a straightforward step to specify atoms in terms of their atomic structure and so count the atoms of an isotope of an element as a natural kind, and this leads to recognising the elements themselves and their compounds as natural kinds. How much further we can go is a matter of debate. Ellis himself doubts that animal species, the traditional exemplars of natural kinds since Aristotle, are such in the light of Darwinian evolution. Whatever the facts of such matters are, I, for one, am prepared to concede to Ellis that there are some natural kinds.

Ellis exploits the notion of a natural kind to construct an account of physical necessity. The essential properties associated with a natural kind will be the properties members of that kind must have in order to qualify as a member of that kind. A specified mass, charge and spin are essential properties of an electron, and anything that lacks any one of them is not an electron. Fundamental particles such as electrons do not have intrinsic properties other than their essential ones. Members of more complicated natural kinds have accidental as well as essential properties. A sample of sodium chloride can be crystalline or powdered, hot or cold and remain sodium chloride, but it must essentially be comprised of sodium chloride molecules with their characteristic electron structure if it is to be admitted into the natural kind.

Essential properties give Ellis the basis for his account of natural necessity and his attempt to construe it as a species of logical necessity. Electrons must necessarily radiate when they accelerate because, firstly, they must be charged otherwise they would not be electrons, and second, because part of what it is to be charged is to have the capacity to radiate when accelerated.

This brief sketch of some key features of Ellis's essentialism is sufficient for us to see why Musgrave's two main objections to essentialism do not apply to it. There is no reason for Ellis to identify particular essentialist explanations with ultimate ones, or specific essential properties with ontologically basic ones. Explanations can appeal to the essential properties of the chemical elements irrespective of the fact that those properties can be explained at a deeper level by appealing to electron structure, and there is nothing in Ellis's position that rules out the possibility of some future physics explaining how electrons come to have the properties that they do. Musgrave's other charge is that, if essentialist definitions are necessary then they are true by definition, which suggests that knowledge of them can be known a priori. This, of course, conflicts with the idea that science is empirical. Electrons and their properties needed to be discovered not merely defined. 'If anything is true by definition, then a definition is' writes Musgrave (in this volume, p. 330). If definitions are understood as the stipulation of linguistic conventions, as it is often appropriate to do, then Musgrave's observations are to the point. But definitions of essences are not appropriately interpreted in that way. For Ellis, as, arguably, for Aristotle, essentialist definitions are intended to characterise the nature of natural kinds that exist independently of us and our descriptions, and so can be true or false. Bachelors do not constitute a natural kind. We create that kind by defining them. As a consequence, 'bachelors are unmarried' is an analytic truth known a priori. By contrast, electrons form a natural kind whether we know it or like it or not. Our 'definition' of them, that is, our characterisation of their essential properties, may or may not correspond to what they actually are. The adequacy of our essentialist definitions needs to be established empirically. Musgrave may object to calling characterisations of essential properties 'definitions'. However, whatever we call them, the important point remains that they help to sustain Ellis's account of physical necessity.

As well as accounting for physical necessity, Ellis's position can be readily adapted to resolve the related problems that I have raised above. The scientific essentialist will explain outcomes or processes by showing how they are consequences of the essential properties of whatever it is that gives rise to them. In the circumstance in which a stone falls from a cliff, we can explain why it is that it has the velocity that it has when it reaches the ground by appealing to the law of fall, which is built into the characterisation of the stone as a massive body. In a similar way, we can calculate the distance it must have fallen given the time of fall. In the given circumstances, the distance fallen happens to be the height of the cliff. But that is accidental. One could ask how the cliff came to have the height that it has, and our best attempt at this would involve an understanding of the essence of rocks and water waves as mechanical systems. The sense in which explanations are relative to what it is that an explanation is sought for need not embarrass the essentialist nor anyone else. There would be no explanations in a world without humans to seek or construct them. What is crucial, for an essentialist, is that the world is inherently active and that things happen by virtue of this. So when we do seek to explain a situation in science we should seek to identify what it was in the world that brought it about. The logical necessities involved in our explanations must reflect physical necessities present in the world.

5. PROBLEMS WITH MAD DOG ESSENTIALISM

I do not recommend that Alan Musgrave adopt scientific essentialism in the form that Ellis has formulated it. It has problematic features and, in any case, does not give a satisfactory solution to the problems associated with physical necessity that I have identified in this article.

The solution that scientific essentialism offers for our problems associated with natural necessity depends crucially on the existence of natural kinds. Members of natural kinds necessarily behave in the ways that they do by virtue of the properties they necessarily must have to qualify as members of that kind. I have admitted that there are natural kinds that conform to the strict demands that Ellis makes of them. Electrons constitute such a kind. However, such natural kinds constitute too narrow a category to serve as a basis for an adequate account of physical necessity.

According to Ellis the boundaries of natural kinds must be objective, independent of our interests, psychology, perceptual apparatus, languages, practices or choices (2001, p. 19). But Ellis's position is much stronger than that. He insists that the boundaries between natural kinds be sharp, so that there be no borderline cases. 'For natural kinds to exist, there must be discreteness or discontinuity at the most fundamental level' (2001, p. 31). It is undoubtedly the case that there is discreteness in nature. Electrons are members of a natural kind sharply and unambiguously distinct from any other kind of particle because of the strong sense in which electrons are identical. This in part explains why the atoms of the chemical elements also belong to discrete natural kinds. The existence of quantum discreteness is an important fact about the physical world that explains the character of a good deal of modern science, especially chemistry. But, according to modern science, there is continuity in the world also. Whilst the energy spectrum of bound electrons is discrete, the energy spectrum of free electrons is continuous. The frequency of electromagnetic radiation varies continuously across the spectrum. So why make such a big metaphysical meal out of the discreteness side of the story at the expense of continuity? There is nothing subjective about the differences in samples of radiation that differ in wavelength, even though the electromagnetic spectrum is continuous, that is, objectively continuous. Further, the fact that the spectrum is continuous does not prevent radiation from being governed by laws. It is governed by the continuous differential equations named after Maxwell. Planck's law specifies how the energy of black-body radiation is distributed across the electromagnetic spectrum as a continuous function of the frequency. There is a straightforward way in which Ellis can respond to my criticism. He can simply drop the demand that the boundaries between natural kinds must be necessarily discrete. Alternatively, he can drop the demand that natural kinds are a necessary prerequisite for the grounding of laws of nature. I do not see why essentialism should not be able to accommodate objective continuity as well as objective discreteness.

Another problem I have with Ellis's elaboration of his scientific essentialism is what I consider to be an overemphasis on fundamental particle and atomic and molecular physics and chemistry, which certainly provides him with nice examples for some of his claims. It is certainly the case that fundamental particles are ontologically basic, in the sense that there can be no atoms or macroscopic substances if there are no fundamental particles, whereas there can be systems of fundamental particles without there being atoms or macroscopic substances (cosmic rays or plasma, for example). But it does not follow from this that once we have fully grasped the ontology of fundamental particles we have grasped all that is of ontological significance or importance. Metals provide me with a nice example of my point.

Individual metals, treated as chemicals, may well be regarded as natural kinds insofar as their atoms are distinct from the atoms of any other metal or any other substance. But what if we turn from the chemistry to the physics of metals, where by metals I mean the solids that preponderate in our world and which metallurgists spend their professional lives studying and manipulating. Early in the twentieth century, some basic physical properties of metals such as their tensile strength and their ductility posed a problem for physicists. Theoretical calculations of such magnitudes based on the assumption that solid metals are comprised of a regular crystal lattice of metal atoms gave values that differed by several orders of magnitude from the values determined empirically. In the 1930's it was hypothesised that the crystal structures of metals contain disruptions or imperfections that came to be known as dislocations. This enabled the theoretical predictions of tensile strength and the like to be brought into line with measurement. By the late 1950's it was possible to detect the presence of the dislocations directly using electron microscopes. Today it is accepted that the properties of bulk metals, as they exist in our world, depend crucially on dislocations, the density of which will vary with the treatment and method of preparation of the bulk metal but which is always far from zero. I doubt whether metals, complete with dislocations, form natural kinds according to Ellis. Certainly the density of dislocations varies (continuously) from sample to sample. And yet there are certainly many laws governing the behaviour of metals, as there are in solid state physics and fluid dynamics generally. There are emergent properties, not mysteriously emergent in the sense that they float free of the fundamental entities on which they depend ontologically, but emergent nevertheless. And many laws in physics and chemistry involve them.

It would seem that metals, that is, solid metals complete with dislocations and perhaps including steel, are not natural kinds according to Ellis's strict demands. But if they are not, then Ellis's account of physical necessity cannot be applied to them. So what of our intuitions about physical necessity, exemplified in our example involving the expanding bottle top? Insofar as our intuitions link physical necessity with laws, they involve a wide range of law-like behaviour that extends well beyond fundamental laws governing the members of natural kinds such as electrons. It would seem that if Ellis's position is to have the wide application it needs to have it is committed to a strong reductionist thesis which assumes that all laws can be reduced to laws governing fundamental natural kinds. This is not a position I would wish to urge on anyone, and certainly not Alan Musgrave.

A case can be made for the view that the notions of physical necessity and causality apply only at the level of the emergent properties of the macroscopic world which we can manipulate and with which we causally interact. At the level of electrons and protons, that is, the level at which we have unambiguous natural kinds satisfying Ellis's demands, interactions are symmetric with respect to time inversion, so that the distinction between cause and effect loses its significance, all properties of particles are essential properties, and everything that happens does so of necessity.[1] So, it would seem, scientific essentialism tied to a strong notion of natural kinds can remove our problems associated with physical necessity only in the domain where they don't exist!

Aspects of Ellis's position make most appeal in the case of causal laws, the laws that govern the attractive power of masses and charged bodies and so on. He identifies three problems that confront any attempt to give a philosophical account of laws, what he calls the necessity problem, the idealisation problem and the ontological problem. His aim is to show how scientific essentialism can solve these problems better than rival positions. But when Ellis comes to make good this claim and show how his account of laws can solve the three problems (2001, pp. 219-222) he does so only in the case of causal laws, and not for more general, highly abstract, laws such as the conservation of energy (which elsewhere Ellis himself distinguishes from causal laws). The Ellis solutions for causal laws are attractive and have at least a superficial appeal. The behaviour of a member of a natural kind necessarily tends to behave in a lawlike way because the properties it necessarily has as a member of that kind causes it to behave in that way. Laws are idealisations in the sense that the ways in which members of a natural kind tend to behave because of the properties they necessarily have by virtue of being a member of that kind are often disturbed by counteracting tendencies. So a leaf, by virtue of being a member of the natural kind, massive bodies, necessarily has a tendency to be attracted to the earth according to the law of gravitation. But leaves rarely fall in a way that exhibits the law of fall because of winds and air resistance. Finally, the ontological problem, the problem of what in the world grounds laws, the problem of what their truth makers are, is solved fairly straightforwardly for causal laws. It is the essential properties of members of natural kinds that cause them to behave, or tend to behave, in accordance with laws. It is the gravitational power that massive bodies necessarily have as such that causes them to tend to move in accordance with the law of gravitational attraction.

I have two lines of criticism which decrease the appeal of this seemingly persuasive account. The first concerns the treatment of such things as massive bodies and charged bodies as natural kinds. These are rather different from electrons, with their specific mass and charge, that I have conceded to Ellis form a natural kind. The mass kind will include electrons but also sticks and stones and planets. Why are massive bodies picked out as constituting natural kinds rather than say, spherical ones. After all, is it not the case that the distinction between spherical

[1] This latter remark would need to be qualified to accommodate processes that are intrinsically probabilistic. Ellis does make the appropriate qualifications, but I have ignored them throughout this paper.

and non-spherical objects is just as objective as the distinction between massive and non-massive ones? Further, is it not the case that, when an object experiences a collision, the outcome of that collision depends as much on its shape as on its mass? I suppose the answer here is that mass is picked out because there are general laws, of inertia and gravitation, governing the behaviour of massive objects, whereas, in the case of shape, this enters only incidentally as determining the direction of the force on impact. We do not have a law of nature that makes generalisations about shaped bodies. But if we do accept this response, then there is a kind of circularity entering here analogous to the kind that Musgrave encountered when noting the difference between the way in which lawlike and accidental generalisations support counterfactuals, and that Popper acknowledged when attempting to use a distinction between initial conditions and laws to explicate physical necessity. Causal laws arise from the essential properties of the items they govern, whilst the properties that count as essential ones are just those governed by a law.

The second line of criticism becomes relevant when we move away from causal laws to consider very general laws such as the laws of thermodynamics, Newton's second law of motion and Schrödinger's equation. One of the striking features of the laws of thermodynamics is that they constrain systems whatever the underlying causal mechanism might be. That is why it was, and continues to be, possible to predict the depression of the freezing point of water under pressure from the fact that ice is less dense than water using thermodynamics, without any knowledge of the molecular mechanism involved. It seems to me that Ellis's attempt to accommodate laws like the conservation of energy into his essentialism by regarding those laws as essential properties of the world considered as one of a kind is merely a verbal manoeuvre that gives only the appearance of such an accommodation. The law of conservation of energy applies, not just to the world as a whole, but to every individual process that takes place in it, whether it be the operation of a steam engine or electric motor or the extraction of energy from food by the body. It is to such systems that the law is applied, and the observation of such systems that gives us what evidence we have for the law. The first law of thermodynamics amounts to the claim that it is not possible to construct a perpetual motion machine of the first kind, provided we define that latter notion carefully enough. When the law is so expressed, the difficulty of construing it as following from the essential property of anything becomes apparent.

I have gone along with Ellis and accepted electrons, with their specific mass and charge, as constituting a natural kind, and even strengthened his claim a little by stressing the strong sense in which electrons and protons are identical. However, the essential properties of electrons and protons do not give us enough to capture their lawlike behaviour. After all, classical laws applied to electrons and protons in hydrogen atoms would have them spiralling into the nucleus. They do not do that. Rather, the interaction of the electrons with the positively charged protons as nucleus yields atoms with discrete energy levels and a corresponding spectrum. How is that explained? The quantum mechanical explanation assumes that all quantum states are represented by vectors in Hilbert space and that observables such as energy are represented by appropriate Hermitian operators, the eigenvalues associated with the eigenstates of which represent the possible values of measurements of those observables. The time development of a quantum state is determined by the time-dependent Schrödinger equation, whilst the time-independent Schrödinger equation yields the energy levels. I suppose all this could be taken as describing the essential properties of quantum systems, but it is not clear what is gained or clarified by such a move. And we are still far short of describing the hydrogen atom. To do so we need to specify the Hamiltonian operator in the case of the hydrogen atom. In doing so we will indeed make use of the fact that the electron has its characteristic (essential) mass and charge as well as its characteristic 'spin', and once we have done so we will get the hydrogen atom with its characteristic energy levels. We do arrive at a position where we can argue for the chemical elements as natural chemical kinds, but, as we have seen, the explanation involves more than an appeal to merely the essential properties of electrons and protons. It involves the fundamental principles of quantum mechanics including Schrödinger's equations. There is an analogous situation in classical mechanics. Newton's second law of motion, force equals mass times acceleration, is an important truth about the classical world, but to fill it out in any specific situation we need to specify the forces in that situation. In doing so we will use laws, such as the law of gravitational attraction, which can be thought of as causal laws in the way that Ellis does. But Newton's second law is presupposed in any such specification and in any case made for the physical necessity of the outcome of such a situation. In all these cases, the most general laws governing the world, and those we may suppose are most fundamentally involved in physical necessity, defy accommodation into Ellis's essentialism in a straightforward way that is not question-begging.

6. MODEST ESSENTIALISM

For reasons outlined in the previous section, a modest version of essentialism that I am inclined to defend drops the strong reliance on natural kinds involved in Ellis's scientific essentialism and the allied attempt to construe physical necessity as a species of logical necessity. It retains the idea that the world in inherently active by virtue of the dispositional properties possessed by objects and systems in it and that real dispositions underlie physical necessity.

For Ellis, a member of a natural kind must necessarily possess the properties that are essential for it to qualify as a member of the kind. What is more, possessing a property has implications for behaviour. For example, an electron must have the appropriate negative charge, and possessing that charge implies, for example, the disposition to repel other negative charges in accordance with Coulomb's law. Physical necessity is a logical consequence of what is built into essential properties. I reject this account as it stands because it fails to accommodate physical necessity exhibited by such things as expanding metal bottle tops, which are not natural kinds in the strong sense demanded by Ellis, (either as bottle tops or as metals) and because physical behaviour depends on accidental as well as essential properties. But I accept the connection between physical necessity and logical consequence involved in Ellis's position. How is it that physical necessity, which is at work in the world, is connected with logical relations, which apply in the domain of propositions and symbolic expressions constructed by humans? A correspondence notion of truth is at work here. If statements truly describe the properties of a system and the situation it finds itself in then its behaviour will be truly described by some logical consequences of those statements. It is because the properties are dispositional that they have implications for behaviour captured by the statements that describe them.

A key component of Ellis's scientific essentialism that needs to be retained, then, and which provides much of what is needed for a comprehension of physical necessity, is the recognition that the world is inherently active rather than passive. What is needed here is, in effect, an updated version of Aristotle's notion of potential being. Systems or objects are what they are in large part by virtue of what they are capable of doing or becoming, of how they are capable of acting, reacting and interacting. The important part of the being of an acorn is its capacity to grow into an oak tree. What is distinctive of an electrically charged body is its capacity to interact with other charged bodies and electromagnetic fields in definite ways, so any characterisation of a body as charged should include those facts. Such a picture runs counter to the Humean notion that situations can be characterised as states of affairs, occurrences or events defined in terms of their momentary being in such a way that whatever states of affairs, occurrences or events that follow is a totally contingent matter to be specified by appeal to totally contingent laws of nature.

I believe Ellis is correct to recognise that the Humean view gained much of its plausibility from the mechanical philosophy that formed the background for empiricist philosophers such as Locke and Hume. From that point of view the universal matter from which the material world is made is passive. A piece of matter is distinct from space through possessing the property of impenetrability. The state of a system is specified by describing the distribution of matter in it. The subsequent motions or rearrangements of matter are comprehended in terms of contingent laws. It is precisely this point of view that is illustrated in the passage from Boyle quoted at the beginning of section 4. After Newton it became necessary to specify the velocities as well as locations of pieces of matter in the initial conditions without changing the overall character of the picture.

The subsequent development of science has rendered the mechanical philosophy totally inadequate. This is evident from the way in which 'initial conditions' figure in routine scientific practice. Suppose a system in question is a system of charged bodies. A crucial element of the initial conditions that a physicist needs to know is the distribution of masses and charges (not just the distribution of 'matter'). Once this distribution is known the physicist can write down the forces on the system by employing Coulomb's law and the like. The general laws of mechanics and electromagnetism then enable the future behaviour of the system to be deduced. So the traditional distinction between laws and initial conditions employed by philosophers is misleading. Some lawlike behaviour is already implicit in the formulation of the initial conditions. If the identification of a body as charged does not imply something about its ability to interact with other charged bodies, then just what is the import of describing it as charged rather than magnetic or merely massive?

This point is connected with another aspect of Ellis's position that I wish to include in my modest version of essentialism. Dispositional properties and related entities such as powers and capacities can be real, in the sense that they cannot be explained away by appeal to non-dispositional properties, although they may well be explicable in terms of other dispositional properties acting at a deeper level. Once we admit dispositional properties as real and recognise that they imply lawlike behaviour we are well on the way to being able to accommodate physical necessity. If the metal bottle top possesses the property of thermal expandability characteristic of metals then it must expand when held under the hot water tap by virtue of possessing that property and what it entails.

Most of the above was stressed by Popper, in Appendix x added to the 1957 English edition of the *Logic of Scientific Discovery* (1972, pp. 420 – 441). He insisted that mundane singular statements such as 'here is a glass of water' and 'this here is a swan' are informative because they contain general terms such as 'swan' and 'water' which are dispositional insofar as they imply lawlike behaviour. They imply behaviour typical of water and swans. But Popper did not quite round off his discussion with an easy accommodation of physical necessity in the way that he might have. One unfortunate, non-realist, aspect of his discussion may have been responsible for this. We have seen how singular statements involving universals imply lawlike behaviour. Thus such statements have implications that go beyond anything that might be 'given' in any experience that might serve as evidence for their truth. 'For even ordinary singular statements are always *interpretations of the "facts" in the light of theories'* (1972, p. 423, italics in original). This remark, however true, is not quite to the point in an appendix devoted to a discussion of physical necessity. Interpretations are human constructions. There would be no such things in a world without humans. So noting the way in which human interpretations take us beyond evidence for them in some given circumstances cannot be relevant to the problem of physical necessity. To think otherwise smacks of word magic. (That should have Alan Musgrave, now retired, turning in his deck chair!) The crucial point is not the interpretations involved in singular statements, but what it is in the world that makes them true. The point I am stressing is that the instantiation of dispositional properties is part of what makes singular statements true. 'There is a charged body' cannot be true if the body lacks the capacity to interact with other charged bodies in the appropriate way.

Modest essentialism departs from mad dog essentialism insofar as it rejects the linkage the latter makes between physical necessity and the essential properties of natural kinds and it rejects the allied doctrine that physical necessity is a species of logical necessity, with the behaviour of members of natural kinds following from the definitions of the essential properties of members of those kinds together with a specification of the circumstances. Modest essentialism shares with the more extreme version the assumption that dispositional properties, and allied notions such as causal powers and capacities, are real and cannot be explained away. An adequate characterisation of the being (or essence) of an object or system must include a characterisation of how that object or system is disposed to act or develop. Any such characterisation will include some lawlike behaviour, and this gives a basis for an understanding of physical necessity and causation.

7. THE ROOTS OF MODEST ESSENTIALISM IN MUSGRAVE'S WRITING

I have indicated that, because the only clear-cut cases of natural kinds on which Ellis bases his scientific essentialism are fundamental particles such as electrons and protons and elementary groupings of them to form atoms, then his position can only serve as a basis for the solution of general problems associated with physical necessity and causation if a strong reductionist thesis is adopted in addition. It is clear that Musgrave has no more inclination to go along with such a thesis than I do. He insists that explaining the properties of the tables of common sense by appeal to their molecular structure does not have the consequence that there are no common sense tables. I echo his own words (1999, p. 133) when I say 'if the properties of metals cannot be cashed out in terms of electrons and protons, then so much the worse for electrons and protons.' I am confident that he will agree with me that problems about necessity involving expanding bottle tops and silver coins should be confronted in their own terms rather than evaded by shifting to the level of electrons and protons.

Musgrave does not make explicit his acceptance of another view I would like to urge on him, namely, the reality of dispositional properties. However, there are elements of his stated position that bring him close to implying it. Musgrave shares Popper's view that the singular statements that are reports on observations are fallible and testable. But does this not mean that those statements have implications that go beyond what is instantiated in the original observed situation that suggested them? The testability of these statements implies that their truth content goes beyond the original circumstances. That is, it implies the reality of dispositional properties of just the kind that modest essentialism involves.

Once we have dispositional properties that are real, then we have most of what modest essentialism needs. Objects or systems behave in the way that they do because they have the dispositional properties that they have. A precise characterisation of what those dispositional properties amount to will include a specification of the laws governing, or implicit in, those dispositions. Objects made of copper have a disposition to expand when heated but no disposition to change to silver when placed in my pocket. The charge on a body and its obedience to Coulomb's law are not separable, because obeying Coulomb's law is part of what having the dispositional property 'charged' amounts to, although this is not obvious and needed to be empirically discovered.

It might well be questioned whether the modest position that I am attempting to urge on Musgrave warrants the description 'essentialist'. Once we move beyond natural kinds in the strict Ellis sense then the distinction between essential and accidental properties becomes cloudy, as my discussion of mass and shape in section 5 indicated. And once this is acknowledged then we can no longer avail ourselves of clear-cut essentialist definitions that pick out just the essential properties that must be possessed to warrant membership of a kind. There is some justification for the name insofar as modest essentialism shares some of the characteristics of the position that Ellis has defended under the name of essentialism, and also has an affinity with Aristotelian essentialism, defended by Brody (1972), for example. It may be that all that stands in the way of Musgrave accepting modest essentialism is the name, in which case I would be happy to drop it. It does not trouble me a great deal if the position I have arrived at involves no disagreement with Musgrave whatsoever, provided we have learnt something along the way. Agreement with Alan Musgrave is not something which, in itself, causes me any discomfort whatsoever.

REFERENCES

Brody, B. (1972) 'Towards an Aristotelian Theory of Scientific Essentialism.' *Philosophy of Science* 39: 20-31.

Cartwright, N. (1983) *How the Laws of Physics Lie*. Oxford: Oxford University Press.

Ellis, B. (2001) *Scientific Essentialism*. Cambridge: Cambridge University Press.

Musgrave, A. (1999) *Essays on Realism and Rationalism*. Amsterdam: Rodopi.

Popper, K. R. (1972) *The Logic of Scientific Discovery*. London: Hutchinson.

Russell, B. (1912-13) 'On the Notion of Cause.' *Proceedings of the Aristotelian Society*. 13: 1-26.

Stewart, M. (1979) *Selected Philosophical Papers of Robert Boyle*. Manchester: Manchester University Press.

ROBERT NOLA

THE METAPHYSICS OF REALISM AND STRUCTURAL REALISM

Scientific realism is the default position of many philosophers of science and most working scientists, Alan Musgrave having done much to keep the nature of this default position before us and to provide arguments for it.[1] For this, as well as much else, we are all in his debt. But the term 'realism' needs careful handling since not only do the friends of realism offer us definitions that can be misleading, but those who are not its friends offer us tendentious definitions that can also lead us astray. To set matters straight, section 1 offers a definition of (but no argument for) a strong version of ontological (or metaphysical) scientific realism for a number of categories of observable and unobservable items found in science including such particulars as objects, events, processes and tropes, as well as non-particulars such as properties and universals. Given this background it is then possible to characterise structural realism, viz., the view that realists should also admit structures into their ontology, especially mathematical structures expressed by mathematical equations. One task is to give an unproblematic characterisation of such structures.

Contrary to positivists and empiricists, metaphysics plays an important role in our account of science. In this paper a liberal stance will be adopted towards all of the metaphysical categories mentioned above (from particulars to universals) in that all will be admitted without discrimination into the ontology of science. This will be contrasted with a more systematic metaphysics for science in which the liberal stance is eschewed and a more parsimonious approach to ontology is adopted in which some of the categories permitted by the liberal metaphysician are admitted, or not, by the systematic metaphysician (from nominalist to Platonist). What ontology we ought to adopt involves complex arguments found in metaphysics that have little to do with science; these will be mentioned only in passing. The liberal approach adopted here is provisional; it is adopted in order to make clear exactly what the metaphysical commitments of structural realism might be in a non-question-begging manner. Often nominalistically inclined realists pursue their metaphysical agenda alongside their realism, and this can obscure issues.

[1] His papers over the years in defence of scientific realism are collected in Part I 'Realism' of Musgrave (1999).

C. Cheyne & J. Worrall (eds.), Rationality and Reality: Conversations with Alan Musgrave, 183–223.

In section 2 some aspects of structuralism in the ontology of science are discussed.[2] These vary from tropes of structure to laws of nature as relations between universals. Some non-liberal metaphysicians might find this last commitment too high a price to pay; but it is one that some are willing to pay, for example those who follow, say, Armstrong (1983, Part B) in arguing that laws of nature, understood ontologically as part of the furniture of the world, are really higher-order relations of necessitation holding between universals. As will be seen, structural realists do not form a united band. Some would claim: (a) there are structures that can be characterised mathematically; (b) there are particulars, such as objects, which are placeholders in the structures. Given the liberal stance adopted here, both (a) and (b) will be accepted. In addition it will be claimed that we can have knowledge of both (a) and (b). However some structural realists adopt what might be called *epistemic structural realism* in which it is claimed that while we can have knowledge of (a) we cannot have (much or any) knowledge of (b). That is, we can have knowledge of structures but cannot know the items that are placeholders in such structures (such as objects); they are a 'something-we-know-not-what'. Yet others adopt what might be called *Platonistic ontological structural realism* in which (a) holds but not (b). That is, all that exists are mathematical structures, and we can have knowledge of these. But there do not exist placeholders, such as objects, within the structures: so we cannot have knowledge of these at all. On the stance adopted in this paper, one need be neither an epistemic nor a Platonistic structural realist. There are a number of different arguments for these positions, only a few central ones being addressed in this paper.

Structural realists are realists who accept certain arguments, such as inference to the best explanation, to support their realism. But they also take seriously the pessimistic meta-inductive argument that shows that there are significant ontological discontinuities between successive theories. The ontologies of our evolving theories are alleged to change, so that the (kinds of) objects postulated by earlier theories are denied by later theories, which in turn postulate new (kinds of) objects. This tends to undermine any realism about the (kinds of) objects postulated in theories with theory change. So, are there any significant continuities between theories? If not, realism would be undermined. For structural realists scientific realism is saved by an alleged continuity of laws, mathematical equations or 'structures' between successive theories. It is important for realists to recognise that there might well be such structures (however they are to be understood) and that not all continuity need be sought at the level of objects. In fact, if the pessimistic meta-induction is accepted, it had better not be sought only at that level. However for some there is a 'downside' to structural realism in its strong forms. For both the epistemic and Platonistic structuralist we can only have knowledge of the allegedly invariant structures; we cannot have knowledge of the objects standing in these structures. More extremely, for the Platonistic structuralist discontinuities at the level of objects ceases to be a real issue since there are no such objects to countenance in the first place.

[2] Since structuralism has become a many-splendoured thing, there are other aspects of structuralism not discussed in this paper. For some of these, and a critical response, see Psillos (2001) and Psillos (forthcoming).

Are structural realists right in claiming that there are continuities in mathematical structure with theory change? There is a confusion that needs to be avoided in posing this question that is addressed in section 3. A distinction should be drawn between formulations of laws in some language, and the objective worldly structures they attempt to represent. What is argued is that, at best, structural realists can only appeal to law formulations and not the objective structures that they purportedly represent. Also, even though there are some cases in which law formulations have been invariant with theory change, there are other cases in which there is change in law formulation with theory change; we do renovate our knowledge of the mathematical laws that apply in science. So the structural realists' worry about lack of continuity of objects with theory change can also come to infect the very 'mathematical structures', or more correctly law formulations, to which they appeal in order to overcome the realists' problem about lack of continuity with theory change.

Structural realists were led to their position through accepting the pessimistic meta-inductive argument. It is argued in section 4 that this involves an unsound inference. This is not to deny that ontological change, either at the level of postulated objects of law formulations, has not occurred in the history of science. Rather it does not have the dire consequences for scientific realism that structural realists have feared.

Section 5 examines a further argument that structural realists adopt for their 'flight from objects'. Some structural realists have seen, in the use of the Ramsey sentence to formulate a theory, a way of advocating a version of structuralism in science (see Maxwell, 1970); this reason for structuralism will not be discussed in this paper. But in this section David Lewis' version of the Ramsey sentence will be used to show that we can have continuity in objects with theory change in some historical cases where this has been denied to occur. Structural realists claim that there has been a change in ontology, as suggested by Poincaré and others following him such as Worrall, in the theory of light from Fresnel to Maxwell, and that this in turn supports an epistemic version of structural realism. But the Ramsey-Lewis theory of reference fixing can be used to show, appearances to the contrary, that there is continuity in our reference to 'objects' between these two theories; and the strategy can be generalised to other cases (not discussed). That there appears to be ontological incommensurability is at best illusory. Further, even if our knowledge of the intrinsic properties of light might still be slim, there are no grounds for drawing the strong conclusion that knowledge of the intrinsic (not the same as essential) properties of such 'objects' is impossible and that all we can ever know are their extrinsic relational properties. The position advocated here is that of a liberal epistemology in which there can be knowledge of objects, events, processes and properties as well as structures; what is resisted is the idea that we can only have knowledge of structures and not the placeholders within the structures.

In sum, structural realists have pointed to a feature of the ontology of science that nominalistically inclined realists often obscure. But if one does not adopt the systematic metaphysical stance of the nominalist and is more liberal, then there is room to consider structures as well. But there is no need to go overboard in the opposite direction and adopt the systematic metaphysics of a Platonist and deny the existence of particulars such as objects and admit only Platonistic structures. Nor are there good grounds to support only an epistemic version of structuralism. Of course, a systematic metaphysics will be eliminative with respect to some of the ontological categories admitted by the liberal metaphysician. But it is hard to see how matters only relating to science could resolve these issues. Rather they are the province of a systematic metaphysics not addressed here.

1. A CONSPECTUS OF REALISMS

1.1 Common-Sense Realism

The term 'realism' needs to be defined with respect to some category of items rather than realism *tout court*. Thus we need to say that we are realists with respect to broad categories such as common-sense objects, the unobservable objects of science, natural kinds or numbers, but not, say, with respect to universals, or possible worlds, and so on.[3] Such a realism has two dimensions, an existence dimension which says that within each category of items something exists, and an independence dimension which says that they exist mind-independently.

The second clause is needed to exclude virulent forms of human chauvinism in which what exists is said, in some sense, to depend on us; either nothing exists independently of us (commonly called idealism), or if something does, then we can know nothing of it (a Kantian idealism). By the mind-independent existence of some x (such as the broad categories just mentioned, or items in them) is to be understood as the existence of x independently of our perception of x, our thoughts, beliefs or theories about x, and the language we use to talk of x. That is, x would still exist, and do its thing or be the kind of thing it is, even though we humans were not to have existed; or if we do exist, x would still exist and do its thing or be the kind of thing it is even if we were never to perceive x, or think or theorise about x or have a language to talk of x. Central to the definition of realism is that such counterfactuals hold of our actual world for the broad categories of items so far envisaged. Arguments for the truth of such counterfactuals lie outside the definitional aspects of realism addressed here; but some arguments will surface in later sections. Such a conception of realism rules out the claim that xs are, somehow, a construct (logical or not), out of our 'sense data' or 'experience', or a construct out of our cognising activities, or out of our historical-social-cultural circumstance, or the consensus of some scientific community (as many contemporary social constructivists would

[3] Nothing will be said here about the full range of items with respect to which philosophers have been realist. Our concerns will be largely with realism with respect to common-sense and unobservable scientific objects, events, processes, etc. And even though the paper mentions common-sense and unobservable properties (such as having mass, momentum, charge, etc) this should not be taken to commit us necessarily to Platonistic universals, one of the original sites of controversy within realism. In one sense these are issues of a systematic metaphysics which lie beyond the concerns of philosophy of science; in any case, issues of realism in the philosophy of science presuppose the results of a systematic metaphysics, thereby showing that there is no science free of metaphysical presuppositions as positivists and empiricists have sought.

have it). It is in some such way that opponents of common sense realism resist the mind-independence clause.

Only the most rabid of idealists, relativists or postmodernists would deny that anything exists mind-independently. Following the account developed by Devitt (1997, chapter 2), a more contentful, contestable, but contingent and empirical doctrine of realism lists the items that exist in each category. It turns on the core idea that we are mainly (but not wholly) right in our folk beliefs about what exists, but characterises it in a way that is independent of talk of beliefs. More precisely, realism [common-sense observable objects] can be specified by a large list of the things that exist in this category such as: most of the items such as the Sun, Mt Everest, the Adriatic Sea, Napoleon, and so on indefinitely, exist mind-independently. The word 'most'[4] needs to be added in case some non-existent items are erroneously included in the list, such as Cyclops, Atlantis, Vulcan (the intra-Mercurial planet), and so on. We are not realist about these objects; they do not exist mind-independently since they do not exist at all. Realism [common-sense observable tokens of kinds] also gives a large list such as: tokens[5] of most common-sense kinds such as cats, water, human beings, stars, aspects of the experimental apparatus used in science,[6] etc, exist mind-independently. Again the word 'most' is to be included because some kinds might be listed that do not exist, even mind-independently, such as dragons, Loch Ness monsters, flying saucers, witches, etc. Both theses about realism are clearly contingent and are empirically (rather than *a priori*) knowable; and they entail the largely undisputed and much less risky bland ontological claim that *something* exists mind-independently.

There are, of course, some common sense items whose existence is dependent on our minds. Such is the case for ordinary artefacts such screwdrivers or watches, or experimental apparatus in science such as Bunsen burners, microscopes and CAT scanners (but not, say, the materials out of which they are made). Also there are other clearly social objects such as money or birth certificates that exist only mind-dependently. It is not the task of this paper to give an account of this important category of items (see Searle (1995) for an extended account of social kinds).

[4] Devitt (1997) chapter 2 gives an account of the qualification 'most' and how that can be understood as a vast disjunction of conjuncts of items in the list under each category, each conjunct having a minimum 'quorum' from the list. It is then guaranteed that at least one or more of the conjuncts in the vast disjunction will be true.

[5] Devitt's use of the term 'token' indicates that he restricts his position to a form of nominalism that admits only tokens (instances, examples) of kinds of objects. Though we go along with this for the purposes of formulating a definition of scientific realism, such a nominalism is not part of the more ecumenical position adopted in this paper in which realism is not restricted to nominalistic realism. Any argument for nominalism is an issue for systematic metaphysics that is not addressed here.

[6] Here we will not discuss the status of items which are artefacts; if scientists' experimental apparatus uses natural materials then they can serve as good examples of items about which one can be realist in the sense being defined. Thus in Newton's experiments with prisms and light beams, the beams of light and the materials in the glass which constitute the prisms are things about which one can properly be realist; however glass and prisms are made by us and as such are artefacts which, it might be argued, are not items about which one can be realist in a strong sense. For one account of the dependent existence of artefacts see Searle (1995), chapter 1.

We can also be realists about categories of particulars other than ordinary objects, such as events, processes, properties and property instances (tropes). Here science comes into contact with metaphysics in the different categories of particulars to which it can be said to be committed. The liberal metaphysics adopted here allows a separate category of events. In contrast some systematic metaphysicians would do away with such a category in favour of three other items from other categories such as objects, a property and a time (See Kim (1998) in which events are claimed to be property exemplifications). These matters aside, we can define 'realism [observable events]' as 'tokens of most event kinds exist mind-independently, such as: volcanic eruptions occur mind independently, the turning red of litmus paper when placed in acid occurs mind independently, etc'. Again the qualification 'most' is also understood to apply here; if we were to list kinds of event that do not occur, such as alien abductions, etc, then this would not vitiate a substantive realism with respect to observable happenings. There is also realism [common-sense processes] as when it is said that tokens of most process kinds such as tidal movements, the eclipsing of the Sun by the moon, embryonic development, etc, exist (occur) mind-independently. Again the word 'most' allows that we might be wrong about some kind of observable process occurring such as the alleged extraction of (observable) ectoplasm in the course of a séance. The case of properties and tropes is left to section 1.3.

1.2 Scientific Realism

What now of scientific realism? Some express it as a goal of our scientific endeavours, for example, 'we aim to seek the truth not only at the level of the observable but also at the level of the unobservable'. While such goals are not denied by realists, this expresses a semantic version of realism, (to be discussed shortly), or is a methodological principle of realism, or part of realist axiology. What is intended here is an ontological doctrine in which, within each ontological category, there exist, in the mind-independent way specified, a number of unobservable items. But this is a bland, and empty kind of realism to which few would object. More contentfully, scientific realism is the view that scientific theories are largely (but not wholly) right about the unobservables that exist. But again, this is couched in a way that is dependent of talk of theories, or languages or other semantic notions. Expressing this as an ontological thesis, we have the following contingent, but empirically knowable, version of scientific realism, viz., an open-ended list of items, most of which exist mind-independently. Thus scientific realism [kinds of (unobservable) objects] is the doctrine that tokens of a long list of most unobservable kinds of objects, such as electrons, black holes, genes, viruses, forces, tectonic plates, pulsars, etc, exist mind-independently. Again the qualification 'most' allows that we have been wrong about some kinds that can appear in the list such as phlogiston, polywater, caloric, a pervasive electromagnetic ether, and so on.

We can also be scientific realists with respect to unobservable events and processes such as the emission of an electron during neutron decay, or the catalytic process whereby chlorinated fluorocarbon atoms eat up ozone in the ozone layer above the Earth. So far these are standard items with respect to which one can be a scientific realist.

The distinction between the various sorts of common sense and scientific realisms turns on a distinction between what is observable and unobservable in respect of objects, events and processes. Here we will simply assume that such an epistemic distinction can be drawn between items in each ontological category. Conjoining these two doctrines of realism yields what we might simply call 'realism [objects/kinds/events/processes]' regardless of whether these are observable or not. Opponents of this broad realism will resist any autonomous mind-independent existence of each category and items in them, often claiming that they are, somehow, a 'construct', (logical or not) out of our sensory, experience, or out of our cognising activities when we theorise, or out of our socio-historico-cultural circumstance, or the agreements communities reach about their scientific activities. Since only definitions of various kinds of realism are being considered, arguments for and against realism and its rivals will not be considered here.

In subsequent sub-sections we will examine further ontological categories in respect of realism, such as properties, tropes, universals and structures, as a prelude to the ontological claims of structural realism in science. But before this, it is important to see that the kind of ontological scientific realism introduced here is quite distinct from other ways of characterising scientific realism. One already mentioned is that of truth as the aim of science. Ontological realists will not want to eschew truth as an aim of their inquiries, or something like this such as increased verisimilitude, or approximate truth (for those who put store on the claim that all theories are idealisations and can only ever yield approximate truth at either the observable or theoretical level). But such additional claims about scientific realism are 'add-ons' which depend crucially on semantic notions such as that of a theory or the statements (sentences, propositions) being *true*, or the descriptive terms of theory *referring*. The conception of ontological realism, through its mind-independence clause, puts strong emphasis on the existence of categories of item independently of our beliefs or theories about such items, or the language, terms and sentences we use to talk of them. Ontological realism is also distinct from the commonly cited Putnam-Boyd characterisation of 'realism as an over-arching empirical hypothesis by means of two principles: (1) Terms of a mature science typically refer. (2) The laws of a theory belonging to a mature science are typically approximately true' (Putnam 1978, p. 20). While ontological realism and Putnam-Boyd realism, are both empirical, the latter cannot be constitutive of ontological realism because of its talk of semantic items such as the truth of theories or the reference of terms. Of course ontological realists will also want the terms of their mature science to refer and its laws to be approximately true. But this is a matter independent of the constitutive claims of ontological realism.[7]

[7] The account of ontological realism set out above is modelled on that given in Devitt (1997) chapter 2, and Devitt (forthcoming). He provides an account of the ways in which ontological (or metaphysical) realism can, using the disquotational notion of truth as a device for semantic assent or descent, be expressed using semantic notions such as 'refers' and 'true'. Thus there are equivalences such as: 'electron' refers if electrons exist; 'electrons exist mind-independently' is true iff electrons exist mind-independently; and so on. But such formulations are innocent of semantics and leave untouched other

In his excellent conspectus of realisms, Hellman (1983) invites us to broadly distinguish between semantic, ontological and epistemic formulations of realism. The first two have already been mentioned. The third goes beyond the ontological thesis R, viz., that most of the tokens of unobservable kinds in science exist mind-independently; it invites us to add various epistemic operators to R and instead to *believe* that R, or claim that *there is reason or evidence to believe* that R, or that we *know* that R, and so on. Note that such epistemic formulations are in terms of the more fundamental ontological thesis R. Realists do not eschew such formulations. In fact they need to give arguments for them (but such arguments are not canvassed in this paper). The very truth-value of epistemic formulations such 'there is good evidence for R', has a strong bearing on our acceptance (or rejection) of ontological scientific realism. But realists ought not to accept these (and other) epistemic formulations as constitutive of the central aspects of the doctrine of realism.

Advocates of structural realism in science often adopt semantic versions of realism. Thus John Worrall tells us that the default position of realists is such that 'Most of us unreflectingly take it that the *statements* in [the] observation-transcendent part of the theory are attempted descriptions of a reality "behind" the observable phenomena … or at any rate "essentially" or "approximately" accurate' (Worrall 1989, p. 100; emphasis added). Another advocate of structural realism, James Ladyman, characterises Worrall's position more strongly:

> According to Worrall, we should not accept full-blown scientific realism, which asserts that the *nature* of things is correctly described by the metaphysical and physical content of our best theories. Rather we should adopt the structural realist emphasis on the mathematical or *structural* content of our theories. (Ladyman 1998, p. 409)

What is described here as full-blown scientific realism is not ontological realism. It is semantic in that it explicitly makes reference to our theories, and requires, for at least a standard account of this brand of realism, that our theories be correct, i.e., true (or approximately true) not just about what things there are but also what their natures are. But even for semantic realism, this is to add an over-strong, and irrelevant, requirement that our theories not only be right about the *things* that exist, but also be right about the *nature* of those things. We may doubt whether even our best theory of electrons gets anywhere nearly correct about the nature of electrons, even though it tells us a lot about electrons. In contrast a case can be made for chemistry having got right the nature of water as a collection of H_2O molecules (give or take impurities). One can be a semantic realist about our theories being right about the existence of things without also requiring, in order for our theories to be realistically understood, that they also be right about natures. Importantly one can be right in claims about ontological realism with respect to, say, the chemical kinds water, hydrogen, oxygen, etc, without having to mention any theory, or any doctrine about semantic realism (full-blown or not) in the definition of ontological realism.

We have ventured into the territory of structural realism. One way of expressing this kind of realism is to say that our theories are right, or approximately right, about the laws or mathematical equations they propose. However, as will be urged

ways in which substantive semantic notions, such as truth and reference, can impinge upon definitions of realism. Here the task is to define, and not argue for, ontological realism (scientific or otherwise).

subsequently, such a semantic realism is about our laws-like formulations expressed in some language, such as the formulation of mathematical equations. This is quite distinct from an ontological realism with respect to laws of nature, or mathematical structures, understood ontologically as features of the actual world. What this latter kind of realism proposes is that there are in reality structures, some of which may be mathematical in character; alternatively, there are laws of nature, understood as brute facts of nature and not as law-like formulations in some language, that attempt to *describe* these mathematical structures or laws of nature. Liberal-minded ontological realists will wish to admit within their ontology such laws of nature or mathematical structures (however they may be characterised, for example as relations between universals, or whatever); in this way they propose an additional category of entities to the categories of objects, events and processes they also admit. But as already indicated, there are some Platonistic structural realists who wish to adopt a restricted ontology that eschews particulars such as objects, and admits only structures. We will return to the different brands of structural realism in section 2 once some motivation has been given for adopting such structures in one's ontology. This will be done by considering a realism with respect to properties and universals, and with respect to property particulars also known as tropes.

1.3 Realism about Properties, Universals and Tropes

Realists who are also nominalists wish to admit only particulars (such as objects, and perhaps events and processes); they do not wish to admit other items such as kinds, properties or universals. Such a stance is evident in the Devitt-style definition of common sense and scientific realism given above in which it is *tokens* of kinds of objects, events or processes which are said to exist mind-independently, and not the *types* or *kinds* themselves. As far as it goes there is nothing metaphysically unproblematic about this nominalist style of definition. However since the stance of this paper is liberal towards the various metaphysical categories, and sets aside the claims of a more systematic metaphysics, such a strictly nominalist stance will not be adopted.[8] This opens the way to admit categories of properties, universals and tropes into scientific realism.

A scientific realism [kinds of object] admits the mind-independent existence of kinds such as electron, boson, oxygen, common salt, gene, tiger, pulsar, etc. A scientific realism [kinds of property] admits the mind-independent existence of items such as mass, charge, density, volume, geometrical shape, kinetic energy, valency, being recessive or dominant (of a gene), and so on. Some properties are one-place while others are two-, or more, place, i.e., they are relations. Standard

[8] No arguments will be given here for this position. It should be noted that Quine even rejects the label of a nominalist telling us, in a symposium with Devitt and Armstrong that, like Armstrong, he 'espouse[s] rather a realism of universals', and that he 'see[s] no way of meeting the needs of scientific theory, let alone those of everyday discourse, without admitting universals irreducibly into our ontology' (Quine (1980), p. 450). Quine also gives no arguments in his paper but refers to other places where he does. Thus the grounds for the more liberal ontological stance adopted in this paper are not without foundation; but, as indicated, the grounds for a more systematic metaphysics which rejects one or other of the ontological categories of the liberal metaphysician are only hinted at.

examples are: being more massive than, being between, having a velocity or an acceleration in some frame of reference, and so on.

How is such talk of properties to be understood? Some might take a predicate nominalist (and thus metaphysically anti-realist) stance towards them in which some object might have predicated of it 'being a ball', 'being spherical', 'having such-and-such a volume', etc, but not admit that there are properties as such.[9] Others might be quite realist about them and claim that the properties are like universals; each universal admits multiple instantiations, the very same universal being present in each instantiation. A third alternative is that of a trope metaphysician. They are realists about what philosophers have called abstract particulars, or particularised universals, or property instances, all alternative names for tropes. Serious trope metaphysicians are not realists about universals such as greenness or sphericity; nor do they admit substantive objects (which they regard as bundles of tropes (see Armstrong (1989, chapter 6) and Campbell (1990)). In fact they are extreme particularists, each trope being a particular in its own right. Thus they claim that there are distinct items such as the trope which is the greenness of this pea, another trope which is the greenness of that pea, a further trope which is the greenness of this other pea—in fact there are at least as many such tropes as there are green peas.

Systematic trope metaphysicians claim that tropes are the only ontological item that one need admit; there is no need for objects, or events or universals.[10] This aspect of trope metaphysics will not be pursued here. We will admit that there are tropes, but admit other kinds of entity as well. But to illustrate, systematic trope metaphysics claims that there is no need for an independent category of events; tropes can do their work, especially as the items that stand in particular causal relations. Thus it is the whiteness of the wash that advertisers like to say pleases some housewife. Again it is the weightiness of the box (held by a person) that causes muscle strain in the person, or the sphericity of the ball (amongst other things) that is a cause of its rolling so far, and so on. If tropes are the very items that stand in causal relations, then this only reinforces claims about realism with respect to tropes (see Campbell 1990, chapters 1.7 and 5.9-5.15).

For our purposes we will adopt Ellis' characterisation (see Ellis 2001, p. 24) of a trope as the pair < U, a > where the two 'constituents' are U (which is some property or a universal), and a (which is a particular such as an object, an event, or an ordered sequence of objects or events, that instantiate some many-place universal U). < U, a > can be read as: 'the U-ness of a' (or 'the U-hood of a', or 'the U-ship of a', or 'the U-ity of a'). But note that this is not to claim that tropes can be eliminated since they are nothing but such ordered pairs of items from other ontological categories. This would be to adopt the stance of a systematic metaphysics hostile to tropes. For our purposes we will continue to admit tropes; they are a kind of particular and are

[9] This is a claim of the systematic metaphysics of nominalism. Several varieties of nominalism are described and critically evaluated in Armstrong (1978) chapters 2 to 5, such as natural class, resemblance, mereological and conceptual nominalism, as well as the predicate nominalism above.

[10] For a defence of a systematic metaphysics of tropes see Campbell (1990) in which it is argued that no other ontological category is needed other than that of tropes. Armstrong (1989) chapter 6 contains an assessment of the comparative virtues of a systematic metaphysics of tropes *versus* universals in which a trope metaphysics come a close second to his preferred theory of universals.

nothing like an ordered pair, or a set. Though the identity conditions for tropes are not clear, on this characterisation there are different tropes either if the universal U instantiated is different (e.g., the greenness of a pea, the sphericity of the same pea, etc.), or if the object a is different (the greenness of this pea, the greenness of that pea, and so on).

Since structural realism is the main topic of this paper, it is useful to expand on what will be called relational tropes and tropes of structure. In what follows we will consider tropes of the form $< U, < a, b, c, \ldots, n >>$ where U is an n-place universal and a to n are the items (objects, events, etc) all of which are needed to instantiate U. To illustrate, whether or not one is a realist about the universal between-ness, one can be a realist about the trope of the between-ness of the lectern with respect to the speaker and the audience. And this is a different relational trope from the trope that is the between-ness of one's car with respect to this other car and that further car (as when one parks one's car in a parking vacancy in the line of cars at the curbside). And both of these are different from the trope that is the between-ness of the elbow with respect to some individual's shoulder and hand. Though it might appear that the same relational universal of between-ness is instanced in all these cases, for the trope theorist each property instance is really a distinct particular; but a relation of close resemblance can hold between them (in the case of a resemblance nominalism with respect to tropes).

Taking relations further in the case where many different relations hold between some set of particulars, one might wish to speak of tropes of structure. Whether or not one is a realist about spiral-ness, or spiral-hood, one still needs to admit the trope which is the spiral structure of a given snail shell. Again, whether or not one admits relational universals of Fibonacci structure one still has to admit the Fibonacci structure of the arrangement of leaves on the stem of a given plant. Such structural tropes are an important part of the claims of structural realism in science. They can be represented as follows: $< \mathbf{S}, \{a, b, \ldots, n\}>$ where **S** is a set of m universals $\{U_1, U_2, \ldots, U_m,\}$ each having i places (where i ranges from 1 to some number $k \leq n$), and the set of particulars (such as objects, events, or even other tropes) $\{a, b, \ldots, n\}$, is such that some subset, formed as an ordered i-tuple, instantiates each universal in **S**.[11]

Scientific realism can also admit unobservable tropes of relations, or structures. We can be realist about tropes of structure such as the structure of a given water molecule, there being as many different tropes of structure of water molecules as there are molecules of water (though they may all bear some resemblance to one another). Admitting such a multitude of tropes is independent of whether or not we also admit a universal of structure, the water molecule structure. Again there is the trope of the helical structure of each DNA molecule, the trope of the 4-dimensional structure of a given crystal of diamond, the trope of the planetary orbital structure of

[11] Ellis ((2001), p. 25) asks us to note two kinds of instantiation relation, instantiated *in* and *by*. We will say that the universal whiteness is instantiated *by* a cup that is white, but not instantiated *in* the cup which is white. The instantiated *by* relation holds between a universal and an object. In contrast whiteness is instantiated *in* the trope, the whiteness of the cup; it is not merely instantiated *by* the trope, the whiteness of the cup. The instantiated *in* relation holds between a universal and a trope.

the solar system and the trope of the cellular structure of a given human body. The moulding of a piece of clay produces both observable and unobservable tropes. There is the trope of the form of the clay at a given time; as the form is changed there is yet another trope of form at another time. Again there is the trope of the structural relations between the clay molecules at a given time; and as the clay is pressed into different shapes there are yet other tropes at other times. There is also the trope of the distribution of stars and galaxies through the universe at some given time; and at a later time, given the dynamic nature of the cosmos, there will be another trope of structural distribution. And so on. In each case the trope is of the form mentioned above but with a temporal index, t: $< \mathbf{S}, \{a, b, \ldots, n\}, t >$ where **S** is a set of m universals $\{U_1, U_2, \ldots, U_m,\}$ each having i places (where i ranges from 1 to some number $k \leq n$), and the set of objects or events, $\{a, b, \ldots, n\}$ is such that some subset, formed as an ordered i-tuple, instantiates each universal in **S** over, or at, time t.

As indicated, such structural tropes are to be carefully distinguished from the universals of structure, the *very same* universal, or set of universals, being instanced many times over in different water molecules, or different DNA molecules, or different diamond crystals, or different planetary systems, and so on. Such many-placed universals are said, in the talk of those who are advocates of universals, to be the *same* in each of their instantiations. It is *exactly the same* universal of structure that is said to be instantiated in, say, two different water molecules, etc. However the trope metaphysician must say that the trope of structure of *this* water molecule and the trope structure of *that* water molecule *resemble* one another, or *exactly resemble* on another; but they cannot be the same. It is not part of the task here to go into how a 'trope + resemblance' metaphysics is to be compared with a metaphysics of universals. All that will be done here is to indicate the differences between these two metaphysical positions at certain crucial points as we attempt to come to terms with the various kinds of scientific realism that engage with issues to do with structure.[12]

In the light of the above, we need to take into account within scientific realism not just particulars such as objects and events which are either observable of unobservable, but also properties and tropes which are both observable and unobservable, or have their locus in observable or unobservable objects or events. And these tropes may be tropes that involve just a single one-place property, such as the whiteness of the cup, or the charge of an electron. Or the tropes might be structural involving a whole set of relations and properties instantiated in a whole range of objects. In fact the entire cosmos, as well as containing objects and events also contains tropes, from property to relational and structural tropes. Even the very distribution of the matter in the cosmos is one trope of the spatio-temporal structure of stuff at a time.

[12] For Armstrong ((1989), chapter 6) in assessing the best metaphysical theory, an ontology of Armstrongian universals only just wins out; a close second is an ontology of tropes between which there is a resemblance relation as the preferred systematic metaphysics.

2. SOME CLAIMS OF STRUCTURAL REALISM

2.1 Tropes and Universals of Structure

Section 1 set out the case for scientific realism combined with a liberal metaphysics about what ontological categories might be countenanced. In this section we will mainly focus on the specific claims of structural realism with an eye to what metaphysical account can be given of the structures they propose. Here tropes of structure have an important place as one ontological category scientific realists of even a mild structuralist persuasion could adopt. But this is not enough to satisfy many structural realists; they want structures not just as tropes, but also of higher order tropes, or of universals that may or may not be instantiated by particulars. Mathematical equations play an important role in structural realism; they are either the very structures themselves or are expressions of structure. This requires that we keep in mind the distinction between, on the one hand, the linguistic expressions, or the functional formulae, or laws, we use to express purported relations between items and, on the other hand, their ontological counterparts, worldly structures of tropes of various orders, or of universals.

To begin, let us consider the simple trope schema: $< U, < a, b, c, \ldots, n >>$. The a, b, …, n are particulars which, so far, have been taken to be objects or events. But can they also be tropes? There is no reason why not. In the above let us replace each of the a, b, …, n by tropes $t_1, t_2, \ldots, t_n$. We will call these 'lower order tropes' and they will each contain some lower order universal instantiated by, say, k items; such lower order tropes can be expressed as: $t_i = < U_i, <a, b, c, \ldots, k >>$. These tropes are then instantiated in a universal of a relatively 'higher-order' that we can indicate in bold underlined by $\underline{\mathbf{U}}$. Thus higher order universals which are instantiated by tropes (containing universals of lower order) can be express as: $< \underline{\mathbf{U}}, < t_1, t_2, \ldots, t_n >>$.

A simple example of this can be illustrated in the case of particular causation, if we accept the view of trope metaphysicians that tropes stand in causal relations. As examples of lower order tropes which are cause and effect, consider the trope t_1 = the sphericity of the ball, and the trope t_2 = the rolling of the ball. Then these two tropes can be the particulars that stand in a higher order relational trope that has as a constituent, causation, as the higher order relational universal. This is evident in claims of the following sort: the sphericity of this ball causes it's rolling. This we can express as follows: $< \underline{\mathbf{C}}, < t_1, t_2 >>$, where $\underline{\mathbf{C}}$ is a higher order relational universal of causation. Though not standard English, we can understand this to express the trope: the causing of the ball's rolling by the ball's sphericity.

Tropes are quite particular and do not admit of variation. Thus the greenness of this pea is different from the greenness of that pea, even though the greens may exactly resemble one another. Moreover such tropes are qualitative rather than quantitative. We also need to consider tropes that are quantitative in that they have magnitudes, or numerical values, on some scale of units. An important issue about quantitative tropes arises in the following way through a consideration of the identity conditions of tropes. The trope of the temperature, or the trope of the momentum, of a given body at a time are two examples of particular tropes; and the tropes of the temperature, or momentum, of another body either at the same time or at another time, are two further different tropes. The first and the third, or the second and the forth, of these tropes might exactly resemble one another in respect of having the same magnitude as one another. What makes these different tropes is that the locus of the two tropes is in two different bodies. But what do we say about, for example, the trope of the temperature of a given body at a time, which has one magnitude, and the trope of the temperature of the very same body at another time when this trope has a different magnitude? Are these the same or different tropes? Here we need to turn to trope metaphysicians and ask them what are the identity conditions for tropes (not always a straightforward matter).

Considering the one body, we can talk of the trope of the temperature of the body. So, on the face of it, it looks as if there is just one trope. But the magnitude of the temperature can be different at different times. In the light of the strong particularity of tropes it looks as if there are two tropes to consider and not one, simply because of the difference in magnitude. This answer is reinforced when we consider causal affects. If an alleged single trope can have different magnitudes of temperature, then depending on the magnitude quite different causal effects can arise. A metallic device in the thermostat of a heater with low magnitude of temperature may not turn the heater on, but one with a much higher magnitude of temperature does turn the heater on. These considerations suggest that there are two tropes and not one. So, it will be assumed that for tropes that admit quantitative variation, where they differ in their magnitudes at different times, there are different tropes.

A consequence is that, since there can be continuous variation in the magnitude of temperature, there are an uncountably infinite number of tropes of temperature for a given body. Thus for properties like temperature, or momentum, which are quantitative and admit of continuous variation in their quantity, a trope needs to be represented as < U, a, t >, where U is the universal, a is a particular body, and t is a given length, or instant, of time. The best we can say is that, for one trope of temperature of item a at a time t, and of a given magnitude, and another trope of temperature of a at another time t′ with a different magnitude, there is some relation of resemblance which holds between them (they are both temperatures), but one that falls short of exact resemblance in all respects (because of difference in magnitudes).

One might baulk at such a large number of particulars to be countenanced by trope theorists. But the baulking is due to the urgings of a systematic metaphysics unfriendly to tropes. Those who advocate universals may wish to abstract the object a and the time t from the trope and separate out the universal U, which in the example above is temperature. But the friends of universals also have to admit a large number of universals of various orders when quantitative matters come to the fore. Thus there will be the large number of objects that instantiate the universal having temperature of T° Centigrade, as opposed to those objects having temperature of T*° Centigrade, and so on for other amounts of temperature. But all of these will be instances of the universal *temperature* regardless of the magnitude of the units. Resolving such problems for rival systematic theories of metaphysics need not concern us further.

Granted the above discussion, we are now well on the way to regarding laws of nature (considered ontologically and not as linguistic expressions) as structural features of the world. Such laws are not tropes of structure but universals of structure. There is an already made theory of laws of nature, such as that in Armstrong (1983, Part B), in which laws are relations of necessitation between universals, which does capture much that structural realists wish to say about the structures that they understand realistically. But perhaps they have trope-ist inclinations (since it is often unclear whether they posit tropes or universals of structure). Which of these they adopt becomes a matter to be relegated to systematic metaphysics for resolution. In the light of this it is necessary to say more about the difference between the realist conception of laws from law formulations.

2.2 Law Formulations

Consider the historical sequence of equations that have been proposed for gasses. There is the well-known Boyle-Charles Law relating properties of the gas such as pressure P, volume V and temperature T; its common linguistic expression is the mathematical function 'PV = kT'. But it was well known that this does not capture the actual relations between P, V and T of a real gas. A sequence of gas laws were proposed which captured better the actual behaviour of gases, thereby underlining the common metaphysical assumption that in reality there is some relation that P, V and T do stand, whatever that be.

The following sets out a sequence of five of the proposed laws (see Bromberg 1980, sections 2.8 to 2. 14) that govern a gas:

Boyle-Charles: $PV = RT$

However this was known to be accurate for gases only a long way from their liquefaction point; close to it, it is quite inaccurate. The following are attempts to improve on the law by taking into account aspects of the molecules contained in the gas.

Van der Waals: $(P + a/V^2)(V-b) = RT$

This removes the assumption that the molecules take up no room in the gas container (the factor b), and takes into account the attractive forces between the molecules. (Note that this is really a cubic equation in V.) However other equations are based on models of the molecules that are quite different from those that lead to the above two equations.[13] They result in further equations such as:

Dieterci: $P(V - b) = RT(exp\text{-}a/VRT)$

Bertholet: $(P + a/T\ V2)(V - b) + RT$

[13] For a discussion of the different models and equations see Morrison (2000) pp. 47-52.

Virial Equation: $PV/RT = 1 + B/V + C/V2 + D/V3 + \ldots$

(where A, B, etc are functions of T for which further mathematical equations are to be given). [Note: the term 'virial' is Clausius' term, derived from *vis* for force, which has to do with the stresses due to inter-molecular attraction, repulsion and impact.]

The latter equations attempt to deal with the assumption that the molecules are inelastic and to allow for elasticity in inter-molecular collisions. Yet other equations not mentioned attempt to take into account yet other factors such as the charges on the molecules, effects due to quantum considerations, and so on. The sequence of laws illustrates a version of the Correspondence Principle which says: each law at the higher level will, when limiting assumptions are introduced, entail the law at the next lowest level. Thus the Van der Waals equation entails the Boyle-Charles Law when a gas is well away from its liquefaction point so the b is so small it can be neglected, and the attractive forces between the molecules, given by the factor a/V^2, are also negligible.

The above sequence of law formulations are attempts to give linguistic and mathematical expression to the real relations of P, V and T that hold of some real gas. And they graphically illustrate the ways in which we have changed and improved the mathematically expressed functions that have been proposed. These are intended, in the long run, to capture the real relations we suppose hold between P, V and T of a gas. This is the ontological characterisation of laws of nature. At any one time t there will be a real structural higher order trope relating the lower order tropes the pressure of the gas, the temperature of the gas and the volume of the gas. And the magnitudes of these tropes also stand in some mathematical relation. How, for the liberal metaphysician, do such tropes lead us to laws of nature, considered ontologically? This is the topic of the next sub-section.

2.3 Laws as Part of the Furniture of the World and Structuralism

Continuing the example of a gas G with the properties of temperature T, volume V and pressure P, G is the locus of at least the following three tropes (indicated using the notation to express tropes, where the first expression indicates a universal, G is the object the gas, and the last is a time):

The trope of the temperature of G at time $t_1 = < \text{TEMP.}, G, t_1 >$

The trope of the volume of G at time $t_1 = < \text{VOL.}, G, t_1 >$

The trope of the pressure of G at time $t_1 = < \text{PRES.}, G, t_1 >$.

Each of these tropes has a magnitude measured in some units. Thus we may now say, where a, b, c, etc are real numbers:

Magnitude of the trope $< \text{TEMP.}, G, t_1 >$ in degrees Kelvin = a

Magnitude of the trope < VOL., G, t_1 > in cubic centimetres = b

Magnitude of the trope < PRES., G, t_1 > in Pascals = c.

Now the world is such that the magnitudes of these three tropes cannot find their locus in a gas with any arbitrary value assigned to them; they do, as a brute matter of fact, come in a mathematically linked way. If we assume (falsely) that the ontological gas law is expressed by the Boyle Charles Law then the relationship will be, for the above three tropes: cb = ka. In contrast, the body might also have a fourth trope, that of being coloured in some way. As best as we can tell, it does not matter what colour trope the gas has, or whether it changes colour or not; this is a matter independent of the tropes of temperature, pressure and volume at a time. (Such a claim might be false of some gases in that colour might vary with, say, temperature; but the above claim is true at least of the atmosphere as the given gas, when at common temperatures.)

Granted that the magnitudes of the tropes must be related in a certain way, what can we say of the relationships between the tropes themselves? It is a short step to move from the claim that the magnitudes stand in a certain mathematical relation to say that the tropes of which they are magnitudes must also stand in the same kind of relation. This indicates that tropes can also stand in some relation of structure. That the magnitudes must stand in a certain mathematical structural relations is what underpins the claim that their associated tropes must also stand in related real structural relations. And that real structure is (let us suppose falsely again) that indicated by the Boyle-Charles Law. What this shows is that not only are there tropes of properties, relations and structure, but also that these very tropes themselves can be items in further higher order structural relations. What might these further structural relations be?

As indicated above, there is a certain mathematical structure in which stand the values of the magnitudes of the tropes of temperature, pressure and volume of the gas at a time. This we have assumed for the sake of illustration to be: cb = ka. Corresponding to this mathematical structural relationship between the magnitudes will also be a similar structural relation in which the tropes of these magnitudes stand. If we denote the mathematical structure by **MS**, then we can say that the three tropes of temperature, volume and pressure, also stand as lower order tropes in a higher order trope, thus: **MS** << TEMP., G, t_1 >, < VOL., G, t_1 >, < PRES., G, t_1 >>.

What does this last expression indicate? We could bundle together all the tropes of a given gas at a time t_1, including its physical tropes, its colour tropes, and so on. Some of these tropes are independent of one another; yet others stand in a further relation to one another specified by mathematical structural relationships. Such is the case for the tropes of temperature, pressures and volume of the gas at t_1 (but not the colour tropes). And it is this specific structural relation that the expression above indicates. Note also that the expression indicates only the higher order mathematical structural trope at a single time t_1. There are a host of other times, t_2, t_3, t_4, etc, at each of which a different triple of lower order tropes will prevail. These triples will be the constituents of a different higher-order mathematical structural trope, but with the same universal **MS**. Similarly, one can also consider a range of different gases G_j. Thus a general form for the many tropes prevailing for any gas at any time would be: **MS** << TEMP., G_j, t_i >, < VOL., G_j, t_i >, < PRES., G_j, t_i >> (where for each different gas G_j, t_i stands for a range of times).

The next step turns on the particular systematic metaphysics one adopts. If one is a systematic 'trope + resemblance' metaphysician, then one is committed to a particularism with respect to tropes that eschews universals. So one stays with the set of the large number of such higher order tropes (in which for each fixed j, i varies over times). One also admits that there is a strong resemblance relation between them indicated by the common **MS** of mathematical structure; but one does not reify this structure. In contrast, the systematic metaphysician who advocates universals does just this. They claim that the mathematical structure **MS** is a universal; and it is the very same universal, the very same mathematical structure, which is in each of the set of tropes (in which for each fixed gas G_j, i varies over times).

The above is a piece of metaphysical speculation on behalf of structural realists about what are the kinds of structure to which they appeal. It is a metaphysics that is liberal as to what are the particulars it endorses, such as ordinary or microscopic objects, events and tropes. And it allows that any of these particulars instantiate universals. It allows structures that are particular items, such as tropes of structure found in, say, each DNA molecule. And it allows law structures as general features of the world (however this is to be understood). Structural realists also make much of mathematical structures. Like all realists they postulate that there is an independent world about which science tells us something. But they accept (wrongly as will be argued in section 5) that there is no continuity in reference to objects with theory change; so realism is in trouble. What they allege rescues realism from oblivion is continuity in mathematical equations. But it is unclear what this might mean. If it is continuity in the formulation of mathematical equations and laws, then in some cases, as the previous section 2.2 argued, there is no such continuity to be found, unless some version of the Correspondence Principle is invoked to guarantee continuity thereby weakening the notion of continuity. Is there continuity in laws of nature considered ontologically? This is an odd question to ask. It is not as if there could be continuity or discontinuity in such structures; they either exist out there in the world, or they do not exist. If there are really laws of nature, and the world is structured in some way, then the supposition by realists, and structural realists in particular, is correct. Our law formulations will capture these structures with varying degrees of approximation.[14]

2.4 Epistemic and Ontological Structuralism

From the stance of a liberal metaphysics and epistemology for scientific realism, there is no obvious reason why we cannot obtain knowledge of both the particulars,

[14] The above account of laws will have to be modified if one thinks that some law formulations are idealisations, or others have complex ceteris paribus clauses. However the gas laws are apt for the point being made of the relationship between tropes and their instantiation of laws.

or kinds of particulars, that stand in structures, and the law structures themselves. In the table[15] below of possible epistemic theses concerning structuralism, this position is represented in the first line; it is the most initially plausible position to adopt. The second line allows for the odd position that we can only know particulars but not the relations in which they stand; this possibility need not detain us. The third line is the position of Epistemic Structural Realism (ESR) which claims that we cannot obtain knowledge of (kinds of) particulars, especially (kinds of) objects, which populate the world. ESR does not deny that such particulars exist; rather it remains agnostic about them and their properties. The only knowledge we can obtain is of the relations between such particulars. The classic expression of this position comes from Poincaré who says in respect of our theories of light that at best we use 'merely names of the images we substituted for the real objects which Nature will hide for ever from our eyes. The true relations between these real objects are the only reality we can attain' (Poincaré (1952), p. 161). In sections 4 and 5 some reasons for rejecting such claims will be given. The fourth position of all-round scepticism need not detain us. The table shows that both realists and structural realists can agree that we do have knowledge of the mathematical and law-like structures of the world; they disagree over knowledge of particulars. They can also differ over the systematic metaphysics in which such claims are to be embedded.

The bottom half of the diagram lists possible ontological positions with respect to particulars and the structures in which they might stand. The first is the liberal position in which there are both particulars (of whatever category) and structures, which may be a commitment to tropes of structure only, or to the more fully-fledged existence of relations between universals. In the later case, the liberal position allows that the particulars instantiate the universals, and so is consistent with a more Aristotelian rather than Platonic view of universals (in which there are no uninstantiated universals). As the second line suggests realists can be nominalists who accept some category of particulars but who eschew any commitment to universals. Many realists are also nominalists who eschew all talk of structures. The last possibility, which seems not to admit anything, would have no serious adherents. What is of interest is the third possibility that appears to endorse a strong Platonism about structuralism which we may call Platonistic Ontological Structural Realism (POSR). This is a position in which, because of the absence of any instantiating particulars, the universals of structure exist uninstantiated (i.e., a strongly Platonic and non-Aristotelian view of universals is required). Several structural realists gravitate in various ways towards this position. Granted this position epistemic structural realism follows since there are no particulars to know.

[15] The elements of the table are suggested by a number of writers on structuralism, including its critics such as Psillos (2001) where some of the entries arise in section 3 'The Downward Path' of his paper.

		Particulars (objects, (events, processes)	Structure (tropes or universals)
Epistemology	Liberal	Yes	Yes
		Yes	No
(ESR)	Epistemic Structuralism	No	Yes
	Extreme sceptic	No	No
Ontology	Liberal	Yes	Yes
	Nominalism	Yes	No
(POSR)	Platonistic Structuralism	No	Yes
	(weird 'nothing-ism')	No	No

There is a well-known dispute in traditional systematic metaphysics that bears on POSR that can be put briefly in the following way. Objects can be thought of as substances that exist in their own right, and in which properties are instantiated in some way. Such a dualistic ontology of substance and property is to be contrasted with a monism that declares that objects (or substances) are nothing but bundles of properties (or universals). An issue, not canvassed here, is just how the properties clump together in a space-time region. Systematic trope theorists also eschew substances; these are to be thought of as bundles of tropes (there is also an issue of how and why they clump together). In either case the traditional category of objects (or substances) has been eliminated since the work this category does can allegedly be done by the items in some other category (universals, or tropes). Just how well the work can be done is a dispute in systematic metaphysics into which we need not enter (for one account see Armstrong (1989), chapters 4 and 6).

In eschewing objects, do advocates of POSR intend that they be reconstrued along the lines of bundles of universals, or bundles of tropes? What is clear is that in rejecting objects they overlook other categories of particulars, such as events or tropes, which can also instantiate universals of structure. Not to admit that structures are to be instantiated in some particulars that inhabit our actual world (whether objects, events, or tropes,) is to, in effect, reject the actual world altogether and adopt an extreme form of Platonism in which there are just universals of structure, but there are no particular relata in which these relations are instantiated. Structure is all there is to reality; and the structures remain unanchored in any actual world particulars. Such a Platonist account of the laws of nature runs contrary to the highly plausible idea that there is a very evident world, and that it obeys such laws.

What are some of the reasons for adopting a version of POSR, or ESR, that plays down the role of particulars such as objects, or our knowledge of them? Two reasons are discussed, the first in section 4 which deals with the pessimistic meta-induction, the second in section 5 which deals with claims about lack of ontological continuity at the level of (kinds of) objects. Other reasons can only be mentioned here briefly. The distinction between ESR and POSR, which appears in Ladyman (1998), is accompanied by several other considerations in favour of some version of POSR, only two of which will be briefly mentioned.[16]

The first consideration is based in a semantic, as opposed to a syntactic, account of theories, especially the model-theoretic account of van Fraassen and Giere. But these two accounts differ in that for van Fraassen all one can model are the phenomena; the mathematical models used are judged to the extent that they fit the phenomena, and not how they fit some unobserved reality. Though a case can be made for an appeal to certain structures in such models that go beyond modelling the phenomena (see van Fraassen, forthcoming), they need not be those that would be pleasing to the scientific realist. Van Fraassen's 'constructive empiricism' is sceptical of our ability to ever know whether such models fit the unobservable, including not only theoretical objects but also the very laws to which advocates of structural realism wish to appeal.

In contrast Giere is a 'constructive realist' who does require that there be some relation of *fit* holding between models and reality, whether observed or unobserved. Importantly the relations of fit can hold between the 'objects' of the model and the real objects being modelled. As he says of his theoretical models: 'these are abstract objects, imaginary entities whose structure might or might not be similar to aspects of objects and processes in the real world' (Giere 1999, p. 5). Thus for Giere there are mind-independent unobservable objects in the world which we hope to represent in our models. In contrast for Giere the appeal to laws in science is a suspect notion that has its origin in theology; he hopes to present a 'portrait of science that captures our everyday understanding of success without invoking laws of nature understood as true universal generalisations' (Giere 1999, p. 24). Setting aside the conception of laws just mentioned, it is clear that the idea of laws as structural features of the world is to be downplayed. This is not good news for the advocate of POSR. There does not appear to be any secure inference from a model-theoretic account of science (either that of Giere or van Fraassen) to POSR with its central claim that there are structures but no objects. For van Fraassen we are to remain sceptical of both (unobservable) objects and structures, but not about structures understood as the (mathematical) models we construct. For Giere objects play a central role in what is being modelled as well as what does the modelling. Moreover it appears that we can model both the objects and structures of reality (understood as tropes). But an account of science that insists on objects while downplaying any role for laws, hardly give succour to a strong version of structuralism.

A second consideration (see Ladyman 1998, section 3.2, and French and Ladyman, 2003) arises from the ontology that appears to be adopted in some understandings of quantum mechanics. In our ordinary everyday discourse, and also in classical physics, not to mention the metaphysical positions that philosophers have elaborated to accompany these, it is assumed that there are objects with continuing identity and individuality. Thus even if there are different permutations of objects within the same structure it is recognised that these are distinct

[16] As well as Ladyman (1998), there are other papers in which some version of POSR is advocated; see French (1999) and French and Ladyman (2003). Critics of this position include Psillos (2001) and Chakravartty (2003).

arrangements: the same object is at one time here—later there. This is not the case for some understandings of QM; such permutations are not recognised as different. This raises matters that cannot be explored in this paper. But it does raise the question that, even if we do abandon the idea of an object with substantive identity conditions, do we thereby have to give support to POSR? Even if we were to abandon the idea of an object with substantive identity conditions, it does not follow that we would have thereby abandoned an appeal to any particulars whatever, such as events, processes and tropes as items in which structures can be realised. To be left with abstract structures (however understood) with empty placeholders that some might particulars might have filled but in fact do not, is to have lost a grip on the actual world of particulars altogether (as opposed to a quite Platonistic world).

3. THE PROBLEM OF THE SUCCESSION OF LAWS AND THEORIES FOR STRUCTURALISM

When one theory succeeds another in a given science there are a number of important issues to consider from a philosophical point of view. One is methodological and asks if the succession is rational, and if so what methodological principle(s) legitimates the succession. A second issue is ontological and concerns the variance or invariance of items postulated in theories under theory change. A third issue concerns the nature of the models that accompany successive theories and whether successive models can be consistently embedded in one another, or not. A fourth issue concerns the adequacy of the formulation of laws in the language of a given theory. When one considers how laws succeed one another with theory change, all four issues can come to the fore. These issues are important for the versions of structural realism considered here since their core position is that given the (alleged) variance of objects postulated in theories with theory change, something else should remain invariant if realism is to be a viable position at all. And this something else is laws. The versions of structural realism considered here tend to assume that in theory succession the law formulations are invariant under theory succession. They are usually mathematically formulated, and so there is an alleged preservation of hypothesised mathematical structures; and successive theories have the same (family of) models.

But there are a range of cases to consider, not all of which support the structuralists' position. The most promising examples are those in which exactly the same law in one theory is shown to be a strict logical consequence of a law in the successive theory (and if there are different models for the two theories, one is simply a sub-model of the other). The commonly cited example is that of the preservation of Fresnel's laws of optics in Maxwell's theory; Fresnel's laws are an exact consequence of Maxwell's more general laws. (This is discussed further in section 5.)

More often the succeeding laws are inconsistent with one another , but later laws improve upon and correct preceding laws. An example has been given in section 2.2 of the succession of gas laws from Boyle-Charles Law to the Virial Equation in which with each successive theory there is an improvement of the law. (There is also the issue of many models; see Morrison (2000), section 2.3.2.) Other examples can readily be supplied, such as the succession of Galilean free fall laws, or Kepler's planetary motion laws, by Newton's laws, or the succession of Newtonian laws by laws from Special and General Relativity, and so on. In these cases the succession relation has to take into account corrections, additions of missing parameters, convergence under special assumptions, and so on. So at best, there is only succession of the same law if one assumes some version of the Correspondence Principle (see section 2.2) that permits only a very weak condition that hardly counts as the sameness of laws in two successive theories.

In the face of the predominance of the latter sort of succession, it looks as if invariance of law formulations, while not non-existent, is not common. The structural realists' attempt to seek continuity in laws will only apply in some cases and not others that may also be alleged to raise problems for scientific realism because of lack of continuity in objects. If one turns to models that (sets of) laws are often said to define, then there will not always be an isomorphism between the models, though there can be different degrees of similarity between them. Note that model-model similarity relations are quite different from model-reality relations. If we invoke reality, then we can inquire of the similarity relations that hold between successive models and reality and discover whether there is an improvement in degree of fit or similarity between models and reality. Reverting to the linguistic formulation for laws, we could also speak of increasing verisimilitude of successive laws. Moreover realists might want to explain the increased success of the succeeding laws by appeal to their increased verisimilitude. That is, the explanation of the success of the succeeding laws is in terms of a more generalised account of the truth of the laws. This is simply a special case of the kind of explanation that realists have often given of the success of laws and theories; they appeal to their truth or verisimilitude, using inference to the best explanation to explain that success.

In contrast advocates of constructive empiricism, while accepting the idea of succession, reject the realist's way of explaining success. Thus van Fraassen adopts the following methodological prescription: 'Requirement on Succession: The new theory is so related to the old that we can explain the empirical success of the old theory if we accept the new.' (van Fraassen, forthcoming). He sees this as one of our principles of scientific method that imposes a constraint on the acceptance of new theories. As such it also applies across scientific revolutions.[17] What van Fraassen objects to is any explanation of the sort given at the end of the previous paragraph which invokes inference to the best explanation and the truth, or verisimilitude, of our theories.

Instead, what he advocates is something like the following. If the succeeded theory is T, with empirical success es(T), and T′ is the succeeding theory with empirical success es(T′) such that es(T′) > es(T), then what T′ can do is explain, not T itself, but rather why T had the empirical success es(T) it did. Though van Fraassen does not say it, T can also explain the lack of empirical success, or the

[17] This can be challenged as a viable methodological principle by those who are persuaded by Kuhn and Feyerabend that there is often empirical loss with theory succession.

empirical failures of T, as revealed by the greater success of T′; and it does this through revealing the limited assumptions under which T got the empirical success it managed to have. Note the emphasis on empirical success. For van Fraassen there can be quite radical changes at the level of theory, particularly the models adopted, in which there is increased empirical success when T′ succeeds T. The requirement of succession is not so conservative as to further restrict succession at the deeper level of the non-phenomenal aspects of models.

The upshot of the above for constructive empiricists, is that the Requirement on Succession operates only at the level at which empirical success is determined. It does not operate at the level of the unobservable postulates of a theory; these can vary quite radically so that there is no discernible continuity at this level in the succession. Putting this in terms of models, successive models might not have a high degree of overall fit at all. Where the successive models do fit is in some sub-model that models not the whole of reality but only the observable phenomena.

The constructive empiricist accepts only the idea that a succeeding theory T′ can explain the *empirical success* of a succeeded theory T. What they reject is the kind of explanation of success that realists usually give of why the succeeding theory T′ is as successful as it is, and also in explaining the empirical success of the theory T that preceded it. Realists hanker after explanations which appeal to the truth, or verisimilitude, of theories across the board. But constructive empiricists see this as part of the hubris of realists. Such arguments are not evaluated here. Rather the task is to see what bearing the successful succession of theories and laws has on structural realism. What we have shown here is that the Requirement on Succession, something that realists can adopt as well as constructive empiricists, really counts against structural realism since it puts emphasis on empirical success only. No room is allowed for the invariance of law formulations (i.e., structures in the non-phenomenal part of models) that express the same mathematical structures under conditions of theory succession. Constructive empiricists do eschew invariance at the theoretical level.

Though invariance of laws, and thus of formulations of structure, can be a feature of some theory successions, it is not a feature of all of them. As van Fraassen points out, there can be quite radical theory change while a modest version of the Requirement on Succession still holds. So there is room for a variety of realism that does not put too much emphasis on the preservation of structure, but which will opportunistically take advantage of it when it occurs. Here there can be some agreement between ordinary scientific realists and constructive empiricists but not with structural realists. Where scientific realists (either of the ordinary sort or structural realists) and constructive empiricists part company is that realists still think that they can get a handle on the world and use the theory-world connections to explain the success of our theories; but this is just what the constructive empiricist denies one can do.

4. WHO'S AFRAID OF THE BIG BAD PESSIMISTIC META-INDUCTION?

For scientific realism to be a viable doctrine, there had better be some continuities in what our theories postulate despite the fact that they are constantly undergoing renovation allegedly revealing deep discontinuities. But there is an argument that shows that with discontinuity of theory there is also discontinuity in the (kind of) objects postulated in our theories. So realists had better find something other than objects if there is to be continuity. What saves the day for scientific realism, say the structural realists, is continuity of structures. The argument against continuity in (kinds of) objects is the 'pessimistic meta-induction' (PMI). This structural realists support; for some it is their main reason for being structural realists. PMI comes in a number of different forms. The first to be discussed is that of Putnam (who perhaps does not endorse the argument given the context in which he presents it); the second that of Laudan. It will be argued that PMI should not be accepted.

4.1 The Putnam Version of the Pessimistic Meta-Induction

Putnam expresses one form of PMI in the context of discussing Kuhnian and Feyerabendian incommensurability, and illustrates it with the (alleged) example of the early Bohr-Rutherford theory of the electron compared with the later Bohr's theory of the electron. Is it the same item, the electron, that is being referred to in the early theory as in the later theory (i.e., there is referential invariance with theory change)? Or are there two different objects, the Bohr-Rutherford electron and the mature-Bohr electron (i.e., there is referential variance with theory change)?[18] Putnam puts the issue as follows:

> What if all the theoretical entities postulated by one generation (molecules, genes, etc. as well as electrons) invariably 'don't exist' from the standpoint of later science?' … One reason this is a serious worry is that eventually the following meta-induction becomes overwhelmingly compelling: *just as no term used in science of more than fifty* (or whatever) *years ago referred, so it will turn out that no term used now* (except maybe some observation terms) *refers*. (Putnam 1978, pp. 24 -5).

PMI as an induction over the sciences seems to have the following premise:

> For any time t in the past to t + 50 years later, and for all scientific theories T entertained at t, each of the T at t were discovered, by the end of t + 50, to have theoretical terms all of which failed to refer (i.e., the items in the theory's ontology do not exist);

On the basis of this, one can make an inductive prediction concerning the next case of our current theories:

[18] For an argument that there is no referential change with change in theories from those of Bohr-Rutherford to the later Bohr, see Norton (2000). Whether or not Putnam takes seriously the meta-induction he locates in Kuhn's stance, need not detain us. Putnam makes the point, emphasised later in the section, that it would be a desideratum of any theory of reference that the argument to such massive reference failure be blocked.

> The terms of the scientific theories we currently hold at t = now, will at t + 50 be shown not to refer.

Rather more disturbing would be the inference to the inductive generalisation:

> For all theories at any time t, at t + 50 their terms will be shown not to refer; i.e., all our scientific theories are not about anything at all!

How does PMI challenge scientific realism? Take the example given above in which it is alleged: the term 'electron' does not refer. This is clearly equivalent to: electrons do not exist. In section 1.2 scientific realism was defined via a list in which was included the claim 'electrons exist mind-independently'. Not both of 'electrons do not exist' and 'electrons exist mind-independently' can be true. The same contradiction can be generated for all the other unobservable objects listed in the definition of scientific realism. So if one accepts PMI, then scientific realism is false. PMI is a direct challenge to the very definition of ontological scientific realism.

PMI is to be distinguished from a general kind of philosophical scepticism. Though the conclusion of PMI is close to that of a philosophically based scepticism about whether our theories are ever about anything, the kind of argument given, and the considerations invoked based in the history of science, are not the usual sort found in premises for arguments about philosophical scepticism.

The expression of PMI involves semantic assent and says of some term thought to refer at one time that it will be discovered not to refer at a later time. If we semantically descend, then when formerly it was believed that there existed items such as celestial spheres, epicycles and deferents, impetus, phlogiston, caloric, electromagnetic ether, etc, it is now believed that these do not exist. A realist would have to agree, if they accept the received wisdom of historical studies in science, that such entities do not exist. But the non-existence of these entities is something that can be accommodated in the definition of scientific realism with its claim that *most*, but not all, the entities they list do exist mind-independently. What realists need to resist is the claim that none of the items they currently think exist do in fact exist. So, is the inductive inference strong or weak, or even fallacious? Second, is the premise in which it is based correct?

As expressed, the inductive inference (either with a prediction or a generalisation as conclusion) does appear to be strong. So, is the premise true? If one took a proper random statistical sample of theories over a 50 year period, it would appear, on a cursory investigation, that the frequency of cases in which there was no referential loss at the end of a 50 year period would be far greater than those in which there was referential loss. Proceeding differently, it would be open to vary the length of the period, rather than adopt a 50-year period, and take a proper random sample from different lengths of time period, from a few years to centuries. (Note that it took about 1500 years for the theory of epicycles and deferents to be abandoned, assuming that astronomers who used them actually had a realist view of them and did not claim, right from the start, they were only items in a model and were non-existent.). Sampling over varying time periods might hardly alter the verdict just given on the premise of PMI. Thus it would appear that as a generalisation the premise is false; converting it to a statistical claim would only give us a low probability of referential loss. To illustrate, consider the table of elements developed over the last 250 years. Apart from a few classic cases, such as that of phlogiston, there have been a very large number of cases in which an element postulated at an earlier time is still an existent.

Perhaps the premise of PMI can be better expressed as follows with the period of time being left open rather than fixed at 50 years:

> For any time t in the past, and for all scientific theories T entertained at t, there exists some n such that $n \geq 50$ years, for each T at t thought to contain referring expressions, then for T at $t + n$ it is discovered that theoretical terms of T fail to refer (i.e., the items in the theory's ontology do not exist).

Putting the matter this way leaves it open as to when any theory T is found to fail to have referring terms, though in the long run fail it must. There are two problems with this formulation. First it is unfalsifiable in that if some T is not shown to have its terms fail to refer today, then there is always tomorrow or the next day, and so on, at which its terms will have allegedly been shown to have failed to refer. Second, it follows from the last consideration that the very formulation presupposes that the terms of T will fail to refer at some time; but this is the very matter up for discussion and for which evidence is being sought. So the above version of the PMI premise is unacceptable; one has to revert to a time interval of either $t+50$ from which to take one's statistical sample; or one has to take an appropriate sample over $t+n$ (where n can vary but remains reasonably finite). Putting matters this way gives PMI a definite content that is open to test.

PMI has been countered by saying that, at best, the evidence from historical cases, based in proper statistical sampling, only gives its premise a low probability. But the situation is much more complex in a way which does not assist PMI. It assumes that there is a clear way in which we can say that a theoretical term appears to have its reference fixed, and then after some period in which there has been theory change, it is shown not to refer. But this depends very much upon semantic theories about how the meaning and reference of terms could be fixed that avoids some of the implausible aspects of massive amounts of Kuhnian and Feyerabendian incommensurability; the theories of meaning they adopted entail a rapid turn-over of terms with putative reference. Nor should the bare falsity of a theory be sufficient to declare its theoretical terms non-referring; there can be many a false theory about actual theoretical entities. This indicates that PMI cannot be taken at its face value. Lurking in some way in the background are presuppositions about how the reference of terms is to be fixed. Without any account of reference fixing, PMI offers very weak grounds on which to challenge scientific realism. Even supplemented with such an account it might still fail (see the next section).

Structural realists should not accept the verdict of such a version of PMI when it comes to doubts about the objects postulated by theories. For PMI can be applied just as well to the very laws formulations they think are invariant with theory change. Simply replace in PMI talk of the reference of terms in a theory, by talk of the correctness of law formulations. Then just as a law formulation was, at time t, thought to be correct, so at t + n it can be shown that the law formulation is not correct. In fact it can be argued that, on the whole, the renovation of laws over time proceeds at a pace even greater than that of the replacement of objects. So there are not the invariant laws to which structuralists can appeal to rescue realism. However there are also good grounds for rejecting this version of PMI as applied to laws. For one thing it fails to take into account things like the Correspondence Principle, or appeals to the increasing approximate truth of the laws proposed over time.

4.2 The Laudan Version of the Pessimistic Meta-Induction

A different version of PMI can be culled from Laudan's critique of scientific realism (Laudan 1981). Here we can be brief as the argument has been well formulated, and then criticised, by Peter Lewis (2001). Following Lewis (p. 373) the pessimistic argument is:

1. Assume the success of a theory is a reliable test for its truth.
2. Most current scientific theories are successful.
3. So, most current theories are true.
4. Then most past scientific theories are false, since they differ from current theories in significant ways.
5. Many of these false past theories were successful.
6. So, the success of a theory is not a reliable test for its truth.

The (valid) argument from (1) and (2) to (3), which is attributed to the realist, is to be undermined. (4) could be justified in two ways; either past theories say false things about items that do really exist, or past theories say false things about what exists. It is the latter which is intended here. (5) is Laudan's claim, based in historical investigations, that past theories which says false things about what exists can nevertheless be successful. From this (6) follows, which undermines the realists' assumption (1). The argument can be challenged by undermining premises (4) and (5). Thus it can be argued against (5) that success can come in degrees and that past false theories, while having some success, were not as successful as has been alleged. And against (4) it can be urged, on the basis of some account of reference, that it is wrong to claim that the theory is about some non-existents. Or it can be urged that even if (some? most? all?) terms of a theory do not refer then the theory can still obtain a respectable degree of approximate truth, i.e., have some success.[19]

Lewis' challenge is different; it is directed against the assumed correctness of the inference from (5) to (6). Even if we grant that false theories are successful we cannot infer that such success is not a reliable test for the truth; it can still be a reliable test. We can readily observe the success of a theory, but not its truth; so we

[19] Strategies of these sorts have been tried by various authors, amongst whom is Psillos (1996).

can attempt to use success as a test for truth. In providing such a test we can regard as unproblematic cases where a successful theory is true and an unsuccessful theory is false. What needs to be watched in any statistical examination are cases where tests can go wrong, especially the kind of error due to false positives in which a theory is successful but false (and another kind of error which would be due to false negatives in which a theory is unsuccessful but true – but this we can set aside). A reliable test is one in which both rates are low. Lewis' strategy for evaluating PMI turns on the need to avoid the *paradox of false positives* mentioned in many statistics textbooks. Here only a brief summary of Lewis' considerations will be given.

The crux of Lewis' argument can be illustrated in similar reasoning about diseases. Suppose some test is reliable, that is, the rate of false positives and false negatives is very low (say, 1 in 100). But also suppose that the condition is rather rare – only 1 in 10,000 people have it. That is, in a random sample of the population only 1 in 10,000 will be a true positive. But what of the false positives that could arise, i.e., a person does not have the disease but tests positive? This will be 1 in 100. So what is surprising is that for every true positive there should be about 100 false positives. But the test can still be reliable to the extent specified.

Turning now to truth and success, suppose that for the disease we substitute true theories, and for the population successful theories. We can allow, as Laudan does, that true (and so successful) theories are rather rare in the past and now, and that then and now they were far outweighed by false theories that are successful (Laudan suggests that the ratio might be as high as 1 in 6). But that the number of successful but false theories is well in excess of the number of successful and true theories, does not show that success is not a reliable test for truth (by which is meant that the error rate is low). So from the fact of the prevalence of false but successful theories, it cannot be concluded that success is not a reliable test for truth. This, in outline, is the strategy Lewis uses to subvert PMI as a form of statistical inference.

5. THE FLIGHT FROM OBJECTS TO STRUCTURE: WHY?

The previous section shows, in general, why PMI is incorrect with respect to the ontology of successive theories, particularly the (kinds of) objects they propose. Even though there are structures to countenance, the structuralists' flight from objects to structure is unwarranted (either in the form of ESR or POSR). This final section investigates a particular historical episode in which it is alleged that there is a change of (kind of) object with change of theory, e.g., the transition from the Fresnel to the Maxwell theories of the nature of light. It proposes an account of reference fixing which shows how there can be continuity in the (kind of) 'object', light, even in the transition from Fresnel's to Maxwell's theory.

5.1 The Flight From the 'Object' Light—a Little History

One episode in the 19th century history of light has been made familiar in Worrall's (1989) seminal paper on structuralism that illustrates a flight from objects to structure. We can readily admit that there are observable items such as beams of light. Newton experimented with them using a small chink in a wall through which beams of light passed into a darkened room. These beams can be manipulated by us as when we increase or decrease the size of the chink, reflect them off mirrors, interpose various screens along the path of a beam, etc. And they can be more easily seen in dusty rooms in which some of the beams get reflected off dust particles. The question arises: 'What is a beam, or ray, of light?' If this is a question about the very nature or essence of light, then perhaps we have no good answer even now. But at least we know some *properties* of beams, both intrinsic and extrinsic, and some of the laws of reflection, refraction, etc. that light obeys.[20] What is said by advocates of structural realism (see Worrall 1989, pp. 107-8) is that there is no agreed account of the fundamental nature, or even intrinsic properties, of light from theory to theory; but from theory to theory there is continuity of equations governing the properties of light.

To illustrate Worrall points out that in the 18th century it was commonly held, following Newton, that light was something akin to a shower of unobservable particles. Early in the next century this ontology was rejected in favour of the idea that light is certain kind of vibratory motion set up in an all-pervading elastic but mechanical medium, the 'luminiferous ether'. Such a view, proposed by Fesnel, was shortly rejected by Maxwell in favour of the idea that light is a series of 'wave-like changes in a disembodied electromagnetic field'; that is, light is the vibration of electric and magnetic field vectors. The ontologies are different: 'a mechanical vibration and an electric ('displacement') current are surely radically different sorts of thing' (Worrall 1989, p. 108). And in the 20th century there are the views that light exhibits a problematic particle-wave duality, or the view that it is a stream of photons, in both cases obeying a quite different mechanics from the 'shower of particles ontology' of 18th century optics (*loc. cit.*). Though it is alleged that there have been quite revolutionary changes at the level of the fundamental ontology of theories, there has been steady accumulation at the phenomenal level in which there is increasing law-like capture of the phenomena of light from reflection, refraction, interference and diffraction, polarisation, electric and magnetic effects, photoelectric effects, and so on. Even though there is no ontology of (kinds of) objects which remains invariant with theory change, there is invariance of equations; either they are retained as such or they are incorporated in more general equations. Such has been the view of a large number of writers as different as Poincaré, Russell on various occasions, Eddington and others (including even a structuralist interpretation of Wittgenstein's *Tractatus*).

The following brief historical episode, which illustrates the case of Fresnel on light, is taken from Worrall (1989). Consider a ray of light which is both reflected and refracted as it passes from one medium to another, say, air to glass. Of this Worrall says (p. 119):

[20] For any object the class of intrinsic properties are generally broader than the class of properties comprising natures, or essential properties. The notion of intrinsic being used here is that of a property which does not imply the existence of anything else and is compatible with loneliness. See Lewis (1999) chapters 5 and 6 on extrinsic and intrinsic properties.

> Ordinary unpolarised light can be analysed into two components: one polarized in the plane of incidence [the plane containing the incident, reflected, and refracted beams], the other polarized at right angles to it. Let I, R and X be the intensities of the components polarised in the plane of incidence of the incident, reflected and refracted beams respectively; while I′ , R′, and X′ are the components polarized at right angles to the plane of incidence. Finally, let i and r be the angles made by the incident and refracted beams with the normal to a plane, reflecting surface. Fresnel's equations state:
>
> $R/I = \tan(i-r)/\tan(i+r)$
>
> $R'/I' = \sin(i-r)/\sin(i+r)$
>
> $X/I = (2\sin r.\cos i)(\sin(i+r)\cos(i-r))$
>
> $X'/I' = 2\sin r.\cos i)/\sin(i+r)$

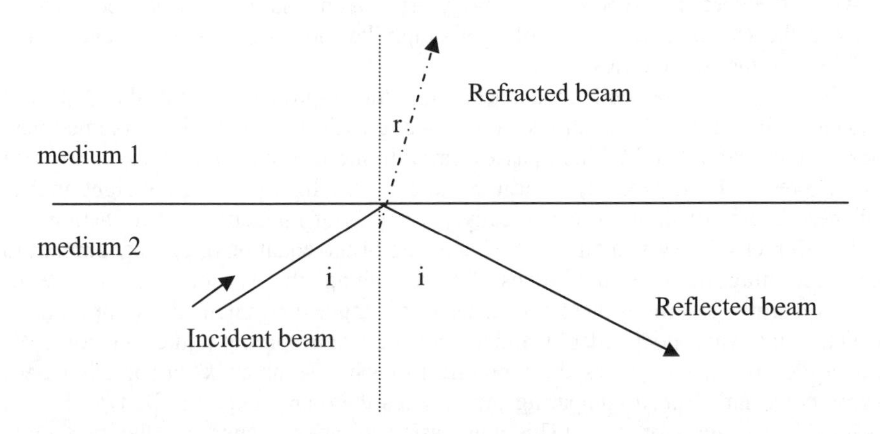

Let us simply refer for convenience to these four Fresnel equations as 'the F-equations' and the properties (often relational) of angles of reflection and refraction, intensities, etc, as 'the F-properties'. More explicitly, they are functions of the following 8 variables that can be written as 'F(R, R′, I, I′, X, X′, i, r)'. Also the intensities I, R and X are the square roots of the magnitudes of the amplitudes of the waves which constitute the vibrations. (If one draws a sinusoidal wavy curve, then the amplitude is the maximum displacement from the x-axis.) Though their magnitudes are related in this direct way, intensity and amplitude are distinct properties, and should be treated as such. The larger the vibration of whatever is vibrating (Fresnel thought it was the ether), the greater the intensity of the light. So what the equations above can be understood to represent are ratios of intensities, or ratios of amplitudes of whatever does the vibrating.

Worrall then continues telling us about the big difference between Fresnel and Maxwell over what does the vibrating:

> Fresnel developed these equations on the basis of the following picture of light. Light consists of vibrations transmitted through a mechanical medium. These vibrations occur at right angles to the direction of the transmission of light through the medium. ... From the vantage point of Maxwell's theory as eventually accepted this account, to repeat, is entirely wrong. How could it be anything else when there is no elastic ether to do any

> vibrating? Nonetheless from this vantage point, Fresnel's theory has exactly the right structure—it's 'just' that what vibrates according to Maxwell's theory, are the electric and magnetic field strengths. And if we in fact interpret I, R, X, etc. as the amplitudes of the 'vibration' of the relevant electric vectors, then Fresnel's equations are directly and fully entailed by Maxwell's theory (Worrall 1989, p. 119).

The 'something' that does the vibrating according to Fresnel is the mechanical medium; but the 'something' that does the vibrating, according to Maxwell, is quite different, the electric and magnetic field strengths expressed as vectors. But both are agreed that there is an intensity, and thus an amplitude, for the 'somethings' that do the vibrating. But the 'somethings' are said not to be the same but are different. And on the face of it this seems to be correct; an elastic mechanical medium is quite unlike a vibrating field vector. Both of these quite different 'somethings' lie at the heart of the quite different models of light advocated by Fresnel and Maxwell. This is where the deep differences on ontology arise. And it is this that has led some to say that the sequence of theories of light simply has no continuity at the level of the ontology of the 'somethings'.

However the Maxwell model contains the F-properties and the Maxwell equations entail the F-equations. What we are dealing with is a (sometimes) observable beam of light which passes through one medium to another, giving rise to reflected and refracted rays; and the same for the light polarised at right angles. Of these beams of light we can specify (1) angles of incidence and refraction, (2) ratios of intensities (given on the left-hand side of the equations), or what amount to the same thing, ratios of amplitudes of a 'something' that is vibrating, (3) a set of four equations governing these properties (viz., angles and intensities (amplitudes)) which cannot vary independently and are constrained according to the equations. We can leave talk of what does the vibrating in each case as an x, 'a something, we know not what'. But in comparing the theories there are alleged to be two distinct 'somethings', and not one. (This impression is strengthened by the two quite different models of light that were heuristically useful in developing the respective theories.) If there is any continuity, then it must be elsewhere. And we are in luck because there is mathematical continuity. There is sameness of mathematical structure from Fresnel's to Maxwell's equations. Structure to the rescue!

The subsequent focus is upon talk of the allegedly different 'somethings' which have the F-properties of amplitude (or intensity), and of angles of incidence and refraction, and obey the F-equations. It will be argued that there is a good case for claiming that there is just one 'something' and not two. Chakravartty makes the right step in this direction by distinguishing between what he calls detection properties, and auxiliary properties: *'detection properties* are those upon which the causal regularities of our detections depend'. In contrast: *'Auxiliary properties* are those associated with the object under consideration, but not essential (in the sense that we do not appeal to them) in establishing existence claims' (Chakravartty 1998, pp. 394-5). As examples of auxiliary properties he cites the claim made by Fresnel that it is a mechanical medium that vibrates. Such a model may have been heuristically useful in developing the F-equations; but it is well known that from false premises true conclusions can follow. However it is the detection properties, viz., in this case the F-properties along with the F-equations, that are open to direct test and so carry the real burden of what it is we are talking about. Chakravartty goes on to say of the F-properties and F-equations: ‘Those properties of light which comprise or give rise to precisely these influences are *detection* properties having to do with causal regularities on the basis of which we infer the existence of the entity possessing them: light.’ (Chakravartty 1998, p. 396). I wish to put this insight to work in a different way using the Lewis sentence[21], which is a small but very significant modification of the Ramsey sentence, and add to it recent work by Papineau (1996) and Kroon and Nola (2001).

5.2 How a Lewis Reference Fixer Rescues Us From the Flight From Objects to Structure.

In general a theory can be given linguistic expression as a conjunction of statements (such as laws, etc). It can be readily expressed as: ‘$T(t_1, t_2, \ldots, t_n, O_1, O_2, \ldots O_m)$’ where the t_i (= 1, …, n) are theoretical terms (T-terms) and the O_j, (j= 1, …, m) are observational terms (O-terms). The distinction between T- and O-terms need not be epistemological. It could be semantic in which the O-terms can be *old* or *original* terms which already have their meaning fixed in some way. What we are interested in is the introduction of some new T-terms on the basis of terms whose meaning has been antecedently settled.

To simplify, let us consider a theory with just one T-term ‘t’ and all the O-terms are simply represented by ‘O’[22]:

(1) $T(t, O)$.

Now the Ramsey sentence is obtained by first replacing all the T-terms by variable, or blanks, thus, giving rise to an open sentence:

(2) $T(-, O)$.

Then the Ramsey Sentence T^R is got by prefixing the existential quantifier:

(3) $T^R = (\exists x)T(x, O)$.

The Lewis sentence, despite the fact that people often refer to the Ramsey-Carnap-Lewis sentence, is a quite different beast and has quite different features. The first thing to note is that the term ‘t’ in the context of ‘T(t, O)’ has it meaning fixed by that context. We can say that the term ‘t’ is implicitly defined by that context, or has its meaning specified by the role it plays in theory T. But this context also specifies

[21] See D. Lewis’ paper ‘How to Define Theoretical Terms’, first published in 1971 and reprinted in Lewis (1983) as chapter 6. Refinements of the view can be found in Lewis (1999) chapter 18.

[22] Not all theoretical terms will be in name position for the subsequently described procedures to take place; so some semantic alternation of the sentences will have to be carried out to make this possible. On this see the beginning of Lewis’ ‘How to define Theoretical Terms’ (in Lewis (1983), chapter 6).

a reference for the term 't' as well. This can be seen as follows. Considering the open sentence (2). Instead of putting the existential quantifier in front, put the definite description operator '¶' instead. And then use the definite description so formed to pick out some entity t in the following manner:

(4) $t = (¶x)T(x, O)$.

This is the Lewis sentence which fixes a reference for 't' *via* a description '(¶x)T(x, O)'. What does the work here is the description; the name can be regarded as a convenient shorthand for the description, both picking out the same item as referent. Generalised descriptions of this sort provide one way in which terms can get introduced, no matter whether they refer to observable or unobservable objects or kinds of object (or events, processes or properties). Note however that Kripke (1980) advocates a causal rather than a descriptivist account for fixing the reference of kind terms in the case of observable kinds. But such an account cannot be used for items which we cannot observe; one must resort to descriptions once more, albeit descriptions which might contain causal and historical elements (as will be seem).

There are some points that need to be noted about whether or not 't' has a reference fixed *via* the reference fixing description (right-hand-side of (4), or alternatively is what satisfies the open sentence (such as (2)). Let us use the open sentence in (2) and say: a reference will have been fixed for 't' if and only if the open sentence is uniquely satisfied, and not otherwise. If the open sentence is not satisfied by any item, then no reference will have been fixed for 't'. What if the open sentence is satisfied by two or more items? In a later modification of his views, Lewis (1999, p. 301) did allow that a name could be so introduced, but it would refer ambiguously to the several items. This is an important modification that can be useful in the context of science. A term can be successfully used, while later scientific development reveals that it has been introduced ambiguously. That is, there is referential indeterminacy for many of our terms that only becomes clear with the advance of science; further advance is then possible through referential refinement. Such is the case when the term 'oxygen' was introduced but it was later recognised that there are isotopes of oxygen that need to be distinguished. Again earlier theories of light supposed that there was just one kind of beam of light; but later it was recognised that some beams are polarised while other are not, thereby leading to a rather harmless referential refinement in our use of the term 'light'. Such referential refinement will not play a significant role in what follows.

Lewis allows not only for perfect satisfaction of an open sentence in order to fix a reference, but also for less than perfect satisfaction. Thus even if there is no perfect satisfier of an open sentence, there might be a unique, nearest and best satisfier; so a reference can be attached to a term, viz., whatever is the best and nearest satisfier. Again, though this is important, it is not a matter that we need to focus on here.

The generalised kinds of descriptions can be used in various ways to fix the reference of terms, or to re-fix the reference of terms with an antecedent use. But one important question is: just how much of the theoretical context of a term needs to be used in a generalised description to fix a reference? It depends; sometimes all, sometimes little. What follows is an attempt to indicate ways in which generalised descriptions can be employed to fix reference. First, there are those which use only observable properties of experimental situations; later fuller theoretical contexts will be illustrated. Good examples of term introductions *via* Lewis-style reference fixers are names for items discovered in experimental situations when little or no theory is known of the items discovered. Thus Röntgen recognised that he had hit upon something which satisfies the following:

> Whatever kind of thing it is, samples of which, in my [Röntgen's] experimental set-up (a) are emitted from a Hittorf tube, (b) cause a screen at the other end of my room treated with barium platinum-cyanide to fluoresce, (c) project an image of the bones in my hand on to the screen when I interpose my hand, (d) 'fogs' up my photographic plates, and (e) are not deflected by an electric field.[23]

Such a description arises from Röntgen's initial experimental investigations. Importantly these investigations took place in the absence of any theory of the 'something' that satisfies the description, or knowledge of the 'nature', or intrinsic properties of the 'something'. The description is unwieldy; so in order to carry on a conversation within the scientific community, a name was introduced *via* the description, viz., 'X-rays'. The name introduced is secondary compared with the identifying description which is uniquely (let us suppose) satisfied by samples of some kind of item involved in the experimental set-up. Further investigation enabled the community of physicists to discover more of the extrinsic properties of X-rays, and later some of the intrinsic properties, and even to rename the item originally so picked out (though the term 'X-ray' is still employed it is now known to refer to electromagnetic radiation of short wavelength).

Other terms can be introduced by means of their causal properties, for example the term 'electron' via the experimental set-up used by J. J. Thompson. (Note that this is not intended to be a historically accurate account of how the term 'electron' got introduced; rather its purpose is to introduce some of the features of reference fixing.[24])

Electrons = whatever kind it is such that samples are (i) emitted from the cathode of cathode ray tubes, (ii) detected when they hit a fluorescent screen at the other end of the cathode tube, and (iii) deflected when passing down the tube by an electric field around the tube.

[23] This is a summary of the experimental identification of X-rays made by Röntgen in the few days of feverous investigation of them after their accidental discovery and which were reported immediately in a paper. For some details see Segrè (1980) pp. 19-25.

[24] In his papers of the time (late 1890s), Thompson talked of 'corpuscles' rather than electrons. How the term 'electron' got introduced to name what Thompson was investigating need not detain us; what is of more significance is the generalised description that arises from Thompson's experimental set-up. A much fuller story need to be told about late 19th century investigations into cathode rays than can be given here, to flesh out the history of, and to illustrate the role that identifying descriptions played in, the manner that we eventually came to talk of electrons.

Such a term introduction tells us nothing much about the nature, or intrinsic properties, of electrons; most of the detection properties used in the reference fixer are extrinsic. But at least the community of scientists can begin to talk about, refer to, and exchange information about, an item without knowing much more about it. This is a prelude to learning more about electrons as the subsequent development of physics reveals from the Millikan experiments, the Bohr-Rutherford model of the electron, and so on.

Thomson believed at one point that electrons were present in an undifferentiated matter much like cherries in a homogeneous cherry-cake. If we add a fourth clause about this alleged feature of the electron in the above, then nothing will satisfy the associated open sentence; so no reference will have been attached to 'electron'. What this suggests is that we need to be cautious in specifying term introducers. We need to steer a course between two undesirable alternatives. The first difficulty is that we list so many properties in the reference fixer that nothing will satisfy it. We need to cut back on the number of properties in the reference fixer so that unique realisation of the associated open sentence occurs. Also we need a principled reason for making the cutback and stopping at some point. The second difficulty is that we mention so few properties that the associated open sentence is all too easily and promiscuously satisfied by too many things. We need to add more properties to reduce the promiscuousness in order to move in the direction of unique realisation. In the case of Thompson, we need to distinguish between (using Chakravartty's distinction) detection properties of the experimental set-up, and auxiliary hypotheses which might not even have had a heuristic role at all, but arise in subsequent theoretical speculation (as in the cherry-cake model Thompson later abandoned).

An important point about this matter is made in Papineau (1996). Let us suppose we do have unique realisation by, say, three reference fixing properties A, B and C. Then a reference will have been fixed by: t = whatever kind x that satisfies (A x & B x & C x). Suppose that we come to learn more about t and discover that it is also D, E and F. Now the reference of t remains invariant when we add D, E and F to our account of what ts are. But while reference remains fixed, what we might call the 'concept of t' is variable and admits of imprecision due to the many 'add-ons' that might be envisaged. But not all imprecision need be bad. We can say that the term 'electron' has had the same reference through a long career of use, perhaps introduced in the way suggested above. That is, we have discovered a lot more about it such as spin, its uncertainty relations, and so on. But if one asks 'what is our current concept of an electron?' the only answer is to get them to understand the latest in electron theory. And it is this that can vary while reference remains invariant and determinate; and during this variation, the concept of electron remains somewhat imprecise.

What the above suggests is that in determining a reference fixer for a term we should not invoke all the theory in which a term is involved. This can lead to Kuhnian and Feyerabendian incommensurability, and to a new argument for PMI. If we use the total theoretical context in which a theory occurs then there is a high probability that reference will fail or vary from one theoretical context to another. Striping down on the contexts in which a term occurs is also a strategy advocated by David Lewis, especially in the case of the reference of the term 'belief'. What does 'belief' refer to? Well, we have a well established 'folk psychology' in which we can readily talk of beliefs without having any idea of what a belief is. We assume that a belief is some kind of internal brain state that has intentional properties, but we do not know very much about what such brain states are like. However it is the folk psychological context that the meaning of the term 'belief' is specified (viz., by its role in all our folk psychological claims), and in which a reference is fixed. In the same way there is a 'folk' physics context of experimental investigation in which we can come to refer to items without knowing much, or anything, of their intrinsic properties (let alone their natures).

There is a controversy about the nature of beliefs that has an interesting parallel to issues concerning the radical version of ontological structural realism (SR) that denies the existence of objects in structures. The use of our folk psychology to fix a reference for beliefs is rejected by some, especially eliminativists who claim that there are no such things as beliefs at all and that any future talk of our mental-brain structure will have to get along without such entities. The parallel to be drawn is between such an eliminativism with respect to beliefs and an eliminativism with respect to objects in favour of an extreme version of SR which denies the existence of objects and admits the existence of structure only. The parallel is between an eliminativism with respect to objects by extreme advocates of SR and an eliminativism with respect to beliefs.

The above suggests that we cannot always find a theoretical context in which a term occurs to fix its reference, because there may be no such context; all there is to hand are experimental situations which allow for detection properties that are largely extrinsic to the objects that give rise to them. However there clearly are some occasions in which a large amount of theory is invoked. Thus some entities have been introduced on the basis of theory alone long before we have any way of even being acquainted with them. Thus Wheeler introduced in the late 1960s the term 'black hole' to refer to something for which there had for some time been a perfectly adequate theoretically loaded generalised description, viz., whatever it is that satisfies the equations that Schwarzchild developed out of Einstein's general Theory of Relativity which showed that there were singularities of a particular sort. Since then even more theory has been developed about black holes without the certainty that one has yet been observed (or detected). Again, the neutrino was a postulate of theory long before there was any experimental detection of their existence; any early reference fixer for 'neutrino' would have to contain much theory.

Such theory-dependent term introductions stand in marked contrast to the cases of 'X-rays' and 'electron' just mentioned. But all have the form of a generalised description. This suggests that the original schema introduced at the beginning of this sub-section needs to be accompanied by a story about what is to be invoked in reference-fixing descriptions to take account of the range of cases from large amounts of theory to little theory but much in the way of knowledge about what is going on in experimental set-ups.[25] The case of light, to which we now turn, is an interesting intermediate case.

[25] For a fuller account of how the original Lewis theory need to be supplemented, see Papineau (1996); also Kroon and Nola (2001) which puts emphasis on the context of belief, and the point of community

The term 'light' already had a use in physics before Fresnel developed his equations about its behaviour. However we can (with a little reconstruction of the historical situation) set this aside and investigate the theoretical context in which the term he uses is embedded. Then the task is to investigate what continuity of reference there may be from Fresnel's theory to Maxwell's. Given the little bit of history set out in the previous section, we can specify an open sentence which is a generalised description, which we may suppose is uniquely satisfied by some kind of thing (viz., light, or beams of light) and which can then be used to determine the reference of a term:

> whatever is the something (kind of object, event, process, etc.) such that the 'something' (i) has the F-properties and (ii) obeys the F-equations.

And that 'something' that satisfies this open sentence is light. Note that it is left open what the 'something' may be. Nothing is determined about the ontological category in which the 'something' falls (whether object, event, process, trope or whatever); and nothing is said about its intrinsic properties, or its nature. However it is well known that Fresnel did employ a model that does tell us something about the nature of light. So an account needs to be given of why elements of the model are left out of the reference-determining description.

Let us take the story further and suppose that we add to conditions (i) and (ii) a further story about the luminiferous ether that goes along with the Fresnel theory. If we do this then such generalised description will have no satisfier; and so the term whose reference it fixes, 'light' does not refer. There is no such thing as the elastic mechanical medium that Fresnel envisaged, and perhaps even used in the construction of the F-equations. But the moral of the above is that we should be more careful and look for a more cautious reference fixer that contains much less in the way of reference specifying properties that would produce reference failure. The question is, how little?

There is a principled answer that asks us to admit into the reference fixer only features that pertain to well established phenomenological laws, such as Fresnel's laws. We do not need to admit much more than this. And it is this, it can be argued, that the scientific community did use in their ongoing conversation about light, whatever different, highly contested theory each member might have also entertained. It is these that Chakravartty calls *detection* properties; it is these properties only, and not what he calls the auxiliary properties, that play the main role in reference fixing. Note also that the F-properties and F-laws may pertain largely to extrinsic properties and relations of light and say very little about what are light's intrinsic properties, and nothing about its nature or its essential properties. But this is not different from the situation of the early reference fixing for 'electron' or 'X-rays' which were identified by mainly extrinsic properties. Only much later did we come to know some of their intrinsic properties. And the same, one may hope, will be the case for the intrinsic properties of light with the advance of science.

communication, at the time of term introduction. None of these proposals need fall back on some account of a fixed meaning for the term or a notion of analyticity.

What needs to be argued now is that the very same thing, light, that has been picked out by the F-properties and the F-equations is also picked out by Maxwell's theory using what we might call the M-properties and the M-equations. In specifying these properties and equations we can set aside anything else that Maxwell might have used in constructing his model of light in much the same way we have set aside the extra and irrelevant baggage of the ether of Fresnel's model. Most of the M-properties are the same as the F-properties; and importantly the M-equations entail the F-equations. From this it follows that whatever unique 'something' that satisfies the F-equations will also satisfy the M-equations. So there is continuity of 'objects' from Fresnel's theory to that of Maxwell, and even beyond.

Given the above account of reference determination for 'light' in the two theories, we can dispense with the claim that there is object discontinuity from Fresnel's theory to that of Maxwell. There is a 'something' that both theories are about. And it is not just structure; it is the kind of particular, light—whatever that kind might be. We have not said very much about the intrinsic properties of light or even it nature or essence, or into which ontological category of kind of particular it falls. All we have is a 'something' which has F-properties and obeys F-equations, and a 'something' which has the M-properties and obeys the M-equations; but there are not really two 'somethings' here but one and the same 'something' with the F- and M-properties, and obeying the M-equations and also the F-equations they entail. None of this forms the basis for an objection to the above account of referential continuity. If the Fresnel and Maxwell equations are correct, then there will be the same 'something' that satisfies them; and even if they are not correct there will be a best approximate satisfier. There is no need for extreme structuralists to deny the existence of 'objects' (i.e., some kind of particular) that stand in the structural relations. So there is no need for ontological flight from objects to structure.

6. CONCLUSION

The aspects of structural realism discussed here take their cue from realists who became convinced, for a number of reasons, that the particulars (kinds of object, event, process, etc,) postulated in their theories did not survive theory change; but structures did. And this saves the day for realism. In pointing to structures, they do an important service to realism in bringing to the fore a category often overlooked, or even denied, by some realists with a special metaphysical agenda (e.g., nominalistic realists). But it is unclear what kind of structures are being advocated. Are they merely law formulations? Are they instantiated structures, instantiated by some category of particular, not necessarily objects? Or are they uninstantiated structures, those of the Platonistic ontological structuralist? Or are they structures which are such that we can never know what instantiates them (epistemic structuralism ESR)? Different considerations of different degrees of persuasiveness are advanced in support of one or other of these kinds of structuralism, only two of which are evaluated here, viz., the argument from PMI and considerations based on the theory of reference. These are found wanting. There is not even an argument that is so strong as to show that our knowledge of particulars must only pertain to their extrinsic and not their intrinsic properties, as ESR would have it. Our normal methods of science have in fact told us something of the intrinsic properties of 'objects' such as electrons (e.g., mass, charge) that undercuts such a strong claim.[26] Whatever else we might want to say about structuralism, there is no need to accept its flight from particulars such as objects and seek refuge for realists in structure alone.

ACKNOWLEDGMENT: I owe a debt of thanks to Jacob Busch for bringing the matter of structuralism to my attention and for his comments.

REFERENCES

Armstrong, D. (1983) *What is a Law of Nature?* Cambridge: Cambridge University Press.
Armstrong, D. (1989) *Universals: An Opinionated Introduction*. Boulder: Westview.
Bromberg, J. (1980) *Physical Chemistry*. Boston: Allyn and Bacon.
Campbell, K. (1990) *Abstract Particulars*. Oxford: Blackwell.
Chakravartty, A. (1998) 'Semirealism.' *Studies in the History and Philosophy of Science* 29: 391-48.
Chakravartty, A. (2003), 'The Structuralist Conception of Objects.' S. Mitchell (ed.) *PSA02*, Proceedings of the Philosophy of Science Association, 867-878.
Devitt, M. (1997) *Realism and Truth*. 2nd edn. Princeton NJ: Princeton University Press.
Devitt, M. (forthcoming) 'Scientific Realism.' in F. Jackson & M. Smith (eds.) *The Oxford Handbook of Contemporary Analytic Philosophy*. Oxford: Oxford University Press.
Ellis, B. (2001) *Scientific Essentialism*. Cambridge: Cambridge University Press.
French, S. (1999) 'On the Withering Away of Physical Objects.' in E. Castellani (ed.) *Interpreting Bodies: Classical and Quantum Objects in Modern Physics*. Princeton: Princeton University Press.
French, S. (unpublished) 'The Dissolution of Objects: Between Platonism and Phenomenalism.'
French S. & Ladyman, J. (2003) 'Remodelling Structural Realism: Quantum Mechanics and the Metaphysics of Structure.' *Synthese* 136: 31-56.
Giere, R. (1999) *Science Without Laws*. Chicago: University of Chicago Press.
Hellman, G. (1983) 'Realist Principles.' *Philosophy of Science* 50: 227-49.
Kim, J. (1998) 'Events as Property Exemplifications.' in S. Laurence & C. Macdonald (eds.) *Contemporary Readings in the Foundations of Metaphysics*. Oxford: Blackwell.
Kripke, S. (1980) *Naming and Necessity*. Oxford: Blackwell.
Kroon, F. & Nola, R. (2001) 'Ramsification, Reference Fixing and Incommensurability.' in P. Hoyningen-Huene & H. Sankey (eds.) *Incommensurability and Related Matters*. Dordrecht: Kluwer Academic Publishers.
Ladyman, J. (1998) 'What is Structural Realism?' *Studies in the History and Philosophy of Science* 29: 409-424.
Langton, R. (1998) *Kantian Humility: Our Ignorance of Things in Themselves*. Oxford: Oxford University Press.
Laudan, L. (1981) 'A Confutation of Convergent Realism.' *Philosophy of Science* 48: 19-29.
Lewis, D. (1983) *Philosophical Papers, Volume 1*. New York: Oxford University Press.
Lewis, D. (1999) *Papers in Metaphysics and Epistemology*. Cambridge: Cambridge University Press.
Lewis, P. (2001) 'Why the Pessimistic Induction is a Fallacy.' *Synthese* 129: 371-380.
Maxwell, G. (1970) 'Theories, Perception and Structural Realism.' in R. G. Colodny (ed.) *The Nature and Function of Scientific Theories*. Pittsburgh: University of Pittsburgh Press.

[26] There are a number of other considerations not explicitly (to my knowledge) advanced by structuralists in the philosophy of science. One such consideration is based in Kant's distinction between noumenal and phenomenal objects. This is understood in Langton (1998) in the sense that we cannot have knowledge of the intrinsic properties of things but only their extrinsic properties. This would suggest a quite different line of support for structuralism.

Morrison, M. (2000) *Unifying Scientific Theories*. Cambridge: Cambridge University Press.
Musgrave, A. (1999) *Essays on Realism and Rationalism*. Amsterdam/Atlanta GA: Rodopi.
Norton, J. (2000) 'How We Know About Electrons.' in R. Nola & H. Sankey (eds.) *After Popper, Kuhn and Feyerabend: Recent Issues in Theories of Scientific Method*. Dordrecht: Kluwer.
Papineau, D. (1996) 'Theory-Dependent Terms.' *Philosophy of Science* 63: 1-20.
Poincaré, H. (1952) *Science and Hypothesis*. New York: Dover; first published in English in 1905.
Psillos, S. (1996) 'Scientific Realism and the "Pessimistic Induction".' L. Darden (ed.) *PSA 1996,* Philosophy of Science Association: S306-S314.
Psillos, S. (2001) 'Is Structural Realism Possible?' in J. Barrett & J. Alexander (eds.) *PSA00*, Supplement to *Philosophy of Science* 68(3), Philosophy of Science Association, S13-S24.
Psillos, S. (forthcoming) '*The* Structure, the *Whole* Structure and Nothing *but* the Structure?', PSA04, Philosophy of Science Association.
Putnam, H. (1978) *Meaning and the Moral Sciences*. London: Routledge & Kegan Paul.
Quine, W. (1980) 'Soft Impeachment Disowned.' *Pacific Philosophical Quarterly* 61: 450-51.
Searle, J. (1995) *The Construction of Social Reality*. London: Penguin.
Segrè, E. (1980) *From X-rays to Quarks*. San Francisco: W. H. Freeman and Company.
van Fraassen, B. (forthcoming) 'Structure: Its Shadow and Substance.' in P. Ehrlich & R. Jones (eds.) *Reverberations of the Shaky Game: Essays in Honor of Arthur Fine*. Chicago: University of Chicago Press.
Worrall, J. (1989) 'Structural Realism: The Best of Both Worlds?' *Dialectica* 43: 99-124.

MARK COLYVAN

SCIENTIFIC REALISM AND MATHEMATICAL NOMINALISM: A MARRIAGE MADE IN HELL

The Quine-Putnam Indispensability argument is the argument for treating mathematical entities on a par with other theoretical entities of our best scientific theories. This argument is usually taken to be an argument for mathematical realism. In this chapter, I will argue that the proper way to understand this argument is as putting pressure on the viability of the marriage of scientific realism and mathematical nominalism. Although such a marriage is a popular option amongst philosophers of science and mathematics, in light of the indispensability argument, the marriage is seen to be very unstable. Unless one is careful about how the Quine-Putnam argument is disarmed, one can be forced to either mathematical realism or, alternatively, scientific instrumentalism.

I will explore the various options: (i) finding a way to reconcile the two partners in the marriage by disarming the indispensability argument (Jody Azzouni (2004), Hartry Field (1980, 1989), Alan Musgrave (1977, 1986), David Papineau (1993)); (ii) embracing mathematical realism (W.V.O. Quine (1981), Michael Resnik (1997), J.J.C. Smart (unpub.)); and (iii) embracing some form of scientific instrumentalism (Otávio Bueno (1999, 2000), Bas van Fraassen (1985)). Elsewhere (Colyvan 2001), I have argued for option (ii) and I won't repeat those arguments here. Instead, I will consider the difficulties for each of the three options just mentioned, with special attention to option (i). In relation to the latter, I will discuss an argument due to Alan Musgrave (1986) as to why option (i) is a plausible and promising approach.

From the discussion of Musgrave's argument, it will emerge that the issue of holist versus separatist theories of confirmation plays a curious role in the realism–antirealism debate in the philosophy of mathematics. I will argue that if you take confirmation to be a holistic matter—it's whole theories (or significant parts thereof) that are confirmed in any experiment—then there's an inclination to opt for (ii) in order to resolve the marital tension outlined above. If, on the other hand, you take it that it's a single hypothesis that's confirmed in a given experiment, then you'll be more inclined towards option (i). As we shall see, Musgrave's argument illuminates, in an interesting and original way, the important role confirmation has to play in realism debates in the philosophy of mathematics.

C. Cheyne & J. Worrall (eds.), Rationality and Reality: Conversations with Alan Musgrave, 225–237.

1. SCIENTIFIC REALISM MEETS MATHEMATICAL NOMINALISM

Scientific realists such as Musgrave (1999) are happy to go beyond what is observable and posit unobservable entities. According to scientific realists, what makes the cloud chamber appear as though there is an electron in it is that *there is an electron in it*. The details of how we go from mere observations, which typically underdetermine the theory, to the positing of unobservable entities vary. Inference to the best explanation is the vehicle of choice amongst most scientific realists. Indeed, it's not stretching things too much to suggest that the scientific realism–antirealism debate can be characterised in terms of the acceptance or rejection (respectively) of inference to the best explanation. In any case, Musgrave, like most scientific realists, accepts this much-discussed form of inference.[1]

Mathematical nominalism is the view that mathematical entities, such as numbers, functions, and sets, do not exist. The opposing view—mathematical realism—holds that at least some mathematical entities exist. One of the primary motivations for mathematical nominalism is that mathematical realism faces a rather daunting epistemological challenge (Benacerraf 1983). The problem is simply that if mathematical entities exist, as the mathematical realist would have it, then we require an adequate account of how we come by knowledge of such entities. After all, mathematical entities, if they exist, do not seem to be the kinds of things that have space-time locations or have causal powers. In short, if they exist, it would seem we cannot have any contact with them and hence we cannot have knowledge of them. Nominalism does not face any vexing epistemological issues, so it seems more reasonable to suppose that mathematical entities do not exist. (Or so the argument goes.)[2]

2. THE TENSION AND THE OPTIONS

At first glance, scientific realism and mathematical nominalism make a handsome couple. There's no need for belief in mysterious abstract mathematical entities, the epistemology is relatively straightforward, and there's a healthy respect for science, taken at face value. No reinterpretation of science in terms of observables or dodgy appeals to the world merely behaving as though there were unobservables. A little thought, however, soon reveals the problems with this union. The problem is that the alliance is very unstable. The scientific realism part of the marriage typically appeals to inference to the best explanation as a reason for belief in unobservable theoretical

[1] Of course scientific realism has its problems. For instance, justifying the clearly invalid inference to the best explanation and dealing with the underdetermination of theory by evidence. (See van Fraassen (1980) for details.) I'll not dwell on such problems in what follows.

[2] Nominalism has some problems too. One we'll look at in the next section, but there is also the problem of supplying a uniform semantics across all natural and scientific language. The problem is simply that scientific sentences such as 'there's a planet closer to the sun than Venus' is true and what makes it true is the existence of Mercury (and the fact that it is closer to the sun than Venus). But nominalists hold that there are no numbers, so it would seem that the nominalist cannot employ the usual semantics to account for the truth of sentences such as 'there is a number smaller than 2' (see Benacerraf (1983) for further details).

entities. But even a cursory glance at any scientific text, from almost any area of science will reveal the crucial role that mathematics plays in science. We have mixed mathematical-empirical statements such as:

(*) The work done in moving a body from a to b is given by $\int_a^b F(s)ds$, where F is the force exerted on the body and s is the body's displacement.

We also have purely mathematical statements such as:

(**) The Gaussian distribution is symmetric about its mean.

Both kinds of statement play important, indeed, indispensable, roles in science. As Quine (1981) and Putnam (1971) have pointed out, if one is to accept such statements as true (as surely we must), then this in turn leads us to accept the existence of real-valued force functions, integrals, displacement functions, Gaussian distributions, and means.

To summarise this line of thought, we ought to count as real any entity that plays an indispensable role in our best scientific theories. As Putnam has stressed, anyone inclined to do otherwise would be guilty of intellectual dishonesty. (This is the sin of 'denying the existence of what one daily presupposes' (Putnam 1971, p. 347).) Following Putnam, let's call this the indispensability argument.[3] This argument is usually construed as an argument for mathematical realism, but since it relies on certain background assumptions (such as naturalism and confirmational holism) it is not going to persuade everyone (at least not without a defence of its background assumptions). But notice that this argument counsels us to accept entities as real, irrespective of whether they are observable or unobservable. All that matters is that the entities in question are indispensable. But what does the latter involve?

One way an entity might play an indispensable role in a scientific theory is that it might be indispensable for explanation. That is, inference to the best explanation is a special case of the indispensability argument (Field 1989, pp. 14–20). Moreover, as has already been noted, this is a style of argument that the scientific realist accepts.[4] In fact that's all we need; we don't really need to consider more general forms of the indispensability argument because mathematical entities surely feature prominently in various explanations.[5] (See (*) and (**) above, for instance, and consider the various scientific explanations such statements feature in.) So here I will take the indispensability argument to be an argument that puts pressure on the marriage of

[3] I lay out this argument in more detail and defend it in Colyvan (2001).

[4] Indeed, this is why Hartry Field (1980, p. 4) suggests that the indispensability argument is the only non-question begging argument for mathematical realism, At least, it is no more question begging than standard arguments for scientific realism.

[5] Joseph Melia (2000) claims that mathematical entities merely allow for more economical statement of theories; they do not simplify the theories in the right kind of way and they do not lend explanatory power to the theories in which they appear. While I think he is wrong about this (see Colyvan (2002)) I agree that there are some interesting issues to be explored here.

scientific realism and nominalism. It does this because the style of argument is one which scientific realists already endorse. Now let's consider the various options facing would-be nominalist scientific realists.

2.1 Marriage Counselling

By far the most popular option for dealing with the tensions I just outlined is to somehow reconcile scientific realism with mathematical nominalism. There are a number of different strategies proposed for this purpose. These divide into what I call 'easy road' and 'hard road' strategies. The easy road strategies involve denying that we ought to have ontological commitment to all the entities that are indispensable to our best scientific theories. That is, we provide some principled demarcation between those parts of our best scientific theories that are to be treated realistically and those which are to be treated instrumentally. And, of course, for this strategy to work, the mathematical entities had better fall on the instrumental side of the divide.[6]

Another way to proceed is to deny that mathematical entities are indispensable to our best scientific theories. This is a hard road since it involves showing how to do science without mathematics. Moreover, an adherent of this approach is also required to offer an explanation of why mathematics, even though dispensable, plays such a prominent role in science. The most influential hard road strategy in recent years has been Hartry Field's fictionalism. According to Field, mathematical sentences such as

$$\text{The unique prime factorisation of 255 is } 17 \times 5 \times 3 \qquad (1)$$

are interpreted at face value. Thus interpreted such sentences imply the existence of mathematical entities[7] and so are literally false. He thus endorses fictionalism about mathematics. Not all of the usually accepted 'truths' of mathematics come out false though. Negative existential claims like 'there is no largest prime' and universally quantified sentences such as 'every natural number has a unique prime factorisation' are true according to the fictionalist. But they are vacuously true; they are true because there are no prime numbers and because there are no natural numbers respectively.

The Field-style fictionalist cannot rest there though. The fictionalist must show how our best scientific theories can be purged of their mathematical content and explain why mathematics can be used in empirical science without (crudely speaking) its falsity infecting the rest of the scientific theories. Field makes significant inroads on the former project by adopting a Hilbert-style geometric approach to Newtonian gravitational theory. On this approach, space-time points are compared with respect to their gravitational potential, for example, and this

[6] Proposals along these lines include Azzouni (2004), Balaguer (1998, chap. 7), Cheyne (2001), and Melia (2000).

[7] Since it follows (in classical logic, at least) that prime numbers such as 17, 5 and 3, and composite numbers such as 255 exist. Of course, in free logic such conclusions do not follow from (1).

eliminates the need for gravitational potential functions (Field 1980). Field then proves a representation theorem which demonstrates the adequacy of the approach. The project of explaining why the falsity of mathematics does not infect the rest of science is tackled by proving (and arguing for the plausibility of) a conservativeness result. The conservativeness result (if correct) shows that a mathematical theory M is conservative in the sense that for any body of nominalistic assertions N and any particular nominalistic assertion A, A is not a consequence of $N + M$ unless it's a consequence of N alone. With this in place, Field-style fictionalism is in a position to resolve the marital difficulties outlined above. One can coherently be both a scientific realist and a mathematical nominalist.[8]

2.2 *Divorce I: Realism Gets the House*

Another way of dealing with the tension outlined above is to move to a more thorough-going realism. One can hold onto one's scientific realist scruples and (perhaps reluctantly) admit that accepting inference to the best explanation and the realist package has some unforeseen consequences: one needs to be realist about a bit more than one initially bargained for. The realism extends to include all entitles indispensable to our best scientific theories, and these include at least some mathematical entities. This option is no reconciliation of scientific realism and mathematical nominalism. On this option, realism wins the day and nominalism is rejected.[9]

No divorce is so neat as this though. The mathematical realist still owes an account of the epistemology of mathematics and perhaps also an account of the nature of mathematical entities that jibes with that epistemology. After all, the indispensability argument, on the face of it at least, does not tell us anything about either mathematical epistemology or the nature of mathematical entities.[10]

2.3 *Divorce II: Instrumentalism Gets the House*

Another option is to hold fast to one's nominalist sensibilities and reject the form of argument that produced the tension in the first place. But as I've already pointed out, rejecting the indispensability argument would seem to undermine a central plank of the scientific realist's platform. 'So be it', you might say. 'If mathematical realism is the price one pays for scientific realism, then the price is too high'. According to this

[8] There are, of course, many further difficulties facing Field's project and many objections. I won't pursue such matters here. (See Burgess and Rosen (1997) for a good discussion of some of these.) At this stage I merely want to outline the various options.

[9] See Colyvan (2001) and Resnik (1997) for defences of this approach.

[10] As we'll see in section 4, I think that the indispensability argument does tell us quite a bit about the epistemology of mathematics. The indispensability argument, after all, does come with a holist epistemology, according to which we have knowledge of mathematical entities by the role they play in our best scientific theories. Moreover, this is no different from how we gain knowledge of other theoretical entities in science. See Cheyne (2001) (for example) for criticism of the holist epistemology that emerges from the indispensability argument and Baker (2003a) for criticism of the indispensability argument's failure to say anything much about the nature of mathematical entities.

line of thought, anti-realism wins the day and it is mathematical realism that is rejected.

The mathematical nominalist is not home free though. It is not enough to simply reject the indispensability argument (and with it inference to the best explanation) and join the anti-realist camp. Consider, for example, what many take to be the most sophisticated anti-realist philosophy of science: Bas van Fraassen's (1980) constructive empiricism. Constructive empiricism makes heavy use of mathematics in both its articulation and defence. Indeed, the crucial notion for constructive empiricism is that of empirical adequacy and this is spelled out in terms of models, structures, and isomorphic mappings—all of which are mathematical entities.[11] This problem for nominalising constructive empiricism has been raised by Michael Resnik (1997, pp. 49–50). Indeed, van Fraassen himself sees the problem and accepts the considerable burden of showing how constructive empiricism might be nominalised:

> I am a nominalist [...] Yet I do not for a moment think that science should eschew the use of mathematics. I have not worked out a nominalist philosophy of mathematics—my trying has not carried me that far. Yet I am clear that it would have to be a fictionalist account, legitimating the use of mathematics and all its intratheoretic distinctions in the course of that use, unaffected by disbelief in the entities mathematical statements purport to be about. (van Fraassen 1985, p. 303)

One option would be to embark on a Field-style nominalisation project but this is not likely to be fruitful. As I pointed out earlier, Field utilises a Hilbert-style geometric approach to space-time. This involves quantification over space-time points and these are thus treated as real entities. This is something that many nominalists are unhappy about. But for constructive empiricists, realism about space-time points is out of the question. What other options are there then? Our constructive empiricist might employ one of the easy road strategies of section 2.1. But then it's not clear what constructive empiricism is bringing to the party. After all, if one of the easy road strategies of section 2.1 can be made to work, there was no need to retreat to constructive empiricism in the first place. Another option would be to reformulate the crucial notions of constructive empiricism—empirical adequacy and so on—in such a way as not to involve quantification over mathematical entities. This may well be possible but I leave the pursuit of such options for those with more sympathy for constructive empiricism.[12] In any case, this is most definitely not an option that a robust realist such as Musgrave is likely to find attractive!

3. MUSGRAVE'S ARGUMENT FOR NOMINALISM

Musgrave has entertained a couple of different approaches to the tension between scientific realism and mathematical nominalism. His first shot at a philosophy of mathematics was a version of if-thenism (Musgrave 1977). According to this view, mathematics consists of conditional statements such as 'If the

[11] A theory is empirically adequate if it has a model such that all appearances are isomorphic to empirical sub-structures of that model (van Fraassen 1980, p. 64).

[12] Otávio Bueno (1999, 2000) has been doing some interesting work in this direction.

conjunction of the Peano axioms, then there are infinitely many prime numbers'. Later, Musgrave defended a Field-style fictionalism (Musgrave 1986). Although these two approaches are rather different in detail they are similar in spirit. They are both nominalist philosophies of mathematics that only accept the truth of mathematical sentences once they are imbedded in a suitable construction such as a conditional ('if the conjunction of the Peano axioms then...') or a fictional operator ('in the story of mathematics...'). Rather than discuss the details of Musgrave's philosophy of mathematics, I want to consider his motivation for treating mathematical entities differently from other unobservable entities.

A central intuition that many nominalists have is that because mathematical entities are non-causal, they cannot make a difference to the way the physical world is.[13] If the existence of mathematical entities doesn't make a difference—that is, the physical world would be the same with or without mathematical entities—then there would seem to be no reason to believe in them. In his paper, 'Arithmetical Platonism: Is Wright Wrong or Must Field Yield?', Musgrave (1986) explores this line of thought in an interesting and original way.[14] Instead of focussing on whether mathematical entities make a difference to the physical world and what bearing this has on the epistemology of mathematics, Musgrave shifts the focus to the question of how we might falsify the hypothesis that there are mathematical entities.

> Imagine that all the evidence that induces scientists to believe (tentatively) in electrons had turned out differently. Imagine that electron-theory turned out to be wrong and electrons went the way of phlogiston or the heavenly spheres. Popperians think this might happen to any of the theoretical posits of science. But can we imagine natural numbers going the way of phlogiston, can we imagine evidence piling up to the effect that there are no natural numbers? This must be possible, if the indispensability argument is right and natural numbers are a theoretical posit in the same epistemological boat as electrons.
>
> But surely, if natural numbers do exist, they exist of necessity, in all possible worlds. If so, no empirical evidence concerning the nature of the actual world can tell against them. If so, no empirical evidence can tell in favour of them either. The indispensability argument for natural numbers is mistaken. (Musgrave 1986, pp. 90–91).

Musgrave, the scientific realist, argues that electrons make a difference to the way the physical world is. This means that the existence of electrons can be confirmed by crucial experiments such as the Millikan oil-drop experiment (and others).[15] But in the case of mathematical entities, Musgrave argues, it is difficult to see how any experiment could provide confirmation of their existence, and, we might add, their properties. The reason Musgrave gives is that mathematical entities, if they exist, exist of necessity, so their presence or absence cannot be established by

[13] See, for example, Azzouni (1998) for an articulation of this line of thought, and Baker (2003b) for a reply.

[14] Musgrave (1986, pp. 90–91) suggests that the argument I'm about to outline is just another way of making Field's (1980, pp. 11–12) point about conservativeness: mathematics does not need to be true to be good, it just needs to be conservative. I think Musgrave's argument is significantly different from Field's. At the very least, Musgrave's argument is different enough to warrant separate attention. I discuss Musgrave's take on Field's argument in Colyvan (2001, chap. 6).

[15] Indeed, experiments such as Millikan's yield crucial confirmations of not only the existence of electrons but also their mass.

appeal to crucial experiments. In short, the hypothesis that electrons exist can be falsified whereas the hypothesis that mathematical objects exist cannot be falsified.

I take it that Musgrave's objection presents serious difficulties for any defender of the indispensability argument who takes mathematical entities to exist of necessity. There is, however, another position for the defender of the indispensability argument to adopt: the position that affords *contingent* existence to mathematical entities.[16] It might seem that this position isn't touched by the Musgrave objection, but I think Musgrave's concerns here run a little deeper. Musgrave may be seen to be challenging the defender of contingently existing mathematical entities to provide the details of possible crucial experiments that might give us reason to accept or reject the existence of mathematical entities. I think this challenge can be met, though perhaps not in a fashion that will satisfy Musgrave. If I'm right about this, we have reached the source of Musgrave's nominalist sympathies and identified an important point of contention between nominalists and mathematical realists. So let me sketch how I take it that the challenge to provide crucial experiments for the existence of mathematical entities might be met.

According to the most plausible reading of the indispensability argument, mathematical entities exist contingently and the evidence for their existence comes from the confirmation of our best scientific theories (and the indispensable role mathematical entities play in those theories). Mathematical entities do not need to play causal roles in those theories (indeed, it is generally agreed that they do not play such roles). But if they do not play causal roles, what roles are left? Asking after a crucial experiment for the existence of an entity is akin to identifying a crucial causal connection of the entity in question. But at this point the Quinean simply digs her heals in and insists that there need not be any crucial experiment in the sense that Musgrave seeks. There will not be any experiment that directly confirms the existence of mathematical entities. This is not to say, however, that mathematical entities are without empirical support. According to the Quinean, mathematical entities are *indirectly* confirmed by whatever confirmation our best scientific theories enjoy.

But this doesn't address the issue of specifying the role of mathematical entities in these theories? Elsewhere (Colyvan 2001, 2002) I've argued that mathematics may contribute to the unificatory power and other theoretical virtues of scientific theories. We need to think of these theories holistically though. We need to resist any demand for crucial experiments—not just for mathematical entities, but for any entity. The thorough-going holist would deny that even the Millikan oil-drop experiment is a crucial experiment. This experiment, after all, had auxiliary assumptions about the behaviour of oil drops in gravitational fields and the behaviour of charged particles in electric fields, for instance. Such assumptions are not particularly controversial—that's not the confirmational holist's point. Their point is simply that a great deal more than the hypothesis in question is being tested and confirmed, even in so-called *crucial* experiments.

[16] This is the position I endorse in Colyvan (2001, section 6.4). Hartry Field (1993) also accepts that the existence or non-existence of mathematical entities is a contingent matter, though he takes mathematical entities to contingently fail to exist.

So, I claim that the source of Musgrave's inclination for trying to salvage the marriage of scientific realism and mathematical nominalism lies in his separatist (Popperian) confirmation theory. Separatist confirmational theories demand more than merely stating that some entity 'plays a role in our best theory'. The separatist wants a crucial experiment that identifies the causal roles of the entities in question. The confirmational holist, on the other hand, sees this latter request as simply unreasonable—at least in the context of establishing the existence of the entities in question.

Identifying the source of the disagreement is one thing, the real issue is surely that of determining which theory of confirmation is to be preferred. Here, however, we have something of a nil-all draw. It's fair to say that both Popperian falsification and Quinean confirmational holism find few supporters these days—most philosophers of science would not see either as a viable theory of confirmation. Be that as it may, the issues we've been concerned with do not depend so much on the fine details of these two theories of confirmation. After all, although Musgrave is a Popperian on such matters, the full details of Popper's philosophy of science were never invoked nor called into question. Musgrave's argument, it would seem, could be advanced on any separatist theory of confirmation.[17] Indeed, Elliot Sober (1993) pursues much the same line of attack on the indispensability argument via another separatist theory of confirmation, namely, his contrastive empiricism. And likewise, as I've argued elsewhere (Colyvan 2001, chap. 2), the confirmational holist need not endorse the more radical holism of Quine. Still, there is a substantial issue sorting out whether any particular separatist or, alternatively, holist theory of confirmation can be made to fly. Obviously that is a large task and one I cannot do justice to here. I'm content to identify a significant intuition that drives Musgrave (and others) in the exploration of issues concerning realism and nominalism in the philosophy of mathematics.[18]

I can't resist mentioning, however, the delicacy of the position the separatist finds himself in. The separatist needs to be able to avoid a couple of nearby slippery slopes. After all, if a crucial experiment must be devised for every kind of entity to which we are to be ontologically committed, care needs to be taken about certain problem cases—those involving entities that scientific realists are committed to but which seem to lack crucial experiments. Some of the problem cases obviously involve unobservable entities such as electrons, quarks, black holes and the like. Typically the scientific realist is able to invoke the causal powers of such entities to design a crucial experiment. This is how the Millikan oil drop experiment worked. But what of unobservable entities that have causal powers but with which we have no contact? Consider, for example, stars and planets outside our own light cone. What are the crucial experiments that establish the existence of these entities? Of course there are responses to such problem cases, but the response must not license a slide to scientific instrumentalism or, alternatively, a slide to mathematical realism. Either slide would be to give the game away. For instance, I take it that the following response to the problem cases is illegitimate: we accept the existence of

[17] Which is why I think, for present purposes, there is little point in criticising falsification.

[18] See Resnik (1997, chap. 7) for a nice discussion of holism and its relevance to the realist–anti-realist debate in the philosophy of mathematics.

stars and planets outside our light cone because they play an indispensable role in our best cosmological theories. This won't do because (a) it violates the separatist criterion of providing a crucial experiment and (b) the appeal to playing an indispensable role (without further qualification) in a best scientific theory would also seem to license the acceptance of mathematical entities. In short, this response amounts to giving the game away to the mathematical realist. A similar unacceptable slide to anti-realism beckons if our nominalist scientific realist decides to deny the existence of stars and planets outside our light cone. As I suggested above, there are options available here, but those wishing to salvage the marriage of scientific realism and mathematical nominalism need to be very careful about their treatment of such problem cases.

I now turn to a well-known epistemic argument used as a motivation for mathematical nominalism. I'll show how similar holist and separatist considerations impact on the ensuing debate in very similar ways as those outlined above.

4. SEPARATISM AND HOLISM ABOUT JUSTIFICATION

A great deal of the literature on the realism–antirealism debate in the philosophy of mathematics focuses on epistemology. In particular, nominalists typically take Paul Benacerraf's (1983) epistemic challenge to mathematical realism as a challenge that cannot be met. Although Benacerraf originally presented his challenge in terms of the causal theory of knowledge, the essence of his argument can be captured without recourse to this now unpopular epistemology. Hartry Field puts the challenge as follows (emphasis in the original):

> Benacerraf's challenge—or at least, the challenge which his paper suggests to me—is to provide an account of the mechanisms that explain how our beliefs about these remote entities can so well reflect the facts about them. The idea is that *if it appears in principle impossible to explain this*, then that tends to *undermine* the belief in mathematical entities, *despite* whatever reasons we might have for believing in them. (Field 1989, p. 26)

As I've already mentioned, Field too is in favour of saving the union of mathematical nominalism and scientific realism. And from the above quotation we see that one of Field's motivations for defending nominalism is the epistemic problem for mathematical realism.[19] What's interesting here is that both Field and Musgrave agree that we should save the marriage in question, and they even agree on the best way to go about this: Field-style fictionalism. The difference is that Field is motivated (in part) by an epistemic problem for mathematical realism, whereas, on the face of it at least, Musgrave is motivated by something else. In the last section I argued that Musgrave's motivation arises from separatist (as opposed to holist) sympathies about theory confirmation. But I think Musgrave's motivation has some interesting points of contact with Field's.

The usual construal of the Benacerraf-Field epistemic challenge is a challenge for the platonist to explain the reliability of mathematicians' beliefs. But implicit in

[19] Field has other motivations as well: the quest for intrinsic explanations and the elimination of arbitrariness from scientific theories. See Field (1980, p. xi) for more on these issues.

this challenge is that the mathematical beliefs be taken *one at a time*. That is, the platonist must account for the reliability of the inference from 'mathematicians believe that *P*' to *P*. But put thus, the challenge assumes a separatist epistemology, according to which beliefs are justified one at a time. The epistemological holist will argue that this is wrong-headed; beliefs are justified as packages. How does this help answer the Benacerraf-Field challenge? Well, if we drop the demand for justification of beliefs one belief at a time, then the mathematical realist can appeal to a holist epistemology to meet the challenge in question: we justify our mathematical beliefs by the role they play in broader systems of beliefs (namely, our best scientific theories).[20] Any dissatisfaction with such a holist response would seem to arise from separatist sympathies with regard to justification.

So Field is right that the main thrust of Benacerraf's epistemic challenge does not rely on the causal theory of knowledge. But by stating the challenge in reliabilist terms, Field still assumes a separatist epistemology. Moreover, it is precisely here that the holist will object. So once again we see that holist sympathies push towards 'realism getting the house' whereas separatist sympathies push for 'saving the marriage'. This time the separatist and holist sympathies concern justification not confirmation, but clearly these two notions are closely related.

5. CONCLUSION

I've outlined some of the problems associated with reconciling scientific realism with mathematical nominalism. In the light of these problems it might be wondered why anyone would want to save this marriage. Well, one reason is that divorces are difficult: both of the divorce options I presented face substantial philosophical problems. This much is well known. Less appreciated, I think, is the role played by separatism and holism about both confirmation and justification. I've argued that, with respect to both justification and confirmation, separatist sympathies push for mathematical nominalism and holist sympathies push for mathematical realism. If this is right, it seems that we have found a fruitful and appropriate place to focus our attention in attacking the realism–antirealism debate in the philosophy of mathematics.[21] I take this to be one of Musgrave's most significant contributions to the philosophy of mathematics. While my sympathies are with holism and mathematical realism, his with separatism and mathematical nominalism, we agree, I think, on how to approach the matter in question and on what some of the broader underlying issues are. In the philosophy of mathematics, at least, such agreement is non-trivial. But at the end of the day, I disagree with Musgrave about the prospects for saving the marriage of

[20] See Colyvan (2001, p. 154), Rosen (1992, chap. 3), and Smart (unpub.) for presentations of this response to the Benacerraf-Field challenge.

[21] Indeed, a great deal of contemporary work in the philosophy of mathematics is directed at questioning confirmational holism. See, for example, Maddy (1997), Sober (1993), and, of course, Musgrave's (1986) contribution.

scientific realism and mathematical nominalism. I'm of the view that this marriage was never meant to be.[22]

REFERENCES

Azzouni, J. (1998) 'On "On What There Is."' *Pacific Philosophical Quarterly* 79: 1–18.

Azzouni, J. (2004) *Deflating Existential Consequence: A Case for Nominalism.* New York: Oxford University Press.

Baker, A. (2003a) 'The Indispensability Argument and Multiple Foundations for Mathematics.' *Philosophical Quarterly* 53: 49–67.

Baker, A. (2003b) 'Does the Existence of Mathematical Objects Make a Difference?' *Australasian Journal of Philosophy* 81: 246–264.

Balaguer, M. (1998) *Platonism and Anti-Platonism in Mathematics.* New York: Oxford University Press.

Benacerraf, P. (1983) 'Mathematical Truth.' in P. Benacerraf and H. Putnam (eds.) *Philosophy of Mathematics Selected Readings* 2nd edn. Cambridge: Cambridge University Press: 403–420.

Bueno, O. (1999) 'Empiricism, Conservativeness and Quasi-Truth.' *Philosophy of Science* 66: S474–S485.

Bueno, O. (2000) 'Empiricism, Scientific Change and Mathematical Change.' *Studies in History and Philosophy of Science* 31: 269–296.

Burgess, J.P. and Rosen, G. (1997) *A Subject with No Object: Strategies for Nominalistic Interpretation of Mathematics.* Oxford: Oxford University Press.

Cheyne, C. (2001) *Knowledge, Cause, and Abstract Objects: Causal Objections to Platonism.* Dordrecht: Kluwer.

Colyvan, M. (2001) *The Indispensability of Mathematics.* New York: Oxford University Press.

Colyvan, M. (2002) 'Mathematics and Aesthetic Considerations in Science.' *Mind* 111: 69–74.

Field, H. (1980) *Science Without Numbers: A Defence of Nominalism.* Blackwell: Oxford.

Field, H. (1989) *Realism, Mathematics and Modality.* Oxford: Blackwell Publishers.

Field, H. (1993) 'The Conceptual Contingency of Mathematical Objects.' *Mind* 102: 285–299.

Maddy, P. (1997) *Naturalism in Mathematics.* Oxford: Clarendon.

Melia, J. (2000) 'Weaseling Away the Indispensability Argument.' *Mind* 109: 455–479.

Musgrave, A. (1977) 'Logicism Revisited.' *British Journal for the Philosophy of Science* 28: 99–127.

Musgrave, A. (1986) 'Arithmetical Platonism: Is Wright Wrong or Must Field Yield?' in M. Fricke (ed.) *Essays in Honour of Bob Durrant.* Dunedin: Otago University Philosophy Department: 90–110.

Musgrave, A. (1999) *Essays on Realism and Rationalism.* Amsterdam: Rodopi.

Papineau, D. (1993) *Philosophical Naturalism.* Oxford: Blackwell Publishers.

[22] I'd like to acknowledge a considerable debt to Alan Musgrave. Early in my graduate student days at the Australian National University, Alan was a visiting fellow for several months. During this time we had many engaging discussions on realism and anti-realism. Although we agreed on a great deal, there was also considerable disagreement. The realism–antirealism debate in the philosophy of mathematics was one issue on which we disagreed. As always, Alan argued for his position vigorously and provided a formidable target for any opponent. Alan forced me to sharpen my thoughts on the realism debate in the philosophy of mathematics, and his example of how to pursue philosophy with both enthusiasm and rigour remains with me to this day. While I don't expect Alan to agree with everything I have to say in this chapter (indeed, I'd be disappointed if he did!), I hope that he recognises his considerable influence on my thinking about the issues in question.

This chapter has benefited from comments from audience members at the University of Cambridge and the 2004 meeting of the British Society for the Philosophy of Science at the University of Kent in Canterbury. I'd also like to thank Otávio Bueno, Colin Cheyne and Mary Leng with whom I've had many fruitful conversations on issues that bear on this chapter. Finally, I'd like to thank the Center for Philosophy of Science at the University of Pittsburgh where I held a Visiting Research Fellowship in the winter term of 2004 and where some of the work on this chapter was carried out. Work on this chapter was funded by an Australian Research Council Discovery Grant (grant number DP0209896).

Putnam, H. (1971) *Philosophy of Logic*. New York: Harper.
Quine, W.V. (1981) 'Success and Limits of Mathematization.' in *Theories and Things*. Cambridge, Mass.: Harvard University Press: 148–155.
Resnik, M.D. (1997) *Mathematics as a Science of Patterns*. Oxford: Clarendon Press.
Rosen, G. 1992. 'Remarks on Modern Nominalism.' Princeton: PhD dissertation.
Smart, J.J.C. (unpub.) 'Prospects for the Philosophy of Mathematics.' Unpublished manuscript.
Sober, E. (1993) 'Mathematics and Indispensability.' *Philosophical Review* 102: 35–57.
van Fraassen, B.C. (1980) *The Scientific Image*. Oxford: Clarendon Press.
van Fraassen, B.C. (1985) 'Empiricism in the Philosophy of Science.' in P.M. Churchland and C.A. Hooker (eds.) *Images of Science: Essays on Realism and Empiricism*. Chicago, Ill: University of Chicago Press: 245–308.

NORETTA KOERTGE

A METHODOLOGICAL CRITIQUE OF THE SEMANTIC CONCEPTION OF THEORIES

A new PhD slated to teach a beginning undergraduate course on scientific reasoning recently asked me to recommend topics. I launched into a description of my 'baby-Popper-plus-statistics' class—give them enough deductive logic to understand the Duhemian problem, do the Galileo case study, use the notion of severe test to introduce a bit of probability theory, then segue to the problem of testing statistical hypotheses.... My interlocutor was looking impatient. 'But I'm a strong adherent of the Semantic Conception of theories,' he said. 'I can't teach all that stuff about trying to falsify bold conjectures.' This was not a moment for proselytizing, so I loaned him a copy of Giere's textbook, which is based on the Semantic Conception, and sent him happily on his way. However, this episode raises an interesting question, one that takes on some urgency as the Semantic Conception of scientific theories (SC) seems well on its way to becoming the new received view: What accounts of scientific method, confirmation and explanation does the SC support?

A major motivation of the Semantic Conception for philosophers was to replace the awkward syntactic account of theories proposed by the positivists, who favoured a formal axiomatic system accompanied by 'correspondence rules' or meaning postulates. But since Popper never gave an account of meaning and was never worried about the problem of how to interpret theoretical terms, it might seem that there should be no inherent tension between Popperian methodology and an account of science that views theories as sets of models. Because the Semantic Conception liberates the content of a theory from any particular linguistic formulation, this move might appear congenial to those in agreement with the Popperian dictum, 'Words don't matter'. Furthermore, the Semantic Conception's emphasis on mapping structures that reside in the world also seems to mesh with Popper's anti-essentialism.

Yet I will argue that the overall approach to scientific inquiry that accompanies the SC approach is antithetical to a Popperian account of scientific methodology, which is intended to maximize the role of criticism. Moreover the methodological glosses that commonly accompany expositions of the Semantic Conception are either antithetical to commonly accepted norms of scientific inquiry or hopelessly ad hoc. The points I wish to make are concerned neither with the technical details of the

C. Cheyne & J. Worrall (eds.), Rationality and Reality: Conversations with Alan Musgrave, 239–253.

SC nor the more idiosyncratic aspects of Popperian methodology, such as his views on induction. Furthermore, for the purposes of this paper I will not worry about whether we should view our best scientific theories as being true, approximately true or merely empirically adequate. My intention is rather to dramatize the *methodological* differences between an approach to science that focuses on model building and one that construes scientific inquiry as a search for true or approximately true or empirically adequate generalisations.

Most of my claims are straightforward and do not hinge on the particular variant of the Semantic Conception one adopts as long as that conception views scientific theories as a class of models of relational structures, not as universal generalisations. Glymour gave this characterization of the SC:

> (T)he product of science in this view is not so much knowledge of general propositions … as an understanding of systems of models and how to embed various classes of phenomena within these models. (Salmon, p. 122)

Thus for my purposes, van Fraassen is not a target simply because his SC account includes generalisations as part of the theory. He says a theory is empirically adequate exactly when '*all* appearances are isomorphic to empirical substructures in at least one of its models' (Boyd, p. 192*)*. Neither am I concerned here about the SC approach to a theory such as cosmology which has only one intended application. The da Costa and French book arrived too late for me to evaluate it carefully, but it may also evade the criticisms that follow.

As these exceptions make clear, my quarrel here is less with the SC *per se* than with the methodology associated with it especially by writers who take the disunity of science to be a virtue. Ron Giere is the SC theorist who has paid most attention to methodological issues. Therefore I will generally illustrate my argument using examples from his writings, which have the additional virtues of being clear and extensive. To my mind, his is an especially interesting version of the Semantic Conception, in part because he takes seriously the goal of accounting for our best examples of scientific practice. Nevertheless, I find that it does not hang together as well as a methodology that construes theories as statements.

1. WHAT DOES THE SC SAY ABOUT REFUTATION?

To remind ourselves of how the Semantic Conception represents science, let us begin with Giere's take on Halley's Comet. What Halley did was fit a 'Newtonian model for two bodies in an elliptical orbit attracting one another by the force of gravity' (1991, p. 72) to Halley's observational data concerning the 1682 comet and show that the same model fit observations of comets sighted at 76 year intervals before. Halley also predicted that the comet would return in 1758 (which it did). Thus the hypothesis that the comet of 1682 fit the Newtonian two-body model was confirmed. Giere does not say that Newton's theory taken as a world system or Newton's Laws were confirmed. Rather he points out that Newtonian models were applied so successfully to both terrestrial and celestial phenomena that Newtonian science became an inspiration for the Age of Enlightenment (1991, p. 69).

Exponents of the SC all make this point quite explicit—a theory defines a set of models. When we find some of these models fit discrete portions of the world we may decide to search for more instantiations in situations that appear to be analogous in some way. But we neither confirm nor refute theories. Rather we show that one model of the theory does or does not fit some real world situation. It is the limited hypothesis that one of the models fits one aspect of the world that is confirmed or refuted.

Giere's account of the failure of the phlogiston theory makes this move explicit. After summarizing the phlogiston theory and Lavoisier's experiments with mercury, Giere concludes: 'The data, therefore, provide evidence that the phlogiston model fails to represent the controlled combustion of mercury as carried out in this experiment' (1991, p. 76). So far, so good. What the Semantic Conception does not sanction is the further claim that Lavoisier's experiment gave evidence that the phlogiston account of other chemical reactions was also thereby discredited.

Of course Lakatos and others have emphasised that we should not advocate a naïve view of refutation. Giere's discussion of the discovery of Neptune illustrates nicely the sort of tinkering with initial conditions that is characteristic of science: When a simple Newtonian model failed to fit the observed orbit of Uranus, scientists were 'forced to conclude that their current models were not correct' (1991, p. 90). Around 1843 Adams and Leverrier proposed a more complicated model that posited a new planet. The new model did fit the orbit and Neptune was observed in 1846. But exactly why did Adams and Leverrier keep tinkering with Newtonian models? Did they think that Newton's Laws were true of all celestial bodies or that the whole universe was a Newtonian system? Giere's explanation is ambiguously worded:

> By that time, there had been so many successful predictions using Newtonian models that they were reluctant to conclude that the general theory could be wrong. (1991, p. 90)

Unless Giere has unconsciously lapsed into thinking of theories as making claims, he probably should replace talk of a theory 'being wrong' with something like 'failing to provide models that fit all situations.'

Giere does not discuss the problem with Mercury's perihelion or it's role in the replacement of Newtonian mechanics and it is difficult to come up with a Semantic Conception account of the motivation for this transition. Perhaps it would go something like this: Research showed that Newtonian models fit an enormous variety of physical phenomena, but, after repeated attempts, scientists failed to find a Newtonian model for the trajectory of Mercury. However they could fit that data with an Einsteinian model. For that reason (and other reasons as well) they decided to start to advocate that in principle Einsteinian models should be used everywhere, even in the situations where the old Newtonian models fitted to within the limits of experimental error.[1]

On the Statement View, Einstein proves Newton false although Newtonian models do fit most data sets pretty well. However, since scientists seek true theories, they prefer Einstein to Newton. On the Semantic Conception Einsteinian models fit

[1] Of course, using the more sophisticated model is not necessary for practical purposes in most cases.

more data sets more accurately than do Newtonian models. To explain the fact that scientists prefer Einstein to Newton, Semantic Conception adherents must also posit that scientists value theories whose models fit more phenomena. Appeals to the truth of the theory have been replaced by appeals to a wide domain of applicability.

The metaphysical differences between these accounts may exercise philosophers, but would it make any difference to actual scientific inquiry if the Semantic Conception were to prevail? As Giere persuasively argues in his 1988 monograph *Explaining Science*, both physics textbooks and the practice of working physicists provide numerous examples of scientists reasoning just as described by the Semantic Conception, so it might seem that there would be no practical consequences to which stance one adopts. Nevertheless, let's probe the matter further by continuing our attempts to translate the traditional maxims of good scientific inquiry into Semantic Conception talk. Although the influence of philosophising on the practice of science may be tenuous, the impact on science education is more direct. In this arena at least, philosophical ideas do have consequences!

2. WHAT DOES THE SC SAY ABOUT VARIETY OF EVIDENCE?

What about the familiar exhortation of Bacon and Mill to test theories against a wide variety of evidence? From the perspective of the Statement View this advice makes good sense. If a theory makes claims about a wide variety of phenomena and we want to find out whether it is true or not, we are well advised to look for its weak spots and to test it against unfamiliar cases, especially ones that were not contemplated when the theory was formulated. In order to capture this aspect of good scientific practice, the Semantic Conception must once again invoke the importance of wide applicability.

But note that this addendum stands in tension with the SC account of how we apply the theory. Remember that on the Semantic Conception Newton's theory does not make universal generalisations about any domain. Rather, it defines a set of models that one can match up against bits of the world. The Semantic Conception *per se* does not tell us where to apply these models.[2] The only systematic advice that the Semantic Conception provides is to look for similarities between cases where the model works and new examples. The analogies may be expressed in ordinary language and describe similarities in the physical systems themselves (if a Newtonian model fits Earth's moon, perhaps it will fit the Jupiter's moons) or they may focus on formal properties of the model (e.g., let's try a harmonic oscillator again in this new case).

The concomitant methodological advice is to extend the domain of applicability of the theory's models cautiously, first looking for new phenomena that are quite similar to the ones already modelled. Deploying models in radically new domains may eventually be necessary, however, to ensure the auxiliary desideratum of wide applicability. Once again the role played by the central value of truth in the Statement View is carried by the secondary desideratum of wide applicability in the

[2] Newton did, of course, but these are *obiter dicta*.

Semantic Conception. Yet we often find scientists going out of their way to test a theory in novel situations where there are no strong analogues to previous successes. Stock examples include Eddington's eclipse expedition and the experiments on conical refraction. Since no one had tried to model the bending of light in a gravitational field before, might not the SC suggest that it might be a better use of Eddington's time and money to simply work out the Einsteinian model of Mercury's perihelion in more detail!

According to Salmon (p. 119), when Poisson showed that Fresnel's wave theory implied that light shone on a circular disk should produce a shadow with a bright spot in the middle, he thought the result was absurd. But Arago performed the experiment and confirmed Fresnel's theory. Such scientific behaviour is what would be expected on the Statement View, but it appears unmotivated from the SC perspective. If Fresnelian models work elsewhere, on the SC why should we care whether or not there is one model predicting an obscure phenomenon whereby a bright spot appears in the middle of the shadow of a circular disk? And why should we be so impressed when this wave theory model is successful?

3. WHAT DOES THE SC SAY ABOUT CRUCIAL EXPERIMENTS?

I also find the SC account of crucial experiments between theories (as opposed to cases where we are choosing between models within a particular theory) a bit contrived. Giere gives a clear summary of the inability of the Ptolemaic model to account for the phases of Venus and draws this conclusion:

> The data would be impossible on the Ptolemaic model, and no other plausible models have been mentioned. So the data must be taken as positive evidence that the Copernican model provides a good fit to the actual universe. (1991, p. 68)

My question is this: At the time, many astronomers considered their accounts of celestial movements to be mathematical devices that 'saved the phenomena' but did not describe what the heavens were really like. In the spirit of the SC why should we not say that the Ptolemaic system was never intended to model the sources of illumination of the planets, only their relative positions? If some other model speaks to that issue, fine, but on the SC why should that fact discredit Ptolemy? Since models are only intended to map certain aspects of the world, why try to stretch them to account for phenomena outside their original scope?

On the Statement View Ptolemy and Copernicus are inconsistent, not both can be true. On the Semantic Conception, we can easily claim that the Ptolemaic model fits the observed positions of planets perfectly well and the fact that navigators continue to use geocentric models reflects this interpretation. We can use the Ptolemaic model for some purposes, the Copernican for others, such as predicting the phases of Venus. On the SC there is no reason to attribute epistemic value to this crucial experiment. A similar gloss can be made on the import of the Eddington eclipse crucial experiment between Newton and Einstein.[3]

[3] I am ignoring the issue of how well Eddington's data actually confirmed either account.

If the Newtonian model does not fit the data, why shouldn't the SC adherent simply say that Newtonian models were never designed to talk about the path of light rays in intense gravitational fields and had never been applied in analogous situations? On the SC, this experiment helps us delimit the scope of application of Newtonian models, but it gives us no grounds for saying that Newtonian theory itself is false.

In *Science Without Laws* Giere recommends as a methodological rule that when two models posit conflicting accounts of some aspect of the world this is 'an indication that one or both types of models fail to fit the world as they might. It is an invitation to further inquiry to find models that eliminate the conflict...' (p. 83). He notes that proceeding as if the world had 'a single structure' has been fruitful in the history of science. So Giere's instincts agree with those who think the goal of science is good comprehensive theories but he has to gerrymander the SC in order to accommodate this salient feature of past science.

We seem to end up with this result. In order for the Semantic Conception of theories to provide a normative analysis of the famous episodes in the history of physical science which philosophers of science use to illustrate basic methodological maxims, it must rely heavily on imperatives that have a pragmatic, but not an epistemic status. To explain scientists' rejection of phlogiston or Ptolemy, or the fact that they consider Einstein better than Newton, the SC must advise students to prefer theories with the most workable models. This is not bad advice, but it rather sounds like Microsoft exhorting us to buy newer versions of their word processing programs—even though they are memory hogs and run slower, they do have additional features that might come in handy! Unlike the Statement View, the SC account does not distinguish the use of theories as instruments and the use of theories as global descriptive claims.

4. WHAT DOES THE SC SAY ABOUT SEVERE TESTING?

One of the most basic maxims of scientific method is nicely illustrated by Bacon's account of those who used the votive offerings of rescued sailors as evidence of the powers of the gods. As Bacon put it in his *New Organon*, 'But where are the offerings of those who were drowned at sea? Such is the way of all superstition.' Running around collecting positive instances of a generalisation or finding places where a template fits the world is only half of the story—we also need to actively look for situations where the theory fails. Thus Popperian and Bayesian accounts place special emphasis on conducting tests in situations where the evidence is unlikely to be positive unless the theory under test is true—neither background knowledge nor plausible competing theories make the positive outcome probable.

Giere adopts the language of severe testing but not surprisingly he restricts it to the testing of singular hypotheses. Thus when discussing what he calls 'marginal science' he points out that Freud's model of Little Hans makes no predictions that go beyond our common-sense accounts of the child's behaviour (1991, p. 103). In the Halley's Comet case, by contrast, one can argue that it is highly unlikely that the

Comet of 1682 would return in 1758 unless the proposed Newtonian model was correct (1991, p. 74).

Once again the SC account works *locally*—it picks out a crucial difference between the model of Little Hans and the model of Halley's Comet. What it fails to provide are global evaluations of Freudian and Newtonian theory. Does the weakness of the Freudian account of Little Hans give us reason to be dubious about future applications of that theory? Does Halley's impressive success give us reason to expect other Newtonian models will work? In particular, the SC cannot discourage the sort of cherry picking of favourable cases so vividly described by Bacon; instead, as we will we see in the next section, it actually encourages us to stick close to favourable cases.

5. WHAT DOES THE SC SAY ABOUT PREDICTION?

Any account of testing in science will talk about deriving predictions from a theory and comparing them to the results of experiment. So when Giere speaks of predicting the return of Halley's Comet, it is natural for someone accustomed to thinking of theories as if-then statements to miss the novelties of the SC account. Perhaps this toy example will highlight some of these distinctive features. Suppose I want to model the shape of my handkerchief as a square. I might proceed as follows. First I determine that my hanky has four sides. Based on my hypothesis that it is a square, I now predict that the sides are of equal length. If this is confirmed I go on to predict that the four corners are congruent. If either prediction fails, my hypothesis is refuted. But all this talk of prediction is really somewhat odd. By definition a square has four equal sides. My hanky either falls under the definition or it doesn't. It makes no sense to take some of the defining characteristics as initial conditions and others as 'predictions'.

Giere's account of how scientists made predictions about the return of Halley's Comet should be read as exactly parallel to the example of my piece-meal efforts to determine whether my hanky fits the model of a square. Certainly scientists performed a mathematical derivation, the conclusion of which was 'The Comet will return in 1758'. But the proposition that was tested is 'Halley's Comet is a Newtonian two-body system'. On the SC there can never be a 'Duhemian' dilemma where one is wondering whether a failed prediction should be blamed on the theory or on the initial conditions. As Rosenberg emphasizes, on the Semantic Conception theories are true by definition (pp. 96ff). So on the SC it is the choice of model which must always be faulted. Yet in the history of science there have been many cases in which scientists treated prediction failures as dilemmas and argued about which premise to fault. A standard example is the failure to observe stellar parallax at the time of Galileo. Did this finding refute Copernicus or simply mean that the distance to the stars was much greater than had been estimated?

Prediction also occurs in science when we apply a successful theory to new domains. On the statement account of theories, the scope of application is an integral part of the theory. The kinetic theory of gases should apply to all gases; Mendeleev's periodic law of the elements was intended to apply to all elements; etc.

On the Semantic Conception, however, theories *per se* simply supply models. Theories do not specify the domain of application. The SC account of extending a theory to new phenomena, therefore, depends crucially on notions such as similarity, resemblance, and analogy. If a theory fits a chloride, then we might try it out next on a bromide, *not* because the theory claims to describe all halogen salts or all chemical compounds, but simply because background knowledge leads us to think bromides may behave similarly to chlorides.

Expositors of the Semantic Conception give few details of the kind of informal inductive reasoning that is required to apply theoretical models to new situations. Giere's naturalistic version, however, suggests that psychological accounts of the structure of concepts may help describe how scientists jump from one application of a theory to another one. He also distinguishes between the similarity relations operating across folk categories and the more abstract structural resemblances that inform scientists' classification schemes. (1999, pp. 100-106)

To sum up, on the SC our standard locutions about the predictive power of successful scientific theories have to be dropped or reconstrued. The theories that scientists admire are, as Cartwright puts it, 'toolkits' for building successful models. The theories themselves do not describe where they apply, but through experience in modelling similar situations the scientific community develops informal guidelines for their deployment.

6. WHAT DOES THE SC SAY ABOUT EXPLANATION?

Let us now turn to current philosophical accounts of explanation and ask how competing models of explanation fit in with the SC. As long as we focus our attention on a single model of a theory, it would seem that the SC can speak of covering law explanations in a very limited sense. In his textbook Giere continues to talk about Newton's Laws. They are what animate the Newtonian models and tell us, for example, when Halley's Comet will return. So the Newtonian model of Halley's Comet both predicts and explains the observed positions of this celestial body by deriving them from general statements about the Comet-sun system. But since refutation on Giere's account logically impacts only the particular model, not the theory as a whole, then by a parallel argument explanation would subsume the motion of Halley's Comet only under a specific two-body model, not under a general Newtonian account of the motions of celestial and terrestrial bodies.

In his more sophisticated philosophical account *Science Without Laws*, Giere argues that Newton's Laws are more accurately denominated as 'equations' or perhaps the more honorific 'principles', in the light of their ubiquitous role in Newtonian models (1999, p. 94). He emphasizes that they do not function as universal generalisations, not even if we add provisos. Giere does believe that there is a relationship of physical necessity between the length and period of a simple pendulum clock, one that even supports counterfactuals, but this is again an addendum to the SC, not part of the core account.

If we look at causal accounts of explanation, the SC would appear to lend itself most naturally to talk of singular causes. If we like, we can trace causal trajectories within one model, but, as was the case with prediction, generalizing the causal account to other systems works only through analogy. The contrast with the Statement View emerges if we look at how each approach handles narrative explanations.

Consider, for example, a so-called Rube Goldberg machine. Here is a description of one of his classic cartoons:

> *Picture Snapping Machine*: As you sit on pneumatic cushion (A), you force air through (B) which starts ice boat (C), causing lighted cigar butt (D) to explode balloon (E). Dictator (F), hearing loud report, thinks he has been shot and falls over backward on bulb (G), snapping picture.(http://www.rubegoldberg.com/html/picture%20snapping.htm)

On the Statement View each step of the machine's operation would be explained using the appropriate causal law from a wide variety of scientific theories—weight on a confined gas increases its pressure, objects move more rapidly when the friction is low, animals are startled by unexpected noises, etc. Our understanding of each step inherits the epistemic and explanatory weight accruing from the fact that it falls under a covering law that has been confirmed in a wide variety of cases.

The Semantic Conception, by contrast, will provide a composite model constructed by referring to models that have worked well in closely analogous situations. So the reference model for the first step might refer to previous experience with pooh-pooh cushions, not general gas laws, the ice-boat model will be analogous to models of moving hockey pucks, etc. Understanding on the SC approach comes from similarity to the nearest neighbouring cases, which may in turn be connected to less similar cases. Of course, there are scientific situations in which the only kind of understanding that is available to us is through a network of family resemblances. But it seems cognitively preferable to have global connections made salient when they exist and this is exactly what the Statement View provides.

The unification account of explanation, though usually presented in terms of statements, incorporates some of the spirit of the SC approach to science. For example, Kitcher presents early genetics as a series of deductive schemata. To account for the distribution of phenotypes in what he calls 'pedigree problems' one fills in slots for genotypes, traits, dominance, etc. and then derives the expected distribution. If the given biological system does not fit the Mendel schema, one can move to the Morgan schema or the Watson-Crick derivation pattern. As in the SC approach, one does not test an overall theory; rather one hunts around until one finds a schema that fits. Kitcher also emphasizes that some theories may not contain universal laws; instead they consist of a variety of 'mini-laws' each figuring in their own deductive schema (1989, p. 447).

However, there is this basic philosophical difference between the two approaches. For Kitcher, unification, namely presenting a small number of schemata that will cover a large number of phenomena, is the defining feature of scientific explanation. But on the Semantic Conception, as we saw above, scientists' preference for an economical set of models appears to be a question of pragmatic convenience, not a central cognitive requirement.

Here is my evaluation so far of how well the Semantic Conception of theories fares when it is taken as the foundation of a general theory of scientific inquiry: It gives no direct rationale for the scientific practice of seeking a variety of evidence and novel test cases—in fact its resemblance account of how scientists apply old models to new instances points in just the opposite direction. It gives only a weak rationale for why crucial experiments between theories (as opposed to between competing models within a single theory) should have the epistemic weight that scientists attribute to them. To say that Einstein is preferred to Newton, or oxygen to phlogiston, or evolution to creationism simply because their models are more widely applicable is too faint praise. The localism of the SC approach that limits the scope of refutation and turns prediction into a search for analogues also yields a feeble account of scientific explanation although I have suggested that it might be strengthened by incorporating some sort of unification view.

7. A CRITICAL LOOK AT THE PRIMA FACIE ADVANTAGES OF THE SC

I have argued above that the Semantic Conception of theories is hard pressed to account for some of the most basic features of scientific inquiry such as the role of attempted refutation, variety of evidence, and crucial experiments in the search for theories with great predictive and explanatory power. Nevertheless, there are aspects of scientific practice where the Semantic Conception on the face of it has the upper hand. One of these can be dealt with quickly. It is it in fact true that scientists spend lots of time tinkering with models and generally blame them instead of the overall theory when discrepancies between data and prediction occur. This phenomenon can be rationalised in a variety of ways—Lakatos' Methodology of Scientific Research Programmes was motivated by this very issue. But the core of any Statement View response will rest on the distinction between models of initial conditions (what Lakatos has in mind) and the sorts of theoretical models invoked by the Semantic Conception.

The Semantic Conception also gives a smooth account of the fact that scientific representations rarely fit the data perfectly and that scientists often promiscuously flit back and forth between models according to what is convenient—sometimes earth-centered, sometimes heliocentric. Sometimes chemists use a valence-bond approach, other times they postulate molecular orbitals. There is a quick Statement View response to the first example—scientists use analyses they know to be false as convenient calculating devices, but they rely on the theory they consider to be true (or closer to the truth) to tell them when the old theory can be employed as an instrument. However, the VB/MO clash is more problematic and it leads us to the biggest challenge that the SC poses: What can the Statement View really say about fields that lack a single, detailed overarching theory? Does not the SC give better advice in such cases? A prime example would be Evolutionary Theory, which Popper at one time construed as unfalsifiable and hence falling on the metaphysical side of his Demarcation Principle. It is no accident that there is strong adherence to the Semantic Conception among philosophers of biology.

I will come back to biology later. But let us first begin with the case of chemists' apparently promiscuous use of Valence Bond descriptions of molecules (e.g., the configuration of methane is tetrahedral because the bonding electrons of carbon are in sp3 orbitals) versus Molecular Orbital accounts (e.g., the spectrum of benzene is best explained in terms of a sea of electrons on each side of the flat ring). Is this not a clear example of model fitting where there are no theoretical statements underwriting the activity?

Well, this is a sort of messy mixed case. It is true that chemists do not simply model initial conditions and derive predictions from Quantum Mechanics—this is *not* a chemical equivalent of the Halley's Comet problem. Nevertheless, there does exist an overarching theory that guides chemical explanations even when it does not figure directly in derivations. Thus it is the Pauli Exclusion Principle which tells chemists the properties of the available bonding electrons. Theory thus puts enormous constraints on the VB or MO models that chemists build. Theory also tells us that the electrons in every molecule are differentially localized. Valence Bond approaches work best when there is a high degree of localization; Molecular Orbitals reflect the other extreme. There are independent theoretical considerations that allow chemists to predict for a new molecule which model will give the best fit. The choice is not just one of trial and error or reasoning by analogy.

So what at first appears to be opportunistic modelling turns out to incorporate elements of the conception of theories as generalisations about a domain. The Statement View illuminates more scientific practice than might at first appear. I will now argue that the wider applicability of the Semantic Conception is actually a mark against it because it allows too much to count as scientific inquiry.

A common mode of analysis in literature is to point out structural similarities between plots. These can range from the simple 'boy meets/loses/gets girl' recipe for B-movies to Vladimir Propp's elaborate structural theory of folktales. Recalling the emphasis on polarities and analogies in Greek science and the proliferation of oppositions that are 'good to think' in Levi-Strauss' structural anthropology, it is evident that the human mind delights in mapping abstract models onto the variegated manifold of experience. In this respect the SC resonates well with very common cognitive strategies. It places Propp's theory of fairy tales on one end of a continuum that has Newtonian models on the other.

But is this a mark in favour of the SC? I think not. Any philosophy of science worthy of study, even a thorough-going naturalism, needs to criticise modes of thinking that lead us astray. From a Popperian point of view, if there is no falsifiable claim about the structure of folktales, then merely pointing to formal similarities where they can be found (thus ignoring stories which don't fit the pattern or imposing the pattern on them in an *ad hoc* way) has no explanatory power. We should be prepared to mount similar criticisms of rational choice theories. It is easy to look around at human behaviour and pick out examples of people maximizing expected utility. On the SC this is a perfectly acceptable scientific activity. But on the Statement View we would urge scientists to specify the domains where they expect rational choice to work as well as to theorize about what sorts of factors tend to vitiate the rational choice approach. Although the SC offers social scientists a veneer of respectability—just like physical scientists they are finding places where models fit the world—it comes with a loss of cognitive clout.

But now it's time to talk about the SC's poster child, biological science. How well does biology fare if we require tougher standards than those inherent to the SC? I have neither the space available nor the expertise to whip out a detailed Popperian analysis but I will record a couple of suggestions. My first take on the adaptationist research program would be identical to the above gloss on rational choice theory. One needs to provide some sort of general account, even if it starts out with only a laundry list of problems that organisms have to 'solve' in order to survive, that gives at least some guidance to which traits are likely to be adaptations.

The Statement View can be tolerant of weak, incomplete, or vague theories as long as one is trying to improve them. What is not acceptable is the opportunistic matching of templates to phenomena where only successes are recorded.

Much of biological research can be viewed as a highly ramified and generalised version of the narrative explanations discussed above. For example, the various steps of the Krebs Cycle fall under different laws of biochemistry; one early step is a case of oxidative decarboxylation. But unlike the Rube Goldberg machine discussed above, the metabolic mechanism described by Krebs operates in most higher animals. So one can also attempt generalisations about where the Krebs chain of reactions will be found. Current work in evolutionary biology, known as 'evo-devo', also attempts to analyse complicated causal mechanisms and make generalisations about where they occur.

There is no doubt that explanatory practices in biology appear more complicated than that which is presented in textbook physics. And, as we have seen, in such cases it is tempting to fall back on the SC because it makes any sort of modelling activity count as full-fledged science. But when we look more carefully at the achievements of biology and why they are considered great, I think we find the tougher methodological values associated with the Statement View operating everywhere. Biologists do not need to espouse the SC in order to be counted as first-rate scientists! Both the form and the content of scientific theories will depend on the nature of the phenomena they are trying to describe and explain.

But perhaps we should give the SC points for providing us with a congenial way to talk about pre-paradigm science. In a bookstore I once leafed through a hefty volume that presented 27 different theories of personality.[4] I suppose one could view this largess either as a testimony to the creativity of psychologists or as a reflection of the complexity of the human psyche. However to me it was a classic indication of the fact that we don't yet have a good scientific understanding of personality. When scientists first explore phenomena they often resort to curve fitting and sometimes develop models with quite limited applicability—one recalls the two equations for Black Body radiation, those of Wien and Rayleigh-Jeans (Kuhn, 1978). But it was only with the development of Planck's theory that one had both a single formula for all wave lengths and a picture of the underlying mechanism that explained it. A similar scenario applies to spectra before the Bohr atom.

[4] It may have been Burger & Thompson (1993).

On the Statement View, equations whose domain of application is restricted for no reason except the brute fact that they don't fit elsewhere are at best viewed as stepping stones to a general, unified account. On the Semantic Conception, however, curve fitting is a paradigm case of model building; mapping limited aspects of a discrete physical system is not only what scientists do best—it's all that they do! Recent advocates of the disunity of science argue that the ideal of a unified science waxes and wanes according to local circumstances, including political considerations.[5] Here is not the place to review those arguments. I would merely say that the SC meshes nicely with a picture of science as a pluralistic, patchwork quilt of partial perspectives.

The nexus is not a necessary one, however. Giere, unlike the more extreme partisans of disunity, advocates that scientists adopt what he calls a one-world methodological rule: 'Proceed as if the world has a single structure. In light of this rule, the existence of conflicting models is an indication that one or both types of models fail to fit the world as well as they might' (1999, p. 83). Giere's justification of this rule does not rest on metaphysics, he claims. Instead he would point to episodes in the history of science where following such a rule was fruitful. Once again, we see that the SC must be supplemented in order to bring it into line with exemplary scientific practice.

And I submit that it is extremely important that one adhere to the one-world methodology. One of the most important sources of deep problems in the history of science has been clashes between theories, each of which are eminently successful in their own domains, but which give incompatible accounts in areas where they overlap. Standard examples include the conflict between the Copernican astronomy and Aristotelian physics, wave vs. corpuscular theories of light, the early Bohr atom and Maxwell's equations. But on the SC none of these inter-theoretic inconsistencies need give scientists pause—it simply means that they must judiciously pick one kind of model to account for spectra and another to account for the behaviour of moving charges. Again, we see that the SC needs to be supplemented if it is to serve as a guide to good scientific practice.

8. IN CONCLUSION

Whatever its flaws (and they were many!) the old logical empiricist account gave an integrated account of the structure of science. The 'layer-cake' model, as Feyerabend mockingly called it, provided an excellent jumping off point for the analysis of prediction, confirmation and explanation, and the interplay of theory and observation. Popper dropped the logical empiricist account of meaning and criticized the account of induction that often accompanied it but kept the idea of a layered deductive structure of statements intact.

The Semantic Conception pretty much started over again from scratch. Its proponents were very concerned to give a radically different approach to the problem of how abstract scientific theories are related to the natural world. The early

[5] For a guide to this literature see Cat (2006).

attempts to formalize scientific theories using a set-theoretic approach used as examples universal theories from physics. At this point there was no reason to think that the SC would lead to a radically different picture of science—it was simply clearing up the vexed positivist problem of meaning. But people quickly realised that the SC might be adapted to Kuhn's philosophy of science as paradigms—in fact Stegmüller enthusiastically discussed the prospect of 'Sneedifying Kuhn'. By viewing theories as tools for building maps, one could bypass many of the worries about truth and verisimilitude. The SC also lent itself naturally to the analysis of sciences such as meteorology where the central projects are the development of computer simulations and models of huge sets of data.

By and large Semantic Conception philosophers have neglected giving a systematic account of scientific methodology. Their intellectual forbears such as early 20th century conventionalists always placed a high value on economy of thought and this allowed them to stress the importance of inter-theoretic as well as intra-theoretic consistency and the search for comprehensive theories that made precise predictions over a wide domain. Giere's account of science as models tries to include these desiderata. However, in the discussion above I have amassed multiple instances where his version of the SC fails to give an adequate account of scientific inquiry unless it is supplemented with methodological rules that are often in tension with the central tenets of the SC approach. Many current SC philosophers see science as much more fragmented and in some cases actively oppose attempts to unify science.[6]

The basic issue, however, is not unification or pluralism *per se*. The key problem for any account of scientific methodology is to tell us how to maximize the critical appraisal of scientific claims. As my first logic teacher Alan Musgrave used to emphasize, logical inconsistency is the engine of criticism. Sometimes the inconsistency is between theoretical predictions and experimental results. Even more challenging are inconsistencies between successful scientific theories. On the Semantic Conception one is invited to take mapping or modelling as the basic activity of science and this very starting point makes it much more difficult to talk about the critical processes that are so distinctive of science—severe testing, clashes between theories, the search for deep explanations. We may someday see an account of science that integrates the good features of model-theoretic devices with a robust, critical methodology, but so far the Semantic Conception approaches do not look promising.

REFERENCES

Boyd, R., Gaspar, P., and Trout, J.D. eds. (1991) *The Philosophy of Science* (MIT Press).

Burger, J.M. and Thompson, K.L. (1993) *Personality* 3rd edn. (Brooks/Cole Publishing Company).

Cat, J. (2006) 'Unity and Disunity of Science' in *The Philosophy of Science: An Encyclopedia* edited by S. Sarkar & J. Pfeifer (Routledge) pp. 842-47.

Da Costa, N.C.A. and French, S. (2003) *Science and Partial Truth: A Unitary Approach to Models and Scientific Reasoning* (Oxford University Press).

[6] See the discussions in Galison (1996).

Galison, P. and Stump, D.J. (eds) (1996) *The Disunity of Science: Boundaries, Contexts, and Power* (Stanford University Press).
Giere, R.N. (1988) *Explaining Science* (The University of Chicago Press).
Giere, R.N. (1991) *Understanding Scientific Reasoning* 3rd edn. (Harcourt Brace Jovanovich College Publishers).
Giere, R.N. (1999) *Science Without Laws* (The University of Chicago Press).
Kitcher, P. (1989) "Explanatory Unification and the Casual Structure of the World" in *Scientific Explanation* (1989) pp. 410-505.
Kuhn, T. (1978) *Black-Body Theory and the Quantum Discontinuity:1894-1912* (Oxford University Press).
Rosenberg, A. (2000) *Philosophy of Science: a contemporary introduction* (Routledge Taylor & Francis Group).
Salmon, M.H. et al. (1992) *Introduction to the Philosophy of Science* (Prentice Hall).
Salmon, W.C. (1966) *The Foundations of Scientific Inference* (University of Pittsburg Press).
Stegmüller, W. (1979) *The Structuralist View of Theories* (Berlin).

GRAHAM ODDIE

A REFUTATION OF PEIRCEAN IDEALISM

Some years ago I arrived at Otago University as a freshman, intending to study law and to become a lawyer. My Hall of Residence thoughtfully assigned me a roommate whose intention it also was to study law and become a lawyer. My roommate had already spent a year working in a lawyer's office, and so he was better informed than I was about the daily practice of the law. The evening before registration he told me all about the Fencing Act—not the laws governing duelling, which might have been interesting, but the laws governing the boundaries between suburban neighbours. The idea of spending my life thinking about such matters didn't thrill me, and I resolved to withdraw from law and sign up as a philosophy major the following day. To make the change I had to visit the Head of the majoring department and ask him to sign the necessary papers. That is when I first met the young Alan Musgrave, Professor of Philosophy.

I have to confess that at that meeting Professor Musgrave did not inspire me, or even encourage me to pursue philosophy. He intimated I was doing something a bit foolish—passing up the chance for a degree (and a lucrative career) in law, for a degree (and almost certain unemployment) in philosophy. Professor Musgrave's reaction must have given me pause for thought—why else would I remember it more than thirty years later? But my next encounter with the Professor was in the first lecture of his Introduction to Philosophy course—and that was a very different experience. I can still see him striding up and down at the front of the lecture hall, talking loudly without lecture notes in his accent from Manchester, punctuating sundry claims with a belligerent 'Yes?' that more or less demanded agreement.

Alan's excellent lectures were highly entertaining, enormously informative, amazingly clear, totally lacking in obfuscation, and bracingly partisan. Unlike many of his contemporaries in the profession, he was adamant that the discipline of philosophy had made real progress, that philosophical conjectures could be, and often were, refuted. It was obvious when Alan thought a philosophical conjecture had been successfully refuted. It was also obvious that he was a passionate advocate

C. Cheyne & J. Worrall (eds.), Rationality and Reality: Conversations with Alan Musgrave, 255–261.

for a brand of common-sense realism mixed in with a healthy dose of scepticism (which he called fallibilism). This was heady stuff for a seventeen year old who knew absolutely nothing. By the time that first lecture was over I was hooked on whatever it was the absurdly young professor was doing up there—philosophy—and I have been ever since.

Musgrave's introduction to philosophy had a powerful and lasting effect on me. But more than the content of his philosophy, the way Alan did philosophy is what intrigued me. I admire, perhaps now more than ever, the virtues I saw dramatically exhibited in that first series of lectures. I was reminded of these virtues while re-reading two of Musgrave's sorties against his bête noir—contemporary idealism (Musgrave 1997 and 1999). And the fruit of that reminder is this paper—a rather ridiculously simple argument against a class of theories that Musgrave finds particularly irksome. These accounts all embrace, in one form or another, the Peircean idea that truth is whatever our best scientific theories endorse in the limit of inquiry.

My argument has a number of famous predecessors, one of which Musgrave himself outlines in his attack on epistemic theories of truth (Musgrave 1997). He attributes the argument to Timothy Williamson, but in fact it has a long and illustrious lineage that can be traced back to a 1963 article in the *Journal of Symbolic Logic* by Fitch. (For a good summary of this history see the on-line encyclopedia article by Brogaard and Salerno. For an application of a Fitch-style argument to the issue of traditional idealism, see Oddie 2000.)

Many philosophers are, as Musgrave puts it, 'certainty freaks'. They cannot tolerate the possibility that we can't really be justifiably certain about anything much of interest, or that most of what passes for knowledge is fallible. Musgrave has argued that idealism, in all of its various guises, is at bottom an attempt to close the certainty gap; an attempt to guarantee that what we experience, or what we believe, or what science proclaims, is the way the world really is. But the gap is closed by metaphysical fiat—the world is declared to be nothing more than the sum total of what we experience, or believe, or what science puts its stamp of approval on.

The range of idealist theories that Musgrave is most exercised about are those in the last category—those according to which endorsement by our best scientific theory is not merely an indicator, but is rather *constitutive*, of the real. The true is what our best science endorses.

Science changes, of course, and our best scientific theory at one particular time might be different from our best scientific theory at some other time. Thus what our best science endorses will also change. Because of this we cannot identify truth with what is endorsed by our best scientific theories *simpliciter*. For a start, no one thinks that science has had its final say, and for another it would render time-independent propositions (like the laws of nature) temporally fickle affairs. This is where the Peircean limit theory of truth comes to the rescue: the truth is not necessarily what we find to be epistemically acceptable in the light of the current crop of theories, but rather what we will (or would) find epistemically acceptable in the light of theories we will (or would) hold at the *limit* of the scientific endeavour.

I say 'theories' rather than 'theory' for two reasons. The first is that there may be two rival theories in a domain, both of which endorse a large number of common propositions, but which also disagree over others. Even if we cannot rank one of the theories above the other then the propositions they agree on should still be deemed epistemically acceptable. The second is that different theories might cover different domains, and there might be some propositions that are the consequences of the best theories, taken together, governing different domains. In that case we would also want such propositions to be deemed epistemically acceptable.

Musgrave marshals a range of refutations of Peircean idealism. In particular, he gives a summary of a Fitch-style argument against the positivist thesis that if a proposition is true then it is knowable. With a couple of very weak principles governing possibility and knowledge we can show that this entails that *if a proposition is true then it is known*. And that is clearly absurd.

It will be useful to sketch the Fitch-style argument and consider its relevance to Peircean idealism. Where Kp is short for *p is known*, and Pp is short for *p is possible*, then we can abbreviate *p is knowable* (that is, it is possible for p to be known) to PKp. The positivist thesis that all truths are knowable is, as Musgrave notes, not supposed to be merely contingently true. It is supposed to be a matter of necessity. Of necessity, any true proposition is knowable. We can state the thesis succinctly thus (where $\Rightarrow$ is logical entailment).

Positivist thesis: For all p, $p \Rightarrow PKp$.

A proposition cannot be known unless it is true. Again, this principle (call it *facticity for K*) is not a contingent feature of knowledge. That p is known entails that p is true.

Facticity for K: For all p, $Kp \Rightarrow p$.

Finally, we assume that knowledge of a conjunction implies knowledge of each conjunct. Call this Distribution for K.

Distribution for K: For all p, q, $K(p \,\&\, q) \Rightarrow Kp \,\&\, Kq$.

Finally, to make all our assumptions explicit we need the modal principle that if a proposition logically entails a contradiction then it isn't possible for it to be true. Note that any concept of possibility, even logical possibility, will deliver this principle.

Impossibility principle: For all p and q, $(p \Rightarrow (q \,\&\, {\sim}q)) \Rightarrow {\sim}Pp$.

Here is one way of running the Fitch-style argument. For the sake of a *reductio* we start with an intriguing assumption. Assume that for some arbitrary proposition p the following is known: that p is true and p is not known to be so, (i.e. $K(p \,\&\, {\sim}Kp)$). We then show that this assumption entails a contradiction. Consequently it is not possible that $K(p \,\&\, {\sim}Kp)$. So ${\sim}PK(p \,\&\, {\sim}Kp))$ for any proposition p. According to the positivist thesis, if $(p \,\&\, {\sim}Kp)$ were true then so too would $PK(p \,\&\, {\sim}Kp)$. But we have just shown that PK(p & ~Kp) is false and so (p & ~Kp) must be too. ~(p & ~Kp) is however, logically equivalent to the thesis we want to prove: that for any p, if p is true then p is known.

It may be helpful to set out this proof formally. Where a step is a matter of undisputed logical inferences I cite *logic* as the justification.

1.	K(p & ~Kp)	Assumption
2.	Kp & K~Kp	1, Distribution
3.	Kp	2, logic
4.	K~Kp	2, logic
5.	~Kp	4, Facticity
6.	K(p & ~Kp) ⇒ Kp & ~Kp	1—5, logic
7.	~PK(p & ~Kp)	6, Impossibility
8.	p & ~Kp ⇒ PK(p & ~Kp)	Positivist thesis
9.	~(p & ~Kp)	7, 8, logic

As noted, step 9 is tantamount to the claim that if p is true then it is known to be so. But since the principles in the derivation up to and including 9 are purely logical, we have established more than the mere truth of ~(p & ~Kp). We have established its logical truth. (The only non-logical assumption is discharged at step 6.) Thus we have established that, of necessity, if p is true then p is known. That is to say, that a proposition p is true logically entails that p is known (p ⇒ Kp). And that is *logically* absurd—unless an omniscient being exists of logical necessity. (A new ontological argument might be culled from positivism!)

If the positivist doctrine were the weaker claim that all truths are knowable as a matter of contingent fact, then all we could derive is the weaker conclusion that if a proposition is true then it is known in fact. But that is also absurd—unless, perhaps, an omniscient being happens to exist. (Positivists would not rejoice to find they have provided an argument for the merely contingent truth of theism.)

Can this Fitch-style strategy be turned against the Peircean idealist directly? The Peircean idealist holds that in the limit science will (or would) reveal all truths, and that does seem to entail that all truths are knowable. However, I would like a more direct refutation of Peircean idealism, one that uses only principles that the Peircean explicitly affirms or which everyone would accept. One reason for seeking out another argument is that even if we can show that the Peircean is committed to the positivist principle, the Fitch *reductio* is by no means uncontroversial. Many of these criticisms are without merit (see Oddie 2000). However, the distribution principle for knowledge does seem a little bit suspect. For together with just one other equally plausible principle of knowledge, distribution entails the closure of K under entailment. Closure of K under entailment is the claim that if a proposition is known then all of its logical consequences are also known.

Closure for K: (Kp & (p⇒q)) ⇒ Kq.

The extra principle that in conjunction with distribution entails closure is the principle of equivalence. (Let $\Leftrightarrow$ be mutual entailment, or equivalence.)

Equivalence for K: $(p \Leftrightarrow q) \Rightarrow (Kp \Leftrightarrow Kq)$.

If we think of propositions as individuated by their truth conditions, then to know a proposition is to know that a certain truth condition is satisfied. Thus if p is known to be true, then any other proposition with the very same truth condition is also known to be true (even if we don't know we know it).

We can easily prove closure from distribution and equivalence:

1. $Kp \,\&\, (p \Rightarrow q)$	Assumption
2. Kp	1, logic
3. $p \Rightarrow q$	1, logic
4. $p \Leftrightarrow (p \,\&\, q)$	3, logic
5. $K(p \,\&\, q)$	2,4 Equivalence
6. Kq	5, Distribution
7. $(Kp \,\&\, (p \Rightarrow q)) \Rightarrow q$	1—6, logic

Closure for knowledge does seem a bit dubious. The fact that a proposition is *known* does not seem to entail that all its logical consequences are known. Since closure is a consequence of distribution and equivalence, that doubt spreads over the conjunction of the latter two principles.

So, it would be nice to have an argument against Peircean idealism that did not get us bogged down in debates about closure for knowledge. But as we will see, we cannot simply take over the Fitch argument utilising the preferred Peircean concepts (the concepts of *truth in the limit* and of *epistemic acceptability*), because these do not necessarily obey the same principles as knowledge and possibility.

Let us abbreviate the claim *proposition p is epistemically acceptable according to our best scientific theories* to Ep. I take it that propositions are not eternal, that they can change their truth values over time. And it is clear that the proposition Ep in particular is not time-independent. Ep can be false at one moment, true at a later moment, and perhaps false again later on. Further, Ep might be true at one moment while $E \sim p$ is true at another later moment. (We can even entertain the possibility that our best theories are jointly contradictory (perhaps unbeknownst to us), so that Ep and $E \sim p$ may both be true at a single moment.) Most importantly for our purposes, however, is that there is no principle of facticity for E. Ep clearly does *not* entail p. That is the main reason we cannot simply apply the Fitch strategy here.

Peircean theories make use of the idea of Ep being true 'in the limit'. The idea of limiting truth does not, however, have to be confined to propositions about epistemic acceptability. It is a quite general notion. But what exactly does it mean to say of a possibly changing proposition that it is true 'in the limit'? There is an obvious answer to this, as well as a slightly less obvious one. First, the obvious answer. Suppose a proposition p routinely changes its truth value. At t, p is true; at t+, p switches its truth value to false; at t++, p switches back to true … and so on, forever, at either regular or irregular intervals. It would seem problematic to say in this case that p is either true or false in the limit. Suppose, on the other hand, that after switching like this for some time, p becomes true and remains true thereafter. Then it does seem entirely appropriate to say that p is true in the limit. So if a proposition becomes true and stays true permanently, then it is true in the limit.

There is another type of case in which we may be tempted to say that p is true in the limit, even though p's truth value keeps switching, and never settles down. Suppose the intervals during which p is false get relatively shorter and shorter, so that the ratio of the sum of the lengths of intervals during which p is false to total elapsed time (measuring both from some given moment), tends to 0 as time passes. Is p true 'in the limit'? There is clearly a sense in which p tends towards permanent truth even though it never quite *makes* it to permanent truth. Satisfying the more obvious notion of limiting truth entails satisfying the less obvious one. For our purposes we can either work with the more obvious idea that p eventually settles on truth permanently, or we can work with the less obvious notion, that p tends towards permanent truth. In any case, let's abbreviate *p is true in the limit* to Lp.

The Peircean says that p is true if and only if in the limit p is epistemically acceptable, or endorsed by our best scientific theories. But, as Musgrave notes, the Peircean is not simply committed to this theory as a contingent truth. He is not just asserting that, as a matter of contingent fact, science will unearth all and only the truths. Rather, this is an account of the *nature* of truth, it is supposed to be true of necessity. In other words, where $\Leftrightarrow$ is logical equivalence Peircean idealism can be summarised thus:

Peircean idealism: $p \Leftrightarrow LEp$

Suppose p is true in the limit, and p entails q. Then whenever p is true q is also true. So, on the simple and obvious account of limiting truth—as eventual permanent truth—q must also be true in the limit. That is to say, L is closed under entailment. But this will also hold on the less obvious account of limiting truth as tendency to permanence. Indeed, whatever account we give of limiting truth, it should be closed under entailment ($\Rightarrow$).

Closure for L: $(Lp \; \& \; (p \Rightarrow q)) \Rightarrow Lq$.

Suppose that at some stage of scientific inquiry, p is epistemically acceptable according to the best theories at the time: Ep is true. Is the proposition Ep itself epistemically acceptable at that time? If Ep is true our best scientific theories tell us we should accept p. But then, in telling us *that*, they thereby endorse Ep. So, in the light of those theories, we should accept not only p itself, we should also accept Ep. But to say that we should accept Ep in the light of those theories is to say that EEp is true. Briefly, if Ep then EEp. Again, there is nothing contingent about this — it is a matter of necessity. So we have the EE principle.

EE principle: $Ep \Rightarrow EEp$.

Given these principles we can demonstrate that every truth is not merely epistemically acceptable in the limit of scientific inquiry, but rather that it is epistemically acceptable now, period:

For any proposition p, $p \Rightarrow Ep$.

Proof:

1. p	Assumption
2. $p \Rightarrow LEp$	Peircean idealism
3. LEp	1,2 logic
4. $Ep \Rightarrow EEp$	EE Principle
5. LEEp	3,4 Closure
6. $LEEp \Rightarrow Ep$	Peircean idealism
7. Ep	5, 6 logic
8. $p \Rightarrow Ep$	1—7, logic

Suppose we weaken Peircean idealism to the contingent claim that *in fact* whatever is true will be epistemically acceptable in the light of scientific inquiry in the limit. Then, our conclusion is merely the contingent claim that if p is true then it is now epistemically acceptable according to scientific theory. Or; conversely, if a proposition is not endorsed by current science then it is false. That's implausible.

You might feel that this *cannot* be right, even if you cannot pinpoint an error. You might feel that a philosophical position as subtle and widespread as Peircean idealism *cannot* be refuted as simply as this. You may be right that there is an error lurking in my refutation of Peircean idealism. But ever since my second encounter with Alan Musgrave I have been confident that refutations in philosophy are indeed *possible*, and that some are actual. Maybe this is one of those.

REFERENCES

Brogaard, B. and Salerno, J. 'Fitch's Paradox of Knowability.' *Stanford On-Line Encyclopedia of Philosophy*, http://plato.stanford.edu/entries/fitch-paradox/

Fitch, F. (1963) 'A Logical Analysis of Some Value Concepts.' *The Journal of Symbolic Logic* 28: 135-42.

Musgrave, A. (1997) 'The T-Scheme plus Epistemic Truth equals Idealism.' *Australasian Journal of Philosophy* 75: 490-496.

Musgrave, A. (1999) 'Conceptual Idealism and Stove's Gem.' in M.L. Dalla Chiara et al. (eds) *Language, Quantum, Music.* Dordrecht: Kluwer.

Oddie, G. (2000) 'Permanent Possibilities of Sensation.' *Philosophical Studies* 98: 345-359.

HANS ALBERT

HISTORIOGRAPHY AS A HYPOTHETICO-DEDUCTIVE SCIENCE: A CRITICISM OF METHODOLOGICAL HISTORISM

I have known Alan Musgrave for a long time, both as a friend and as a partner in discussion. From reading his works, and from conversations with him, I have learned more than my contribution to this volume could possibly show. Therefore, I would like to express my deeply felt gratitude to him at this point.

For two hundred years or more we have been faced with the thesis that there is a radical difference between historiography and the natural sciences. It is claimed that the methodology which is valid for historical research is, in general, completely different from scientific method. In particular, causal laws play no role in historical narratives—historicity and causal regularity are incompatible. I call this thesis methodological historism.[1]

I shall try to show that historism is not true and even that the representatives of historism themselves can be taken as the chief witnesses for my view, against their intentions, of course. First I shall say something about research programmes and problem-situations. Then I shall make some remarks about the research programmes of historism and of naturalism. Next I shall characterise methodological historism and the aim of historiography. Then I shall analyse the three main questions of Droysen, a prominent representative of historism, and his answers to these questions. Next I shall examine the relationship between sources and facts. And finally I shall criticise narrativism.

[1] Methodological historism is to be distinguished from historicism, as it has been analysed in Popper (1957).

C. Cheyne & J. Worrall (eds.), Rationality and Reality: Conversations with Alan Musgrave, 263–272.

1. RESEARCH PROGRAMMES AND PROBLEM-SITUATIONS

That research programmes play an important role in the history of science is a thesis which has been stressed especially in the famous Popper-Kuhn controversy. As is well known, Kuhn claimed that problems and problem solutions in the natural sciences come about within the framework of certain substantial and methodical presuppositions.[2] In this context he used his concept of paradigm and formulated a theory about the development of science in which he distinguished between normal phases and revolutionary phases. In normal phases, he claimed, research develops within a paradigm, that is, within a framework for the solution of problems which is authoritative for all scientists. Only in a revolutionary phase is it possible to develop alternative frameworks, one of which is destined to become the paradigm of the next normal phase. Because successive paradigms are incommensurable, no rational decision with respect to their preferability is possible.

It has turned out that Kuhn's views are not even valid for physics and that the methodological consequences he has drawn are unacceptable.[3] But, of course, his views include a minimal thesis which is acceptable to most of his critics. This is the thesis that certain problems can only arise within a framework of assumptions, and that this framework also delimits the set of possible solutions to those problems.

In this context Karl Popper speaks of problem-situations. At the same time he tries to show that for the solution of a problem it is often necessary to arrive at a new interpretation of the problem-situation. Logically speaking, that means that certain assumptions which belong to the old framework have to be dropped or replaced by other assumptions. Now, a research programme is a combination of such more or less general substantial and methodical framing conditions which are conducive to the formulation of problems. And in the history of knowledge there are enduring programmes of this kind, the influence of which can be followed for long periods. Popper has even tried to show that research programmes which first arose in presocratic thinking have influenced the whole development of knowledge up to modern science.

2. TWO RESEARCH PROGRAMMES: NATURALISM AND HISTORISM

History as a scientific enterprise is a Greek invention. Since Herodotus and Thucydides historical thought has developed more or less continuously into modern historiography. Most of this development took place without historistic tendencies. Up to the enlightenment the old Greek view was dominant, that historical processes are founded on the nature of man and are embedded in cosmic regularities. Therefore, it seemed plausible to the Scottish moral philosophers of the eighteenth century to try to extend the extraordinarily successful research programme of the theoretical natural sciences into the realm of social reality. The alternative historist research programme first arose in the nineteenth century, and was extended to all

[2] Kuhn (1970).

[3] See Musgrave (1999, pp. 193-228) and Andersson (1994).

social and cultural sciences. The controversies in this realm are mainly a result of the clash between these two research programmes, naturalism and historism. So far the history of my problem.

As to the difference between naturalism and historism, two main questions are involved. First there is the ontological question: Are there laws or regularities in the realm of humanity which can play a role in the explanation of occurrences? And second there is the methodological question: Is there a methodological difference between historiography and the factual sciences?

The first question refers to the realm of objects of historical research, the second to the procedure or cognitive practice of historiography, its research programme and the aim of historical thinking. It is plausible to assume that there is a connection between the answers to these two questions. Therefore our central problem can be formulated in the following way: Is it possible to apply the research programme of naturalism to historiography or are there serious objections against this idea?

Historism implies that the naturalistic research programme cannot be applied to historiography, for ontological and for methodological reasons. History has another structure than nature, and historiography has other aims and methods than natural science. As Droysen has stated: history concerns the 'moral cosmos' where 'the activity of volitions' rather than the 'mechanics of atoms' determines events; therefore it is not accessible to the method of the natural sciences but only to the method of understanding.[4]

I shall argue that both statements are mistaken and that the practice of historical research can be interpreted adequately in the framework of the research programme of naturalism. But it must be conceded that historism has made positive contributions to the development of historical thinking and that the representatives of historism have formulated interesting questions, which deserve answers from their critics.

I believe that the modern discussion about historism has gone awry partly because of the influence of hermeneutic and also of analytic philosophy. This debate concerns all sciences of man, the cultural and moral sciences and the social sciences including economics. In all of these sciences there are influences of historism and methodological controversies connected with them.

3. METHODOLOGICAL HISTORISM AND THE AIM OF HISTORIOGRAPHY

Often the contrast between the natural sciences and historiography is stated in the following way. In the natural sciences the aim is to explain natural phenomena on the basis of laws, which can be codified in general theories. In historiography or in the historical sciences the aim is to represent historical events and developments which are unique and thus cannot be explained in this way. Such unique events can only be understood and then narrated. In these sciences, instead of general theories we have narratives or stories. That may sound plausible if one takes only into account certain literary end products of historical research. But is it really true?

[4] Droysen (1869/1960, pp. 11ff. and p. 26).

In fact, there are also natural sciences which analyse historical developments and apply laws and theories to do so, for instance, geology, cosmology and biology. And there are cultural and social sciences which strive after laws and explanations, for instance, economics, sociology, linguistics and psychology. Thus it might be that historiography depends upon laws in a similar way as the historical parts of the natural sciences do.

To find this out, it is not enough to analyse logically the representations of events by historians, that is, to analyse the 'logical grammar' of their sentences and texts as is proposed in analytic philosophy of history.[5] Rather, one has to examine the structure of the cognitive practice of historians, not just the structure of end-products of this practice, for instance, the narrations which they produce. As to the cognitive interest of the historians, one can admit that it is not primarily oriented to the making of explanatory theories of nomological character. Max Weber drew attention to this point long ago. Yet Gadamer, in his criticism of naturalism, thought that it must impute such an aim to historians.[6]

The real question is whether the interest in the clarification of historical facts makes it necessary to use nomological knowledge somewhere in the analysis. That could be the case even if this knowledge is not explicitly represented, because, for instance, it consists of trivial common sense knowledge. In this case the historian is engaging in everyday explanatory habits. Again, this was already pointed out by Max Weber. But, as we shall see, there is more to be said in this connection.

4. THE THREE CENTRAL QUESTIONS OF DROYSEN AND AN ANSWER TO THEM

One of the best representatives of historism is the above mentioned historian Johann Gustav Droysen. Droysen's *Historik* sets out the research programme of historism in its purest form. Droysen formulates in this book his answers to three fundamental questions of historical research, which can be taken as representative for the historist research programm. I shall try to show that they have other consequences than he has stated.[7]

The three things which he refers to are the character of 'understanding' as the historical method, the significance of sources for historical research, and the role of generality in historical thinking. In my view, if one looks more closely into these three points, then surprisingly three arguments against historism and in favour of naturalism are the result.

As to the first of these problems, Droysen states that the method of the historian is to understand by research.[8] He says explicitly that the understanding of the historian is the same as the understanding of language in everyday communication.

[5] For a thorough criticism of this practice see Goldstein (1976) and other works of this author. I shall come back to this problem in the last part of my paper.

[6] See Gadamer (1960/1965, S.2) and also my criticism in Albert (1994, pp. 42ff).

[7] It must be admitted that in later parts of his book Droysen weakened the radical historist view of the first part, and made considerable concessions to naturalism. See the analysis in Spieler (1970).

[8] Droysen (1869/1960, p. 22: 'Unsere Methode ist forschend zu verstehen').

This raises the question, how we can characterise this linguistic understanding methodologically? To answer this question we can refer to classical hermeneutics. This possibility has never been doubted in historistic thinking. On the other hand, neither has it ever been pursued by the representatives of historism.

In fact, classical hermeneutics developed long before the rise of historism, in the eighteenth century. At that time it was conceived of as a rational 'Kunstlehre' (Art) or technology, for instance, by Friedrich Georg Meierr[9] and later by Schleiermacher and Dilthey. This doctrine was conceived of as being based on general knowledge about the functioning of communication by language.

Classical hermeneutics assumed the same relationship between technology and its theoretical basis as is assumed in modern philosophy of science with respect to the practical utilization of the results of the natural sciences. A technological doctrine is only possible in an area where there are laws to be applied in that area. But then law-governed explanations of the usual kind are possible in that area, too.

Now the understanding of the historian is, Droysen stated, a practice of the same kind as can be found in the area of reality to which the historical method refers. That means that the historian understands in the same way as the persons whose actions are being analysed. Therefore we can assume that the laws which are presupposed in hermeneutics are also to be found in the parts of reality to which historians refer in their cognitive practice.

We have to conclude that the possibility of a hermeneutics of the above mentioned kind is incompatible with methodological historism, which excludes laws from the realm of application of historical method. My argument can be extended to the understanding of all meaningful behaviour. The result of this analysis is that with respect to understanding, methodological historism is incompatible with its own ontological presuppositions.

As is well known, Gadamer attacked classical hermeneutics in his book 'Wahrheit und Methode' and offered a philosophical hermeneutics which aims to set forth the conditions of possibility of understanding. I have considered his views elsewhere.[10] Here I will only say that they are full of contradictions and contribute nothing to the solution of our problems, despite their influence in many areas.

Now I turn to the second argument for historism. It refers to the role of sources in historical research. Obviously, historians try to reconstruct past events and developments on the basis of sources, that is, of residues of past occurrences which have been found in the present. This is stressed by Droysen in his analysis.[11] It is not, of course, a specific historistic thesis. All participants in the discussion of the method of historiography agree on the importance of sources. Historists emphasised, and this is to their credit, the need always to be critical of historical sources, and this has been very important for the development of historical research since the nineteenth century.

But how can the aim of reconstructing past events be achieved? Elton tells us, 'historical method is no more than a recognised and tested way of extracting from

[9] Meier (1756/1996).

[10] For a critical examination of the views of Gadamer see Albert (1994).

[11] Droysen (1869/1960, pp. 20ff).

what the past has left the true facts and events of that past.'[12] But how can we characterise this method of extraction more precisely?

It may sound as if extraction is a logical process, involving some kind of logical inference from present sources to past facts. But it is clear that this immediately raises the problem of induction. Hume's famous criticism of inductive reasoning is interesting not only for natural science but also, as we see here, for historiography. Here the inductive inference is from singular facts of today—the sources—to singular facts of the past. But there are no valid inferences of this kind, as Hume pointed out. Therefore we have to find another solution of the problem.

Hints towards a plausible solution are to be found in the work of Leon Goldstein, who has however connected with his proposal an antirealistic position which we need not accept.[13] Goldstein suggests that we interpret historical method as an attempt to reconstruct past events or developments hypothetically in order to explain the existence and the peculiarity of the present sources. The sources are, as I have mentioned, traces of these past occurrences, that is present effects of them.

But to find such explanations we need, of course, nomological statements and therefore laws. We have to find a reconstruction of past events so that from them combined with appropriate laws an explanation of the present sources is possible. The possibility of such an explanation is, as it were, a test for the adequacy of the reconstruction concerned. The better present sources can be explained in this way, the more adequate is the reconstruction of the past events.

It follows that even the fulfilment of the most basic historical task, the mere reconstruction of past events, involves an explanatory performance. This performance cannot be made explicit by a mere narration of these events. One has to know how this narration explains the sources which prompted it. It also follows that mere analysis of the propositions of historical representations—for instance, of narrative sentences—does not sufficiently reveal the explanatory character of historical knowledge.

Furthermore, one can see immediately that such reconstructions of historical facts can in certain circumstances be tested further by finding and analysing new sources, for such new sources may be incompatible with an existing explanation. Of course, a new source cannot without further ado be taken to contradict a reconstruction of past events—it is only a present fact. It has to be interpreted adequately for this purpose, that is, it has to be embedded in an explanatory context. This means that the criticism of sources, to which historians often refer, is not a mere understanding of the above mentioned kind, in spite of the fact that also such an understanding is involved.

So we see again that an adequate solution of the problem which we have analysed is not compatible with methodological historism, because of historism's ontological presupposition that there are no laws in this part of reality. Of course, the laws which are needed for the reconstructions concerned can be found in other sciences or in common sense. No specific 'historical laws' are needed.

[12] Elton (1972, p. 86).
[13] Goldstein (1976).

Now to Droysen's third argument about the role of generality in historical thinking. Representatives of historism, like Droysen,[14] sometimes admit that historiography needs not only the knowledge of singular facts but also some general insights. Insights of this kind cannot consist in the knowledge of institutional facts, because these are historically variable. Nor can they consist, as Droysen argues, in the knowledge of the continuity of historical developments, for this is itself only a singular fact which would require an explanation. A plausible answer to the question of generality seems to refer us to the possibility of using laws in historical research as far as they are adequate to help in the reconstruction of past facts, again in contradiction to methodological historism.

Thus, my analysis shows that methodological historism is incompatible with the ontological assumptions which its representatives usually accept. The method of understanding and the reconstruction of past facts on the basis of sources presuppose laws, and the general insights which are needed should be interpreted as referring to laws. To achieve the aim of historical research one is dependent on the use of theoretical insights and on the possibility of explanations on the basis of nomological knowledge.

5. FACTS AND SOURCES

As I mentioned before, what historism achieved for historical thinking is the emphasis on the significance of sources and the criticism of sources. Often one forgets that the historian is never immediately confronted with the facts he would like to describe. What he has at his disposal are only sources of different kinds. They are the empirical basis of historiography.

This empirical basis raises the same problem which Karl Popper analysed in the context of the methodology of natural science. This basis always involves a theoretical interpretation, even if this is a matter of everyday theories which are not made explicit. This means that the sources are not unproblematic data which the historian can take as a secure basis for research. They must always be identified and interpreted, and that means that even in making them accessible theoretical viewpoints are indispensable.

It is not even clear from the outset which kind of objects are to be accepted as sources. It has long been assumed that written documents are obvious historical sources. But why is their use acceptable at all? Only because we more or less explicitly make general theoretical assumptions which imply that sources of this kind are produced under certain conditions and preserved under certain conditions.

As mentioned before, historical sources are traces of past historical ocurrences, just as for instance photons which arrive on the earth today are traces of past occurences in the cosmos and are interpreted in this way and used in research. From a naturalist perspective, the historian's 'observation of the past' has the same character as the observations which occur in the natural sciences.[15]

[14] Droysen (1869/1960, pp. 26ff).

[15] See Musgrave (1993, pp. 93ff and p. 277) and also Kosso (1992).

In natural sciences hypothetical reconstructions of past occurences on the basis of such traces and of laws are possible—and on the other side, such traces can test earlier reconstructions. Historians usually test their constructions by searching for new sources, or for sources the significance of which has not be seen before, or even for new kinds of sources.

Apart from written documents, historical research makes use of traces of other kinds. Examples are changes of the state of the surface of the earth, of its flora and fauna, its population, the settlements and customs of this population, their tools, weapons and equipment. Facts of all these kinds can be used as historical sources. One has to think only of the works of the Annales-School and of archaeology.

So we can conclude that for historical research not only psychological, sociological and economic theories and the laws codified in them can be useful, but also in principle theories drawn from all factual sciences. They can be used to find new kinds of sources which may lead to revisions of the reconstruction of past occurrences. But not only is the reconstruction of the past always hypothetical, but also its empirical basis, the so-called sources. Thus, the situation in historiography is the same as that described by Karl Popper for natural science.

6. NARRATION AND EXPLANATION: A CRITICISM OF NARRATIVISM

There is a version of analytic philosophy of history which is, in fact, a particularly questionable version of historism. I refer to narrativism, which conceives of historiography as a science which is satisfied with the production of stories or narratives about past occurrences.

Against this view three objections may be raised. First, it involves an unnecessary restriction of the mode of representation which has already been criticized by Droysen.[16] Second, it is the consequence of an inadequate approach to epistemological problems which is often to be found in analytic philosophy. And third, it involves some highly questionable views about language analysis.

As to the first of these points, it has been shown that narration is only one of the modes of representation which are used in historiography, and that there are good examples of other modes. For instance, there is the structural analysis of the feudal system by Otto Brunner,[17] the analysis of the genesis of capitalism by Jakob Strieder and his examination of a hypothesis of Werner Sombart,[18] or the analysis of the culture of the middle ages by Johannes Bühler.[19]

As to the second point, I have mentioned before that language analysis which only examines the results of research is not sufficient, because in epistemology we have to analyse cognitive practice and what it can achieve. Methodology is not reducible to the logical analysis of propositions.

As to the third point, even if one confines oneself to the analysis of narrations as products of historical research, one will find that these narrative texts include a lot of

[16] See Droysen (1869/1960, pp. 359-366) and also Weber (1906, S. 277ff).
[17] Brunner (1959).
[18] Strieder (1935).
[19] Bühler (1931).

causal statements.[20] Thus, causal connections are involved and one must assume that these connections are underpinned by causal laws. That means that even the result of such an analysis is incompatible with the ontological presupposition of historism. It emerges that the reconstruction of past occurrences, which is the aim of historical research, includes causal analysis and with this an explanation of these occurrences and not just an explanation of the sources involved.

Thus, historiography is not exclusively a narrative mode of cognition, and the narrative mode of representation is only one of the possible modes of representation of the results of historical research. In historical thinking we find an argumentative procedure which aims at the production and testing of hypotheses about past occurrences and their possible explanations. In this procedure, any source is used which can be useful for the purpose, including information sourced from other sciences.

Of course, I do not deny the special character of historiography, that is, its special aim and the special cognitive interests connected with it. Its aim is to know concrete past occurrences in the human realm, or rather, selected aspects of these occurrences which seem to be relevant under certain viewpoints. Our values are typically involved in these viewpoints. Historiography does not aim primarily at the knowledge of laws, let alone at special 'historical' laws. But to find past occurrences and their causal connections we always need nomological knowledge. From a methodological perspective, historiography can be characterised as a hypothetico-deductive science. In this respect it is not different from all other sciences.

REFERENCES

Albert, H. (1994) *Kritik der reinen Hermeneutik: Der Antirealismus und das Problem des Verstehens.* Tübingen: Mohr.

Andersson, G. (1994) *Criticism and the History of Science: Kuhn's, Lakatos's and Feyerabend's Criticisms of Critical Rationalism.* Leiden/New York: E.J. Brill.

Brunner, O. (1959) *Land und Herrschaft: Grundfragen der territorialen Verfassungsgeschichte Österreichs im Mittelalter*, 4th edn. Wien/Wiesbaden.

Bühler, J. (1931) *Die Kultur des Mittelalters.* Stuttgart.

Droysen, J. (1869/1960) *Historik: Vorlesungen über Enzyklopädie und Methodologie der Geschichte*, 4th edn. München.

Elton, R.G. (1972) *The Practice of History.* London/Glasgow: Blackwell.

Gadamer (1960/1965) *Wahrheit und Methode*, 2nd edn. Tübingen: Mohr.

Goldstein, L. (1976) *Historical Knowing.* Austin/London: University of Texas Print.

Kosso, P. (1992) 'Observation of the Past.' *History and Theory* 31: 21-36.

Kuhn, T. (1970) *The Structure of Scientific Revolutions*, 2nd edn. Chicago: University of Chicago Press.

Meier, G.M. (1756/1996) *Versuch einer allgemeinen Auslegungskunst.* Edited by Axel Bühler and Luigi Cataldi Madonna. Hamburg: Meiner.

Munz, P. (1977) *The Shapes of Time: A new look at the philosophy of history.* Middletown, CT: Wesleyan.

Musgrave, A. (1993) *Common Sense, Science and Scepticism. A historical introduction to the theory of knowledge.* Cambridge: Cambridge University Press.

Musgrave, A. (1999) *Essays on Realism and Rationalism.* Amsterdam/Atlanta, GA: Rodopi.

Popper, K. (1957) *The Poverty of Historicism.* London: Routledge & Kegan Paul.

[20] Munz (1977, pp. 39-61).

Spieler K-H. (1970) *Untersuchungen zu Johann Gustav Droysens 'Historik'*. Berlin: Duncker & Humblot.

Strieder, J. (1935) *Zur Genesis des modernen Kapitalismus. Forschungen zur Entstehung der großen bürgerlichen Kapitalvermögen am Ausgang des Mittelalters und zu Beginn der Neuzeit*, 2nd edn. München/Leipzig.

Weber, M. (1906) 'Kritische Studien auf dem Gebiet der kulturwissenschaflichen Logik.' in Weber (1951).

Weber, M. (1951) *Gesammelte Aufsätze zur Wissenschaftslehre*, 2nd edn. Tübingen: Mohr.

ANDREW BARKER

PTOLEMY'S MUSICAL MODELS FOR MIND-MAPS AND STAR-MAPS

In a paper first published in 1981,[1] a remarkable fusion of scholarship and philosophical acumen, Alan Musgrave effectively demolished Duhem's instrumentalist interpretation of the geometrical models deployed by certain Greek scientists, and especially by Ptolemy in the *Almagest*, in their analyses of the movements of the heavenly bodies. My topic here coincides with his in two obvious respects; its focus is also on Ptolemy, and my central concern is with the relation between mathematical models on the one hand, and on the other the reality to whose description these models contribute. At this point, however, our ways divide. Issues in astronomy will fall under my scrutiny only in passing, and the text with which I am primarily concerned is not Ptolemy's *Almagest* but his *Harmonics*, a work which for all its sophistication has played a much smaller role in the subsequent history of the sciences.[2] Further, the questions I shall be raising are not about the competing merits of 'instrumentalist' and 'realist' interpretations of mathematical models (though one might be tempted to bend the discussion in that direction). They arise, instead, out of a contrast between two different ways in which the models appear to be related to the realities whose structures they purport to represent. In one kind of case the model can be mapped directly onto the system it describes, and there is no particular difficulty in identifying, in the target system, straightforward counterparts of the terms and quantitative relations embedded in the model. In the other, these conditions are apparently not met. There seems to be no way—certainly no clear and direct way—in which the structure of the system described can be conceived as conforming, part by part, to that of the model, or as defined by the complex of mathematical relations by which the model is constituted. At the same time, as we shall see, Ptolemy's handling of a model of this latter kind is much too detailed and elaborate to be treated as mere rhetorical flummery, especially in the context of so meticulous and methodologically self-conscious a work as the *Harmonics*. My

[1] Musgrave (1981).

[2] The standard Greek edition is that of Düring (1930). See also his commentary (1934), and Alexanderson (1969). There is an English translation with notes in my (1989, pp. 270-391), and another in Solomon (2000).

C. Cheyne & J. Worrall (eds.), Rationality and Reality: Conversations with Alan Musgrave, 273–291.

central question, then, is how Ptolemy's use of such models is to be interpreted, and how, if at all, they can be conceived as contributing to a scientific understanding of the subject in hand.

1. INTRODUCTION

The Greek science called 'harmonics' was concerned above all with the analysis of the structures underlying musical melody—the patterns of pitches and intervals constituting scales or 'attunements', the substructures out of which these scales were composed, the larger structures of which they themselves formed parts, and so on—and of the ways in which those various structures could be systematically interrelated and transformed. In the tradition of harmonic theorising to which Ptolemy's treatise (with some qualifications) belongs, the terms in which these analyses are set and the principles of order which govern the constructions are mathematical, drawn primarily from the mathematics of ratio. Unlike many of his predecessors, however, Ptolemy insists that the harmonic scientist's task is not just to spin pretty patterns with numbers, regardless of their relation to real musical practice. It is to reveal the intelligible, mathematical form of the structures which the ear *perceives* as musical, and thus to show that the beauty we perceive in them is a reflection of rational order. There must therefore be no conflict between the findings of mathematical reasoning and the aesthetic judgements of the musical ear; and the results of his mathematical analyses and constructions cannot be accepted as correct until they have been empirically tested and confirmed. Hence they must be transposed, in all their detail, into a form in which the ear can assess them. This is done by transferring the mathematical structure onto special 'experimental' instruments, translating arithmetical ratios into ratios between lengths of the instruments' strings. Ptolemy's careful description of these pieces of gadgetry, of the principles on which they work, of procedures for testing their accuracy and for controlling irrelevant variables, of the ways in which they can and cannot reliably be used, and so on, are among the most fascinating parts of his treatise; and the overall methodology which he articulates and pursues, combining mathematically sophisticated modes of 'rational' derivation with rigorous empirical test-procedures, is as impressive and compelling as any to be found in our record of the ancient sciences.[3]

The project I have been lightly sketching occupies only the first 91 of the work's 111 large pages in Düring's edition. At this point Ptolemy announces that the task he had set himself at the outset is complete. 'It seems to me, then, that we have demonstrated accurately and in several ways that the nature of musical attunement possesses its own proper ratios, all the way down to the melodic intervals,[4] and that we have shown which ratio belongs to each of them, in such a way that those who

[3] For a full-dress discussion of these methodological issues see my (2000).

[4] 'Melodic' intervals are those which constitute the individual steps in a musical scale. Ptolemy says 'all the way down', because his method of analysis begins with larger intervals (specifically the concords of the octave, perfect fifth and perfect fourth), and then resolves them, through a complex mathematical procedure, into lesser components of which the melodic intervals are the smallest.

have set themselves, with keen enthusiasm, both to grasp the reasoning involved in the propositions we have set out and to undertake their assessment in practice, according to the methods of using the *kanôn*[5] which we have expounded, will be left in no doubt, since they will recognise through all the species [of scales and similar structures] their agreement with what we accept on the basis of our perceptions' (*Harm.* III.3, 91.22-92.1).[6] The remainder of the *Harmonics* (III.3-16, of which some later parts were lost at an early stage of transmission and were reconstructed by Byzantine editors) is therefore an appendix to its main agenda; but it is with this appendix that our present business lies.

Ptolemy begins by setting the results of his harmonic enquiries in a wider context. The structures he has revealed are to be understood as products of a special sort of 'reason' or 'rationality' which he calls 'harmonic reason'. It is this that 'makes correct the ordering that exists among things that are heard'. It works towards three interconnected goals: the 'theoretical discovery of the proportions by means of intelligence'; their 'practical exhibition by means of skill';[7] and the development, through practice and habituation, of a mode of experience which 'follows' them, that is, in which they are recognised and their status is appreciated (III.3, 92.30-93.4). Ptolemy continues:

> When we consider that reason in general also discovers what is good, establishes in practice what it has understood, and brings the underlying matter into conformity with this by habituation, it is to be expected that the science which embraces all the species of science that rely on reason, and which has the special name 'mathematics', is not limited by only the theoretical grasp of beautiful things, as some people would suppose, but includes at the same time the exhibition of them, and the dedication to them which arises from habituation in 'following' them. (III.3, 93.4-10)

Mathematics, then, is a science whose goal is the analysis and exhibition of beauty, together with the development of a trained capacity to recognise it empirically and a disposition to embrace it. Beauty is a property whose real nature lies in rationally ordered structure, and harmonics is the branch of mathematics which is devoted to the study of beauty in the domain of sound. This inspiring conception of mathematics leaves open the possibility that the forms of order underlying the beauty of things in other domains are different from those at work in music; but Ptolemy's reflections in the remainder of III.3 point in the opposite direction. Sight and hearing, alone among the senses, 'assess their objects not only by the standard of pleasure but also … by that of beauty' (93.13-14). Our eyes and

[5] The *kanôn* is, properly speaking, the 'ruler' laid under the string of a musical instrument designed for the scientist's experimental purposes, marked off to indicate the points at which the string will be divided, by bridges, to produce lengths in the ratios required. Often, however, the word is used to refer to the instrument itself. The simplest of these instruments is the monochord, which as its name indicates has only a single string. For many of Ptolemy's investigations this instrument is inadequate, and in the course of the *Harmonics* he describes a number of others. Some of them are much more sophisticated than the monochord, but the principles on which all of them work are broadly the same.

[6] The *Harmonics* is in three books. Here and elsewhere in this paper, in references of the form 'III.3', the Roman numeral is the book number and the Arabic numeral designates the chapter. References in the form '91.22' or '91.22-92.1', indicate the relevant pages and lines of Düring's edition.

[7] 'Skill' here refers primarily to the 'craftsmanship' involved in the design, construction, regulation and practical use of the scientist's 'laboratory' instruments.

ears, but no other sense-organs, are therefore capable of cooperating with one another and of serving as the allies of reason in pursuit of a single, shared objective, that of 'penetrating progressively into what is beautiful and valuable' (94.12). An astronomer's study of visible beauty in the movements of the heavens and a harmonic scientist's study of audible beauty in music are mutually supporting enterprises whose ultimate subject matter is one and the same. It is the mathematically integrated complex of structures which, as a whole or in one or another of its constituent patterns, underlies the beauty accessible to our two privileged senses wherever it is to be found.

In III.4 Ptolemy specifies the kinds of entity, outside the realm of music itself, which possess the 'faculty of attunement'. In the light of the previous chapter's discussion, this must mean that they have the capacity to grasp, to construct and to conform themselves to harmonious and beautiful patterns of order. Such a faculty must exist to some degree, he says, 'in all things that have in themselves a source of movement,[8] ... but especially and to the greatest extent in those that share in a more complete and rational nature. ... Only in these can this faculty bring to light, and preserve fully and clearly, so far as that is possible, the likeness of the ratios which create appropriateness and attunement' (95.4-10). A few lines later the possessors of such natures are identified. They are those whose movements and changes can be specified as manifestations of intelligible formal structures.

> These movements, as we said, are those of things that are the most perfect and rational in their natures, as among divine things are the movements of the heavenly bodies, and among mortal things those of human souls, most particularly, since it is only to each of these that there belong not only the primary and complete mode of movement, that is, movement in place, but also the characteristic of rationality. The faculty of harmonic reason reveals and displays in them, so far as a human being can grasp it, the pattern of organisation that corresponds to the harmonic ratios of the notes, as we can see if we analyse each of these kinds in turn. (95.20-27)

Ptolemy thus construes the beautiful dynamic ordering of the stars and planets, and of the elements of the human psyche, as arising from the operations of a rational faculty which each of them possesses. Through the operations of this faculty they organise their own movements and components into patterns corresponding to the systems of ratios which define the structures of musical attunement. The project which Ptolemy announces at the end of III.4, that of 'analysing each of these kinds in turn' to show that they do indeed display musical forms of structure, occupies the remainder of Book III. It is here that we finally arrive at the issues about scientific models and their functions which I announced at the start. Ptolemy is proposing to elucidate the patterns of organisation manifested in the travels of the stars and planets and in the workings of the human soul, by mapping them onto the model provided, in the main body of the *Harmonics*, by his complex analyses of musical structures. He tackles human psychology first (III.5-7) and stellar phenomena second (III.8-16). We shall reverse this order, since the application of his musical

[8] This is an expression used by Aristotle (e.g. *Physics* 192b13-14), to explicate the notion of a thing that comes into being 'by nature'. It regularly refers to animals and plants, but is applied also to any other beings which are treated as being alive.

model to the heavens is more straightforward than its psychological counterpart, and will provide a reasonably secure basis for comparison.

2. MUSICAL STRUCTURES IN THE HEAVENS

Of the nine chapters that fall under this heading, the greater part of one (III.14) and the whole of another (III.15) are not from Ptolemy's pen, as I noted earlier, and the origins of III.16 are disputed. I shall consider only one chapter, III.9, which I have chosen because it will entangle us in no arcane musicological technicalities. For our purposes it can be taken as a representative sample of what III.8-16 contains, and most of my comments will apply equally, *mutatis mutandis*, to the rest.

After some brief preliminaries, III.9 sets off with the construction of a simple diagram (see Figure 1). We first draw a circle. In it we draw a diameter AB, whose salient property is that in cutting the circumference at A and B, it divides it into two equal parts. Next we draw another straight line, AC, cutting off an arc which is one third of the circumference, and then AD, which cuts off one quarter; and we join B to C with a line cutting off one sixth. Thus A and B, Ptolemy notes, stand diametrically opposite one another; AC will form one side of an equilateral triangle inscribed in the circle; AD, similarly, will form one side of a square, and CB the side of a regular hexagon (102.4-8).

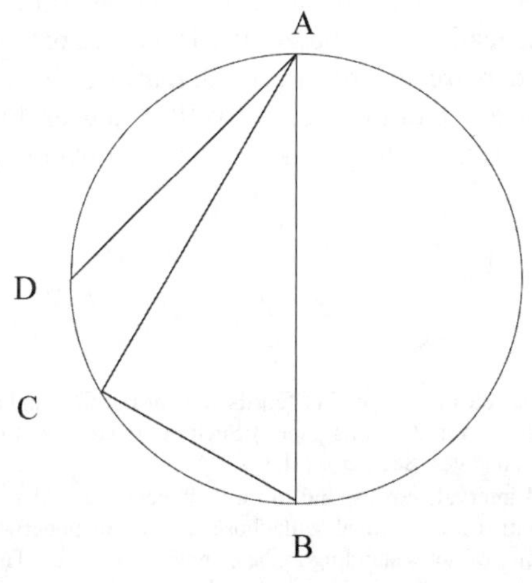

Figure 1

Ptolemy's next step is to correlate the ratios between the arcs defined in this construction with the ratios of the most significant musical relations (that is, the octaves and concords) which are contained in a 'complete' or 'perfect' musical system.[9] These relations are the octave itself (ratio 2:1), the double octave (4:1), the

[9] A complete musical system, in Ptolemy's usage, is a scale which contains all the 'forms' or 'species' of the three primary concords, the octave, the perfect fifth and the perfect fourth. (These 'species' correspond to the different ways in which the 'melodic' intervals within the span of a concord can be

perfect fifth (3:2), the perfect fourth (4:3), the octave plus fifth (3:1) and the octave plus fourth (8:3).[10] One other important relation is also involved, the interval of a tone, which is neither an octave nor a concord, but is defined as the difference between the two smallest concords, the fifth and the fourth.[11] It plays crucial roles in the construction of musical systems. Given its definition, simple calculation shows that its ratio is 9:8. Thus the ratio of the whole circumference to the semicircle AB is 2:1; so is that of arc ABC to arc AC, and that of ACB to AD. The ratio of the whole circumference to AD is 4:1. The whole circumference stands to ABC in the ratio 3:2, as does ABD to AB and AB to AC. The ratio 4:3, similarly, is found in the relations between the circumference and ABD, between ABC and AB, and between AC and AD. The ratio of the circumference to AC and of ABD to AD is 3:1; and that of ABC to AD is 8:3. Arc ABD, finally, stands to arc ABC in the ratio of the tone, 9:8.

What we have here, then, is a simple geometrical construction, analysed in such a way as to reveal the presence within it of all the ratios fundamental to musical attunement. (All other musical ratios are determined through systematic factorisations of these, and the main skeleton of every musical structure is mapped out by reference to octaves, fifths, fourths and tones alone.) The celestial system which it diagrammatically portrays is the circle of the zodiac;[12] and in one obvious but important sense the task of mapping the model onto the target system is simplicity itself. It could be fitted just as easily onto anything else with a circular shape. It connects specifically with the zodiac in an almost equally obvious way, since the smallest arc it demarcates, from C to D, marks off one twelfth of the circle, while all the other arcs are multiples of a twelfth; and as Ptolemy points out at 103.12-13, the zodiac, correspondingly, is divided into twelve parts, each occupied by one of the 'signs'.

rearranged; there are three species of the perfect fourth, four of the fifth and seven of the octave. They will be discussed further in Section 3 of this paper.) Such completeness is reached when the scale is extended to the range of two octaves. See *Harm.* II.3-4.

[10] The discovery that musical intervals correspond to ratios is centuries older than Ptolemy. The values assigned to the principal ratios are identical with those treated in modern physical acoustics as the ratios between the frequencies of notes standing in these musical relations. The Greeks knew nothing of frequencies, but attached the terms of the ratios, in a similar way, to quantitatively variable attributes of the sounds which were held to be responsible for their pitch. More immediately and less speculatively, they were able to match the ratios posited for the relations between pitches themselves with the ratios between the lengths of a stretched string that would emit notes separated by the appropriate intervals. Thus the note sounded by half the length of a string, for instance, is exactly an octave above the note sounded by the whole string. Lengths in the ratio 1:2 give pitches in the ratio 2:1.

[11] Nowadays, of course, we recognise various intervals smaller than the fourth as concordant. The Greeks also found a significant qualitative difference, for example, between the sound of a major third and that of a semitone, but attributed the special property of concordance (*symphônia*) only to fourths, fifths and octaves, and to any larger interval made up of one of these primary concords and one or more octaves.

[12] That the circle in the diagram 'is' the circle of the zodiac has in effect been announced already in Ptolemy's title for this chapter: 'How the concords and discords of attunement are similar to those in the zodiac'

So much is straightforward. It seems much more open to doubt, however, that the various arcs and the ratios between them could be construed by an astronomer as significant elements of celestial structure. Unlike the circles, epicycles, equants and other such constructs which have explanatory roles in the *Almagest*, the relations sketched here seem to explain nothing at all. And indeed that impression is correct; to a project like the *Almagest*'s they have nothing whatever to contribute. As Ptolemy's phrasing suggests in this chapter, and as becomes still clearer in the sequel, the discipline within which the ratios become significant is not what we would call 'scientific astronomy', but astrology. The fact should not prompt us to avert disdainfully our scientific gaze. In Ptolemy's time astrology was not confined to a world of sultry ladies in regrettable head-scarves or brainless prognostications in popular newspapers. Its status was debated, but it was widely recognised as a respectable science, and an extensive work by Ptolemy himself, the *Tetrabiblos*, is a major landmark in its history.[13]

This accounts for several otherwise mysterious statements in III.9, including the one with which it begins.

> Just as the melodic concords involve divisions into four parts and no more, since in the greatest, the double octave, the larger term is four times the smaller, while in the smallest, the perfect fourth, the larger exceeds the smaller by a fourth part of itself, so in the same way divisions of the circle into four parts and no more produce the complete set of configurations in the zodiac that are understood as being concordant and active. (101.27-102.4)

These 'configurations' are those of diametrical opposition and of triangular, square and hexagonal conformation, constructed on the basis of the various arcs; their 'concordance' and 'activity' have nothing to do with the concerns of an astronomer, but are astrological properties which are held to arise when planets or other significant celestial bodies are arranged in these configurations or 'aspects' on the zodiacal circle. Thus we are told in III.16, for example, that 'triangular' configurations of Mars, Venus and the moon are 'bringers of good', while their other configurations are not (111.10-12).

The thesis that these configurations are astrologically meaningful is of course not Ptolemy's own invention. It had long been entrenched in the astrological tradition, and the fact is important. Ptolemy takes the privileged roles of opposition and of triangular, square and hexagonal configuration (or trine, square and sextile aspect) as an established datum, and one that requires explanation. Why is it that these patterns carry meanings and powers while others have none? His strategy is to show that it is precisely these configurations that are determined when a circle is divided into arcs whose ratios correspond to those of fundamental musical relations (the preceding chapter has demonstrated the connection between the circle and the 'complete' two-octave system of harmonic order). Because the configurations constructed by this musical procedure are significant and others are not, the properties of relations on the zodiacal circle are manifestations of the fact that its structure, in astrological respects, is a product of 'harmonic reason' or the 'faculty of attunement'. Meaningful relations between arcs of the zodiacal circle are determined on precisely

[13] For an excellent general survey see Barton (1994).

the same mathematical principles, and by the same modes of derivation, as are relations between the fundamental notes of a musical system. The same point can be made negatively. Points on the zodiac that lie, for example, at the ends of an arc amounting to five twelfths of the circumference, are described, Ptolemy says, as 'uncoordinated' (*asyndeta*), and their 'functions' or 'powers' (*dynameis*) are uncoordinated too. This seems to mean that celestial bodies which are separated on the circle by this fraction of its circumference do not, as it were, 'cooperate' to exert any determinate astrological influence. Correspondingly, the ratios of the circumference to the arcs subtended by such points, 12:5 and 12:7, are not those of any concordant or melodic interval, and do not even exhibit the general form of acceptable musical ratios.[14] In the musical context, as in the astrological, they are meaningless (104.4-5, 8-12).

The astrological theories presupposed by Ptolemy's discussion are nowadays mere relics of an outworn picture of the universe, and no paid-up subscriber to a modern (Western) scientific conception of the world is likely to be convinced by Ptolemy's attempt to recast them on the foundation of mathematical harmonics. Considered in its own terms, however, the attempt is intelligible and potentially fruitful. In the first place, as we have seen, the arithmetical ratios of the model are readily transferred into their new context, as specifications of spatial relations between identifiable points in the target system. Both in the present chapter and in those that follow (though some of the later constructions are less precisely delineated), the task of understanding how the relevant patterns of musical relations are to be projected onto the celestial phenomena presents no serious obstacles. Secondly, the relations defined by the musical model do not become merely arbitrary when transported into a celestial setting; they determine geometrical configurations which had already been established as astrologically significant.

Thirdly, and as a consequence, the musical model serves a recognisably scientific explanatory purpose. By construing the astrologically meaningful configurations as manifestations of the agency of harmonic reason in the celestial domain, Ptolemy not only identifies the cause which determines 'significant structures' in that sphere, but also assimilates it to the wider class of cases in which that agency is at work. He thereby takes a step towards a unified explanation of apparently diverse phenomena, an altogether respectable scientific aspiration. The same principles of order and the same form of agency that are responsible for the harmonious integration of sounds into musical systems are also responsible for the organisation of stars and planets into meaningful and influential configurations.

Correspondingly, the satisfactory fit between the ratios of the model and the significant configurations of the heavenly bodies goes some way towards fulfilling the conditions spelled out in the earlier part of the treatise, whereby a set of 'rational' or mathematical hypotheses is to be submitted to empirical tests. Only if the empirical results match those derived from the hypotheses will the latter be

[14] With certain special exceptions, all such ratios must, according to Ptolemy, be either multiple or epimoric ('superparticular'); in effect they must have either the form mn:n or the form n+1:n. Ptolemy's justification for this rule is complex and we can ignore it for the moment; I shall discuss aspects of it in Section 3 below.

accepted as correct, and appropriate to the subject in hand. Here the data against which the consequences of the hypotheses are checked are 'empirical' in only a rather generous sense. They are those that are standardly accepted as true within the prevailing astrological tradition, rather than being the observed outcomes of the author's or the reader's controlled experiments, as they are in Ptolemy's programme for harmonic science itself. But it is hardly unusual for a scientist to treat the contents of an established consensus as 'data' for such purposes as these; and though the tests that depend on them are much less rigorous than those which turn on direct experimentation, the overall principle is the same. The evident correspondence between these 'significant data' and certain particularly fundamental features of the model provides confirmation, at least in some degree, of the hypothesis that the same principles are responsible for both systems' organisation.

Finally, this application of the model is in a certain sense predictive, and can play a useful heuristic role. Musical structures are replicated in astrologically meaningful structures sufficiently closely and consistently for Ptolemy to conclude that they reflect the same mode of rational order. But no one had hitherto identified significant relations in the heavens which would correspond to every single well-formed musical relation, or to every well-formed complex of them. Many of the details of Ptolemy's harmonic analysis are entirely original, and it would be an astounding coincidence if each of them could be matched precisely to stellar configurations already recognised as meaningful. We can be confident, in fact, even in the absence of adequate records of astrological opinions, that the *Harmonics* specifies many musical relations for which no significant celestial counterpart had yet been identified. Ptolemy's reflections in III.8-16 encourage the prediction that configurations embodying relations of those sorts nevertheless are astrologically important. This prediction would provide a 'research programme' for discovering where these configurations are located, and for reaching an understanding of their powers. (How the astrological hypotheses reached by this method are to be tested in their turn is of course a tricky question, and one that the 'researcher' will need to address. But for present purposes we need not scramble in that thorny territory.) To conclude, the disrepute into which astrology has (no doubt quite properly) fallen in recent times should not blind us to the genuinely scientific merits of Ptolemy's attempt to underpin it with his musical model.

3. MUSICAL MODELS IN HUMAN PSYCHOLOGY

The scientific credentials of psychology, in all its multifarious guises, have been hotly debated over the past century or so, but modern conceptions of it and current assessments of its status are as irrelevant to this enquiry as are twenty-first century attitudes to astrology. Neither should be allowed to affect our evaluation of Ptolemy's treatment of these disciplines in the *Harmonics*. My purpose, in the final part of this paper, is in any case not to represent Ptolemy's approach to psychology as a serious rival to its modern counterparts, but the much more limited one of examining the ways in which, in this connection, he puts his musical model to work. I shall argue that in at least one important respect the manner in which the model is brought into connection with its target, the structure of the human psyche, is different from that involved in his treatment of astrology, and that the model's functions in the psychological context must, as a consequence, be rather differently understood. Once again I shall draw for the most part on material from just one chapter, in this case the first of those devoted to the topic, III.5.

In order to make the contents of this chapter intelligible, I need first to sketch a few of the principles which lie behind Ptolemy's mathematical derivation of the harmonic ratios. He requires, first of all, that the terms of each ratio involved in musical structures should be in a certain sense commensurable with one another. There are two interconnected reasons for this stipulation. The first is that the relation between two notes can be musical only if, in their quantitative guise, they are integrated with one another by being measurable in terms of the same unit. Secondly, if the ear is to perceive this relation as musical, it must be able to compare them by reference to a common 'measure' which is given in the experience of the relation itself; there is no external standard against which they can be assessed. This commensurability may take either of two forms. In one kind of case, the smaller term is a factor of the greater, so that the former itself will serve to 'measure' the latter. Such ratios are 'multiple'; examples are 2:1, 3:1 and so on. In cases of the other sort, the measure is provided by the difference between the terms; each term is an exact multiple of the difference. Ratios of this sort are called 'epimoric' (or 'superparticular' in the more familiar Latinised terminology). They are defined by the fact that the greater term is equal to the smaller plus one integral part of the smaller (one half, one third, and so on); and when taken in their lowest terms each of them has the form n + 1:n. Examples, then, are 3:2, 4:3, 16:15.

Musical relations are characterised by a special kind of coherence. Notes that stand in such relations, as do the notes of a well-formed scale, are integrated with one another in a way in which other collections of pitches are not. The closest kind of integration that can be achieved is that between notes whose pitches are the same; mathematically speaking they stand in the relation of equality, 1:1. Ptolemy therefore grades the 'excellence' of musical intervals by the closeness of their ratios to equality, where this 'closeness' is construed in a very particular sense. He does not mean, for instance, that the ratio 18:17 is 'more musical' than the ratio 4:3, on the grounds that its terms are more nearly equal—on the contrary, in fact. If we restrict ourselves for simplicity's sake to relations within the span of an octave, the 'better' ratios, those closer to equality, are those multiple or epimoric ratios in which the difference between the terms is a larger 'simple part' or factor of the smaller term. Thus the closest to equality is 2:1, in which the difference between the terms is equal to the smaller. The next closest is 3:2, where the difference is half the smaller, and the next is 4:3. These three ratios, as we have seen, are those of the octave, the perfect fifth and the perfect fourth, which are fundamental to attunements of any kind; and just as these ratios are the first three in order of closeness to equality, so the octave, fifth and fourth, in that order, strike the ear as the intervals most similar to a unison. The next ratio in the sequence, 5:4, in modern terms that of a major third, is another 'sweet' or 'smooth' relation, whereas intervals whose ratios involve larger terms (18:17, for example, roughly a semitone) are harsher, more clashing, less well integrated from an aesthetic perspective.

One further point will have some importance here. Ptolemy insists that when we are analysing a musical system—for instance, one of those complexes of four notes, jointly spanning a perfect fourth, which are known in the trade as 'tetrachords'—we must work out the ratios of the lesser intervals by systematic 'divisions' or 'factorisations' of the larger, and not proceed the other way round, by stringing together a series of independently quantified small steps to build up the whole. It is the whole that is fundamental, and from which the smaller, constituent relations must be derived; there is no basis on which they can be established and assigned musical status independently. Thus the ratios of the fifth and the fourth are reached through a simple factorisation of the ratio of the octave (3:2 x 4:3 = 2:1); and in the case of the tetrachord, the various sequences of intervals that may lie within it are quantified through a special procedure, too complex to be explored here, for factorising appropriately the ratio of the perfect fourth.[15]

With these preliminaries behind us, the musicological basis of the opening lines of III.5 becomes readily intelligible.

> There are three primary parts of the soul, the intellectual, the perceptive and the animating, and there are also three primary forms of homophone[16] and concord, the homophone of the octave and the concords of the fifth and fourth. Hence the octave is attuned to the intellectual part, since in each of these there is the greatest degree of simplicity, equality and determinacy; the fifth to the perceptive part; and the fourth to the animating part. For the fifth is closer to the octave than the fourth, since it is more concordant, due to the fact that the difference between its notes is closer to equality; and the perceptive part is closer to the intellectual than the animating part, because it too partakes in a kind of apprehension. (95.28-96.7)

Here the octave is described as 'simple', partly because the two notes bounding it sound almost like one single note, and share the same melodic 'function', as Ptolemy explains in I.6 (13.3-23), and partly because the terms of its ratio are related in the simplest possible way; one is twice the other. Ptolemy implies in the first chapter of the treatise that the ear judges musical relations by a kind of subliminal assessment of the ratios between their terms, and argues that we can assess such ratios more easily and reliably when the amounts by which they differ 'consist in larger parts of the things to which they belong' (4.10-15). The perceived simplicity of a musical relation is thus a reflection of the simplicity of its mathematical form. The octave is also 'closest to equality' in the sense I have outlined; and its 'determinacy' is again a function of the simplicity of the comparison we make when we recognise it. Because it can be assessed accurately with the least degree of difficulty, there is nothing vague or indeterminate about its identity, and there is little room for doubt as to whether a given interval is or is not an octave. The reason why the fifth is 'more concordant' than the fourth is stated in the text of III.5 itself, and the determining factor is again its 'closeness to equality'. From a perceptual point of view, 'more concordant' pairs of notes are ones that are more fully

[15] It may be factorised, for instance, as 4:3 = 9:8 x 8:7 x 28:27, or as 4:3 = 7:6 x 12:11 x 22:21. These and others are worked out in *Harm.* I.15.

[16] Ptolemy uses the term 'homophone' for the octave and its multiples when he wants to distinguish them from concords of other sorts; elsewhere the term 'concord' often covers them all.

integrated with one another; they blend together in our perception of them, creating a 'homogeneous impression' on the hearing (10.25-28).

I shall say nothing yet about the putative psychological correlates of the octave, fifth and fourth, beyond the obvious point that the similarities so far sketched between musical items and 'parts of the soul' seem impressionistic rather than precise. Even if we grant that there are these three parts, and that the intellectual is simpler and more determinate than the perceptive and the perceptive than the animating, we have been given no reason to believe that the grounds of this simplicity and determinacy are the same as in the musical case. Let us postpone this problem and continue with the passage.

> Now things that have animation do not always have perception, and neither do things that have perception always have intellect; things that have perception, conversely, always do have animation, and things that have intellect always have both animation and perception. In just the same way, where there is a fourth there is not always a fifth, and neither is there always an octave where there is a fifth; where there is a fifth, conversely, there is always a fourth too, and where there is an octave there are always both a fifth and a fourth. The reason is that some of them [the smaller ones] are made up of the less perfect melodic intervals and combinations, the others [the greater ones] of the more perfect. (96.7-14)

The line of thought here is simple, if not simple-minded, and the general run of the reasoning is too clear to call for elucidation. The gist of the more abstruse-looking final sentence is only that the octave can be analysed as a combination of two intervals of the 'better' kind, the fifth and the fourth, whereas the fourth, for example, cannot; it can be broken down only into 'melodic' sub-intervals, whose ratios involve greater terms and are therefore further from 'perfection' or 'equality'. But Ptolemy's remarks have a hidden implication. Since the octave is the whole of which the fifth and fourth are parts, rational analysis, according to one of the principles sketched above, will derive the latter from the former, not the other way round. It is the nature and mathematical structure of the octave that explains those of the lesser concords, and harmonic reason will understand and construct them on the basis of its understanding of the octave. If the musical model can properly be applied to the soul, and if its structure too is a product of harmonic reason, a parallel proposition should hold also in its case. Scientific understanding of the soul's perceptive and animating elements will depend on and be derived from understanding of the intellectual part. We shall return to these points later.

The next part of Ptolemy's discussion turns on the notion of the 'forms' or 'species' of the octave, fifth and fourth. The phase of his harmonic theory in which these species play an important role is rather intricate (see II.3-9), but we need not pursue it here. The notion itself is quite straightforward, and can be grasped well enough on the basis of an example. In any fully developed musical scale, the interval of a fourth, whose ratio is 4:3, is broken down into three smaller intervals each of which is normally of a different size and thus has a different ratio. Take the case of Ptolemy's 'tonic diatonic' pattern of attunement, where the three intervals in question have the ratios 9:8, 8:7 and 28:27, reading from the top down. Imagine now that two fourths, each divided in this way, are placed end-on, as they are in certain ranges of a complete scalar series. The ratios of the sequence of intervals (taken again from the top down) will be 9:8, 8:7, 28:27, 9:8, 8:7, 28:27. It can easily be seen that within this sequence, the constituent intervals of the fourth appear in three different orders, as 9:8, 8:7, 28:27, as 8:7, 28:27, 9:8, and as 28:27, 9:8, 8:7. These exhaust the possibilities; other permutations such as 9:8, 28:27, 8:7, which do not appear in the scalar sequence, are musically unacceptable and cannot occur. It is these three orderings that constitute the species of the fourth. The ratios into which the fourth is divided are different in different patterns of attunement, but for any one such pattern there are just three species of the fourth; in the same sense there are four species of the fifth and seven of the octave. Ptolemy defines a musical system or scale as 'perfect' or 'complete' only if it contains within it all the species of the fourth, fifth and octave; and it turns out that this must be a system spanning two octaves. Nothing smaller contains them all, and nothing larger is needed to include all the essential elements of harmonic structure in all their musically proper forms and combinations.

In the second paragraph of III.5, Ptolemy introduces these species into his account of the soul. Just as there are three species or forms of the musical fourth, he says, so there are three forms taken by the animating part of the soul; they are the 'primary powers' of growth, maturity and decline. The perceptive part of the soul, like the musical fifth, takes four forms, sight, hearing, smell and taste (touch being reckoned as an element in all of them). There are seven forms or species of the intellectual part, as of the octave; Ptolemy lists them as imagination (*phantasia*), intellect (*nous*), reflection (*ennoia*), thought (*dianoia*), opinion (*doxa*), reason (*logos*) and knowledge (*epistêmê*) (96.15-27).

Before commenting on this passage, let me outline briefly what happens in the rest of the chapter. Ptolemy announces that there is also another way of dividing the soul, which represents it as a complex of one part that is rational (*logistikos*), one that is 'spirited' (*thumikos*) and one that is appetitive (*epithumêtikos*). This scheme is of course familiar from Plato (notably in the fourth book of the *Republic*). Ptolemy treats this second division in precisely the same way as he did the first. He links the soul's rational part to the octave, its *thumos* to the fifth and its appetitive part to the fourth, and goes on to assert that the virtues belonging to each part fall into the same number of species as does the corresponding concord. The species of virtue proper to the appetitive part are moderation (*sôphrosynê*), self-control (*enkrateia*) and shame (*aidôs*, perhaps more appropriately 'conscience'). Those of the spirited part are gentleness (*praotês*), fearlessness (*aphobia*), courage or manliness (*andreia*) and steadfastness (*karteria*). Those of the rational part, finally, are acuteness (*oxytês*), cleverness (*euphuia*), shrewdness (*anchinoia*), judgement (*euboulia*), wisdom (*sophia*), prudence (*phronêsis*) and experience (*empeiria*, 'experience' conceived in its role as the basis of good dispositions and habits) (96.27-32, 97.9-20).

I have omitted two short passages of discussion and explanation, to which we shall return. The chapter ends with a paragraph on the sovereign virtue of justice, and on the condition of that ideal figure of Platonist and Stoic thought, the 'philosopher' or 'sage'. It is hard to follow and I do not propose to discuss it, but I shall quote it for completeness' sake.

> The best condition of the soul as a whole, justice, is as it were a concord between the parts themselves in their relations to one another, in correspondence with the ratio that presides over greater things, the parts concerned with intelligence and rationality being like the homophones, those concerned with good perception and skill, or with courage and moderation, being like the concords, while those concerned with the things that can produce and the things that participate in attunements are like the species of the melodics. The whole condition of a philosopher is like the whole attunement of the complete system, comparisons between them, part by part, being made by reference to the concords and the virtues, while the most complete comparison is made by reference to what is, as it were, a concord of melodic concords and a virtue of the soul's virtues, constituted out of all the concords and all the virtues. (97.27-98.4)

Let us now consider the uses to which Ptolemy has put his musical models in this chapter, and compare them with those at work in his astrology. The most important and obvious contrast between the two is that whereas the relevant harmonic ratios could be mapped easily and directly onto the geometry of the heavens, there seems to be no comparable way of achieving this in connection with the soul. If the implications of the model were taken literally, the animating and appetitive parts of the soul, for instance, would each be constituted by the relation between two items standing to one another in the ratio 4:3. Between them would lie two others, and the ratios between the four would correspond to the melodic intervals into which a perfect fourth is divided. The three powers of the animating part, and the three virtues of the appetitive, would consist in different orderings of these melodic ratios.

But none of this seems to make sense. We are offered no picture of the soul in which the terms standing in these ratios could be identified, and there is nothing in the philosophical tradition that would fill the gap. We simply have no idea of what the psychological items corresponding to musical notes might be. Hence we do not know in what the relative 'sizes' of the soul's parts consist, or what the components are whose ratios are rearranged to produce the various species. We may well suspect that it is for this reason that Ptolemy avoids any attempt at precise quantification in this chapter; there is nothing in the system that admits of quantification in the manner of a musical structure.

So far so bad; but there are more positive ways of looking at these strange lucubrations. A little earlier I described the first phase of Ptolemy's account as 'impressionistic'. Despite the awkward problem I have noted, the word hardly does justice to the passage as a whole. Though we are left in the dark about important aspects of this 'psychomusicological' system, others, including the identities of the psychic correlates of every one of the species of concords, both in the system of powers and in the system of virtues, twenty-eight of them in all, are spelled out in detail. The list of these powers and virtues is not just a list of names; I have passed over them in my summary, but Ptolemy also appends brief sketches of the natures of each of them. In other Greek scientific treatises (if we leave the special case of astronomy aside), musical models or analogies seem very often to be used with quite vague intent, indicating no more than that certain systems are 'harmoniously' integrated, or that certain items cooperate 'in concord' with one another. Typically, such passages are brief, and enter into no details about supposed correspondences with specific musical relations. None of them approaches the degree of systematic elaboration found in Ptolemy's analysis, here and in the next two chapters.[17] We should at least consider the possibility that the musical model has more substantial functions to perform in his work than in theirs.

We should notice first that Ptolemy's lists of the species of the soul's powers and virtues are not merely arbitrary, as if taken at random from some ancient precursor of Roget's *Thesaurus*.[18] His use of earlier sources is eclectic, to be sure; there are Platonist, Aristotelian, Stoic and perhaps even Epicurean ingredients in the mixture. But all the classifications he presents have respectable roots in the philosophical tradition. All seven of Ptolemy's species of intellectual power, for example, play important roles in Stoic writings (six of them are also common in Plato and Aristotle), and are discussed again in Ptolemy's own essay *On the Criterion*; and all his fourteen species of the virtues proper to the soul's rational, spirited and appetitive parts are listed, with only the most minor variations, in a set of definitions attributed (perhaps a little insecurely) to Speusippus, Plato's immediate successor as head of the Academy.[19] Hence Ptolemy's grounds for adopting these ways of classifying the soul's powers and virtues are much the same as those underlying his specification of significant configurations in astrology. The fourteen powers and fourteen virtues, like the configurations, constitute 'data' established and handed down by Ptolemy's precursors in the intellectual tradition.

Since the lists of powers and virtues were certainly worked out, originally, on a basis wholly independent of musicological principles, the mere fact that their numbers match those of the species of concords so precisely might give legitimate grounds for accepting that there is some real connection between them; the correspondence is not just a construct of Ptolemy's theory itself. If it is only a coincidence that there are exactly three animating powers, three virtues of the appetite and three species of the musical fourth, four forms of perception, four virtues of the spirited part and four species of the fifth, and seven modes of intellection, seven rational virtues and seven species of the octave, then, one might argue, it is a very strange and improbable one. In one of the short, reflective passages of the chapter which I ear-marked for later consideration, Ptolemy offers a reason for one group of these correspondences. 'The more notable distinctions between the virtues proper to each of them [that is, to each of the soul's parts] are equal in number to the distinctions between the species of the primary concords, because melodiousness among notes is a virtue of them, while unmelodiousness is a vice, and conversely virtue among souls is a melodiousness belonging to them, while vice is unmelodiousness. A feature common to both classes is the attunement

[17] In the density of musicological and other detail on which it draws, the treatment closest to Ptolemy's is that in the third book of Aristides Quintilianus *De musica* (the work is translated in my (1989, pp. 392-535). For material comparable to that in Ptolemy III.5, see particularly chapters 14-17. But Aristides' approach is joyfully eclectic and chaotic, governed by no consistent principles, and infected throughout with the mystical numerology of later forms of Pythagoreanism. There is nothing 'scientific' about it, by our standards or by Ptolemy's.

[18] Those who like antiquarian curiosities will be pleased to know that analogues of the *Thesaurus* did in fact exist in the second century AD. One notable surviving specimen is the *Onomastikon* of Julius Pollux, apparently written to increase the word-power of the emperor Commodus.

[19] See Düring (1934, p. 271), and cf. Tarán (1981, pp. 374, 383 n. 196).

of their parts, when they are in a condition conforming to nature, and lack of attunement when they are in a condition contrary to nature' (97.1-8). The point he is stressing is that the property manifested in any virtue of a soul is the same as that which is presented by any well-formed species of a concord. In the musical case we know how many forms the 'melodiousness' of each concord can take; hence it is only to be expected that there will be the same number of species of virtue available to the corresponding part of the soul.

The status of the assumption that each virtue is a form of melodiousness is ambiguous. On the one hand, it is a view that can readily be traced in the philosophical tradition and so treated as a datum; it has in particular the authority of Plato.[20] On the other, it is an ingredient in Ptolemy's own theory, a condition of the applicability of his musical model. If it is straightforwardly accepted on the former basis, it can be called on directly as evidence of the theory's truth. If regarded from the second perspective it cannot serve that purpose, on pain of circularity. Its function will rather be to point to a way in which the theory does not merely describe the phenomena within its domain, but explains them; the species of concord and of virtue correspond *because* they are alternative manifestations of the same formal property.

The second of Ptolemy's reflective passages reads as follows.

> Just as in attunement the accurate construction of the homophones [i.e., the octaves] must take the lead, and those of the concords and the melodic intervals must follow on after it—since a small error in the lesser ratios does not hamper the melody as much as in the larger and more important ones—so also in souls it is natural for the intellectual and rational parts to govern the others, which are subordinate, and more precision is needed in establishing correct ratio in the former [i.e., in the intellectual and rational parts], since it is their errors that are wholly or largely responsible for those of the others. (97.20-27)

The explicit purpose of these remarks is to draw attention to another parallel between musical structures and psychic ones, and so to give further evidence of a connection between them. But here again we see signs of the explanatory role of Ptolemy's theory. The latter part of his statement rehearses a view about the hierarchical ordering of the soul's elements which had been entrenched in philosophical thought since Plato. From the present perspective it is a datum on which any student of psychology could rely. The earlier part reminds us of a principle which I have already mentioned. It is from an understanding of the greater and simpler harmonic relations that the ratios and other features of their smaller components must be derived, since the latter's musical status depends on their role as elements in the larger whole. They have no independent 'musicality'; hence any procedure which begins from them and simply adds them together to produce the larger relations is reversing the proper order, and leaves the musical excellence of the whole unexplained. In that case, if the parallel holds, it is the part of the soul that corresponds to the octave that must take precedence. The virtues of the soul's lesser parts are determined, and constituted as virtues, by their relations to those proper to reason, and their natures and excellences can therefore only be understood in the

[20] See for instance *Republic* 400d-402a.

light of an understanding of the virtues of rationality. This is why failures in 'accuracy' in the sphere of reason are more serious than ones at a lower level, since the former are bound to produce distortions elsewhere in the system, while the latter are not.[21] Ptolemy's overall hypothesis, that what is responsible for good order in the soul is the same harmonic reason that creates well formed musical structures, then recommends itself as an *explanation* of the psychological hierarchy on which earlier philosophers had insisted.

This brings me to the first of two general points on which I shall end. Musical terms and concepts appear very commonly in Greek (and not only Greek) discussions of non-musical topics of every sort, from logic to theology and from embryology to architecture. Usually it is unhelpful to construe them as symptoms of a theoretical position that links music with other subjects in a systematic way, though of course there are exceptions, especially in the Platonist tradition. For the most part they are no more than imaginatively stimulating metaphors (but also no less; the value of such devices in philosophical and scientific writing should not be underestimated). What gives Ptolemy's musical models a more substantial function than that of persuasive analogy is his articulation of an explanatory hypothesis, which represents musical and non-musical structures as products of the same mode of rational agency. This hypothesis is not inferred *post hoc* from the existence of particular, 'observed' analogies such as those listed in III.5-16; it is developed, in the first place, on (roughly speaking) an *a priori* basis, in the reflections of III.3-4 on the nature of mathematics, on the 'beauty' which is the proper subject of its enquiries and the result of its constructions, and on the range of domains in which this beauty can or must exist.

Just as in harmonic science itself, the appropriateness of 'rational hypotheses' and the correctness of the consequences derived from them must be tested against empirical experience, so in psychology and astrology the overall theory must be confronted with independently established data. Unlike some other harmonic theorists, Ptolemy insists that if the consequences of the theory conflict with the data, it is the theory that must be modified or rejected; he has no truck with the time-honoured manoeuvre of appealing to the 'inaccuracy' of perception to evade empirical testimony against the august authority of reason. If the data fit well with results derived from the theory, on the other hand, this fact will not be sufficient to demonstrate its truth conclusively; but what it will do is to show that experience gives no grounds for suspicion about the cogency of the reasoning on which the hypothesis was based, or about its applicability to this particular domain. If the theory is to be acceptable, the data must be consistent with it, if only because they constitute the facts that the theory is supposed to explain. But it is not their job to prove it. It is not because the facts are as they are that the theory is true; it is true because it is 'rational'. Once the data are shown to conform to the theory's

[21] Thus in a musical attunement, if the octave is incorrectly tuned, some or all of the intervals into which it is divided must also be wrongly adjusted; whereas if only a couple of the melodic intervals are incorrectly formed, only localised damage is done.

predictions, on the other hand, our confidence in the reasoning that led us to the theory will be increased, and they can be brought within its explanatory scope; and since musical, astrological and psychological phenomena will all thereby be explained as products of the same rational agency, the structural parallels they display are much more than superficial analogies. They exist because precisely the same principles of mathematical organisation are at work in each.

Ptolemy's model operates, then, within a framework that is designed to provide explanations. It has heuristic functions in psychology too, as it does in astrology; but in psychological contexts it cannot be predictive in quite the same way as in astrological ones. That is, we cannot make scientific use of the prediction that the ratios and patterns of ratios essential to musical structure will reappear elsewhere among the relations between the elements of the soul, since we are in no position to identify the quantifiable psychological 'elements' between which these relations would hold. There are limitations, as Ptolemy recognises at the end of III.4, to the details of mathematical organisation that can be revealed, in such a context as this, by a scientist's analytic use of harmonic reasoning; it reveals such organisation only 'so far as it is possible for a human being to grasp it' (95.25).

Nevertheless, the fact that the psychological data which he identifies are consistent with the model, and that important features of them can be explained in the light of his general hypothesis, allows us to treat the model as 'predictive' and heuristically useful in a looser but still legitimate sense. It suggests that we would do well to consider whether the pattern can be extended to other aspects of the soul and its workings. In III.6-7, Ptolemy identifies psychological counterparts of the three 'genera' of attunement (enharmonic, chromatic and diatonic), and of certain forms of modulation elaborately discussed in Book II; and there is no reason why the search should not be pursued still further. What is achieved by these progressive assimilations of the psychological to the musical is, above all, the systematic integration of otherwise disconnected pieces of data. If we understand, in a musical setting, how the genera are related to one another and to the concords, how the procedures of modulation are related to the species of the octave, and so on, we shall have a sound basis for grasping the way in which a whole array of powers, virtues, emotions, dispositions, motivations and other such items are woven together through the agency of harmonic reason into the single, closely integrated structure that constitutes the soul. This remains true even if the factors in the soul which correspond to musical notes, and serve as the terms of determinate ratios, remain hidden from our scrutiny. The more psychological detail we can accommodate to the model without their aid, the more certain we can be that Ptolemy's overall hypothesis is correct, and hence that such factors must exist, elusive though they are. In the absence of any knowledge of them, the basis of the model's application to the soul will always remain in some degree unstable, lacking the firm anchorage that was available in astrology. Psychomusicological research must proceed *as if* we knew they were present, and *as if* they provided the model with safe moorings; to the extent that it is successful in integrating and explaining its data it increases the probability that the mooring-ropes are indeed secure.

The central difference between the ways in which Ptolemy's musical model can be applied in astrology and in psychology is that in the former case, but not the latter, it is possible to pick out items that correspond directly to the terms of the ratios that define the model. The difficulty infecting the second case might be due merely to the contingencies of intellectual history; such understanding of these matters as Ptolemy and his contemporaries had did not extend so far, but there was no reason why diligent research should not have homed in on the missing items eventually. Alternatively the relevant items might be unidentifiable in principle, for logical, epistemological or metaphysical reasons. Given the history of psychological speculation up to Ptolemy, he would have had every excuse for thinking that they are altogether and in principle beyond our grasp. There is nothing whatever in the intellectual tradition upon which a conception of them could be built, or which could provide the seeds of a method by which one might hope to discover them. I have tried to show that even if we interpret his psychological uses of a musical model in the light (or darkness) of that bleak assumption, his discussion need not be read as a specimen of scientifically empty rhetorical posturing. Even if we are incapable—wholly, inevitably and for ever—of seeing how there can be elements in the target system in whose relations the structural anatomy of the model is replicated, nevertheless the model has genuine explanatory and heuristic power.

REFERENCES

Alexanderson, B. (1969) *Textual Remarks on Ptolemy's Harmonics and Porphyry's Commentary*. Stockholm: Almqvist & Wiksell.

Barker, A. (2000) *Scientific Method in Ptolemy's Harmonics*. Cambridge and New York: Cambridge University Press.

Barker, A. (ed.) (1989) *Greek Musical Writings: Volume 2, Harmonic and Acoustic Theory*. Cambridge: Cambridge University Press.

Barton, T. (1994) *Ancient Astrology*. London and New York: Routledge.

Düring, I. (ed.) (1930) *Die Harmonielehre des Klaudios Ptolemaios*. Göteborg: Elanders Boktryckeri Aktiebolag.

Düring, I. (1934) *Ptolemaios und Pophyrios über die Musik*. Göteborg: Elanders Boktryckeri Aktiebolag.

Musgrave, A. (1981) 'Der Mythos vom Instrumentalismus in der Astronomie' in *Versuchungen Aufsätze zur Philosophie Paul Feyerabends*, edited by Hans Peter Duerr. Frankfurt am Main: Suhrkamp. An English version was published as 'The myth of astronomical instrumentalism' in *Beyond Reason: Essays on the Philosophy of Paul Feyerabend*, edited by G. Munévar. Dordrecht/Boston/London: Kluwer Academic Publishers.

Solomon, J. (2000) *Ptolemy, Harmonics: Translation and Commentary*. Leiden: Brill Academic Publishers.

Tarán, L. (1981) *Speusippus of Athens: a critical study with a collection of the related texts and commentary*. Leiden: Brill Academic Publishers.

ALAN MUSGRAVE

RESPONSES

"Gee, Mum, my name is in lights!" So said David Stove, when I used his name in the title of a paper of mine. Stove deserved the little spotlight I shone on him. I surely do not deserve the floodlights that my friends now shine on me. I am dazzled, like a rabbit transfixed in headlights. Heartfelt thanks to those in the driving seat, for shedding so much light on my worthless carcass!

I have been anxious to defend two pretty commonsensical positions - critical rationalism and critical realism. I learned both from Karl Popper. But most 'Popperians' think I learned badly and got it all wrong. So as to avoid exegetical issues, which are pretty unimportant anyway, I will speak of them as 'my positions'. I take no credit for the good bits, but all the blame for the bad bits.

The papers collected in this volume fall into two groups. There are papers about critical rationalism - the role it gives to observation and testimony (Greg Currie, Colin Cheyne), severe testing (John Worrall, Deborah Mayo), and other critical methods (Volker Gadenne, Howard Sankey, Stathis Psillos), Then there are papers about critical realism - the metaphysics appropriate to it (Michael Redhead, Alan Chalmers, Robert Nola, Mark Colyvan), antirealist views that stand opposed to it (Noretta Koertge, Graham Oddie), its impact on historiography (Hans Albert) and on our understanding of the early history of astronomy (Andrew Barker). I shall organise my responses accordingly, and in that order. My friends will forgive me if, from now on and in deference to academic proprieties, they are usually 'Cheyne' not 'Colin', 'Koertge' not 'Noretta', and so forth.

1. CRITICAL RATIONALISM

Critical rationalism is the view that the best method for trying to understand the world and our place in it is a critical method – propose views and try to criticise them. What do critical methods tell us about truth and belief? If we criticise a view and show it to be false, then obviously we should not believe it. What if we try but fail to show that a view is false? That does not show it to be true. So should we still not believe it? Here critical rationalists distinguish acts of belief (believings) from the things believed (beliefs). They think there are reasons for believings that are not reasons for beliefs. Failing to show that a view is false does not show it to be true, but is a reason to think it true – for the time being anyway. Thus, it may be reasonable to believe a falsehood, if we have sought but failed to find reasons to

C. Cheyne & J. Worrall (eds.), Rationality and Reality: Conversations with Alan Musgrave, 293–333.

think it false. If we later find reason to think a view false, we should no longer believe it. Then we should say that what we previously believed was false – not that it was unreasonable for us ever to have believed it.

This is just common sense. The trouble is that philosophical tradition denies it. Philosophical tradition says that a reason for believing something must also be a reason for what is believed, it must show that what is believed is true, or at least more likely true than not. I call this 'justificationism', and reject it. I think we can justify (give reasons for) believings without justifying the things believed. The chief bone of contention between me and the Popperians concerns this point – they reject all justification, and think critical rationalism has no need of a theory of justified or reasonable believing. I am particularly grateful for Volker Gadenne's support on this point. As he says, critical rationalism bereft of such a theory is really no different from scepticism:

> The rejection of any kind of justification means that, for every proposition P, it is equally justified to believe P as to believe non-P; and this is not rationality, it is Pyrrhonian scepticism. It doesn't help to call criticism rationality as long as one does not make clear how criticism contributes to bringing about situations in which some beliefs turn out to be more acceptable than others with respect to truth. (108)

In fact, philosophical scepticism is underpinned by justificationism. How might we set about establishing the rational credentials of some belief? Justificationism says that we must show that what is believed is true. If we try to do this, by invoking something else that we believe, the sceptic demands that we show this to be true as well. Off we go on an infinite regress, which can only be stopped by invoking certainly true 'first principles' of some kind – 'observation statements' if you are a classical empiricist, 'self-evident axioms' if you are a classical rationalist. The rejection of justificationism enables the critical rationalist to drive a wedge between scepticism about certainty (which is correct) and scepticism about rationality (which is not). Failure to show that a belief is false does not show it to be true, but does show it to be reasonable. But do not sceptics show that our beliefs are false? No, sceptics produce no criticisms of our beliefs – they only produce excellent criticisms of attempts to prove that our beliefs are true.

Justificationism also lies behind inductivism. Given justificationism, empiricists need 'inductive' or 'ampliative' reasoning to show that some evidence-transcending views are true or more likely true than not, and hence reasonably believed. And they need inductive logic to show that inductive reasoning is valid or 'cogent'. Critical rationalists reject justificationism, and hence have no need of inductive reasoning or inductive logic. Deductive reasoning is enough for them, and deductive logic is the only logic they have or need.

Nobody can get by just with reasoning or argument, and critical rationalists are no exception. All arguments, whether deductive or non-deductive, rest upon premises. Not all the premises of our arguments can be conclusions of previous arguments, on pain of infinite regress. Or, putting the same point in terms of beliefs, not all our beliefs can be obtained by inference from previous beliefs, on pain of infinite regress. Our arguments must start somewhere, with premises that are not themselves reached by argument. Or, putting the same point in terms of beliefs, if inquiry is to get started we must have some non-inferential beliefs.

But are any non-inferential beliefs *reasonable* beliefs? If not, and if an inferential belief is reasonable only if the beliefs from which it was obtained are reasonable, then no belief is reasonable. *Logomania*, the view that any reasonable belief must be obtained by reasoning from reasonably-believed premises, is a royal road to wholesale irrationalism.

2. OBSERVATION AND TESTIMONY

These general reflections raise the following questions. What are the sources of non-inferential beliefs? Can non-inferential beliefs be reasonable beliefs, and if so, when? The answer – or part of it – to the first question is obvious. Sense-experience and testimony are obvious sources of non-inferential belief. (A paragraph back I spoke of 'evidence-transcending views' – the evidence that such views transcend is the evidence of the senses.) The answer to the second question – whether beliefs obtained from sense-experience or testimony can be reasonable – is perhaps less obvious.

Greg Currie takes up these questions as regards sense-experience. He is anxious to defend the idea that experience has a role to play in epistemology. I agree. We also agree, I take it, that experience cannot yield an absolutely secure or infallible 'empirical basis' against which theories can be tested, because observation statements transcend the experience that prompts them, and are themselves 'theory-laden'.

Currie rebukes me for having suggested in one place that (as he puts it) "theory-ladenness [is] an essentially linguistic phenomenon" (8). On the contrary, he claims, it makes perfect sense to speak of observation or perception itself as being 'theory-laden' or at least, 'concept-laden'.

I never meant to deny this. I was anxious, first of all, to defend that basic sense of 'see' (more generally, 'perceive') whereby, for example, a cat can see a typewriter without possessing the concept < typewriter >. My cat sees the typewriter, for she does not bump into it when the mouse she is chasing runs under it. Cats (or people) can see an X in that basic sense without possessing the concept < X >, let alone any theory or belief about Xs. There are philosophers who, bemused by Kant, deny this. [PROOF: I once met a German philosopher who said that cats cannot see typewriters because they lack the concept < typewriter >. I said that my cat frequently saw my typewriter. She replied that Musgrave's cat could do impossible things – just like Schrödinger's cat, which manages to be both alive and dead until somebody sees it. She even speculated that Musgrave's cat might become as famous as Schrödinger's cat. I should be so lucky!]

What we need here is, of course, a familiar distinction. There is another sense of 'see', seeing-that, which is clearly conceptual. My cat sees the typewriter when her mouse runs under it, but she cannot see *that* the mouse has run under the typewriter, since she lacks the concept < typewriter >. Seeing-that is propositional, hence conceptual.

In between seeing and seeing-that, there is seeing-as. This is also conceptual – you cannot see X *as a Y* without possessing the concept < Y >. The cat that sees the typewriter cannot see it *as* a typewriter. Perhaps the cat sees the typewriter as something else. Perhaps she sees the mouse as food, and the typewriter as non-food. Perhaps all seeing is conceptual in the sense that whenever A sees B, A sees B *as a C* for some concept C.

Despite my incautious formulations, which I shall not defend here, I never meant to deny the distinctions between seeing, seeing-as, and seeing-that, distinctions that I have used and defended in other places. Nor did I mean to deny that animals bereft of spoken language can see-as and even see-that (more generally, perceive-as or perceive-that). Both of these are 'concept-laden'. But having concepts is one thing, having beliefs or theories that use concepts is another thing, and having words to express those beliefs or theories is yet another thing. Or so I believe.

Which brings me to Greg's own interesting discussion. The old seeing, seeing-as and seeing-that distinctions are notable for their absence from it. He talks of the "content of perception" and says it is "a matter of the way perception represents the world as being" (7). So perception or perceptual experience already has content, already represents the world as being a certain way. As he says: "Believing that owls fly requires that I have the concepts *owl* and *flying,* and having a perception with the content *there is a flying owl* requires this also" (8). Put in terms of the old distinction, seeing-that is the whole focus of his attention. The reason is plain. He focuses on seeing-that because he wants to give the representational content of experiences a justificatory role in epistemology, contrary to what he (and most others) take to be Popper's view of the matter:

> It is the content of experience that matters to epistemology. It is this content which creates the possibility that an experience may provide a rational basis for the assertion of a statement describing some state of affairs. (9)

My view of the matter is this. (I think it Popper's view, too, but I shall not argue the exegetical point here.) Currie speaks of an experience providing a rational basis for *asserting* a statement (or, we might add, for *adopting* or *forming* a belief). Must it, in order to do this, also provide a rational basis for the *statement itself* (or for the *content* of the belief)? Justificationism says YES: a reason for asserting (or believing) that P must be a reason for P itself. We need to reject justificationism in the epistemology of perception, as we do elsewhere. Does Currie reject it?

Suppose (to use his example) that I have a perception with the content *there is a flying owl.* Obviously, the content of my experience is a logically conclusive reason for the (content of the) belief that there is a flying owl, and an equally conclusive reason for the (content of the) assertion that there is a flying owl. After all, the three contents are identical, and *C* logically implies *C*. It is equally obvious that the content C of my experience is no reason at all for forming a perceptual belief with content C, let alone for asserting an observation statement with content C. Forming a belief or asserting a statement is an action that we perform. Reasons for actions are causes of them, and contents or propositions are not causes.

Currie will perhaps agree. At least, he says explicitly that "what matters is not content alone":

> I am claiming that experience is capable of playing a justificatory role in epistemology because of its content, and hence that some particular experiences – namely those with

> the right kinds of contents – do justify some assertions. What matters is not content alone, but the content's being the content of an experience. (9)

An experience with the content *there is a flying owl* presumably has "the right kind of content" for a belief or assertion with exactly the same content. So does the experience justify forming the belief or making the assertion?

The experience does not show that the (content of the) belief or assertion is true. Suppose that what I see is not an owl, but a pigeon – and suppose it is a stuffed pigeon, that is not flying but has been thrown. In this case, the assertion is false, and the perceptual belief is false, and *the experience is false as well*, for the same reasons. Admittedly, the last is linguistically odd. It seems odd to say that I can see that there is a flying owl without there being a flying owl. 'Seeing' is a success-word, like 'knowing'. As ordinarily used, "A sees that P" entails P, just as "A knows that P" entails P. But once we endow perception with content, we must allow that perception might have false content, and we must rid 'seeing-that' of any success connotations that it might carry in ordinary speech. Thus, the fallibility of observation statements or of perceptual beliefs cannot be evaded by endowing experiences with statement-like or belief-like contents. That just makes experiences fallible as well.

Moreover, it is not to be assumed that having an experience with content C invariably issues in a perceptual belief with content C, let alone in an assertion with content C. The latter is obvious – the perceiver may lack spoken language. The former is obvious, too. Seeing is not always believing. I may have an experience with the content *there is a flying owl*, yet not come to believe that there is a flying owl – perhaps because I am also possessed of the mistaken belief that owls are flightless birds. In this sense, perceptual belief is obviously 'cognitively penetrated' (to use Currie's expression).

Currie's discussion of 'cognitive penetration' is puzzling. He seems to have become a 'concept-monger', conflating concept-possession with belief-possession. He runs together the question of whether perceptual content requires the appropriate concepts (it surely does), with the question of whether it requires the appropriate belief. He goes off into a side-issue, conceding that we can *imagine* that there is a flying owl without believing it. He then insists that we can only imagine things if we have a suitable stock of beliefs and belief-generated concepts – "a creature with imaginings but no beliefs is not possible" (11). Not *possible?* As for belief-*generated* concepts, does the belief which *generates* the concept C also contain C, which means that we must already possess C to form the belief? Presumably not. So the belief that generates the concept C does not itself contain C – how then does this generating work? But never mind this. Currie says that "The case where perception and belief have identical contents … is an obvious case where the content of the belief renders intelligible the perception" (11). But identical contents are not required for this. I can believe that there are no black swans and then see one. The content of my perception is rendered intelligible by my belief, if you like, in that both contain the concepts <black> and <swan>. But the contents of belief and perception are not identical – they contradict one another! Currie says that it is hard to specify "the point at which perceptual systems deliver their outputs and belief takes over" (13). But what if belief never 'takes over' the output of the perceptual system? What if I do not accept the 'evidence of my senses', because of other beliefs that I possess? I can have a perception with a certain representational content, yet not form the belief with the same content. I can perceive that there is a flying owl without believing it. To take another example, anybody who is not fooled by the Muller-Lyre illusion is rightly correcting the 'evidence of the senses' in the light of other beliefs.

Where are we? Seeing that P is not always believing that P, let alone saying that P. And seeing that P, believing that P, and saying that P, might all involve a false P. Can no more be said about the epistemological role of experience? Seeing may not always be believing, but it often is. Having an experience with content C often causes a belief with content C. What is caused is not, of course, the content C - contents or propositions have no causes. What is caused is the formation or adoption of a belief with content C. The epistemological question is whether a perceptual cause of a believing is also some kind of reason or justification for that believing. Critical rationalism rejects justficationism and proposes that it is. If reasons for actions are causes, then causes of actions may sometimes be reasons for them. Critical rationalism proposes that when seeing that P causes a believing that P, then the seeing is a (defeasible) reason for the believing. If I have no reason to think P false, then seeing that P is a reason for believing that P. This holds even when P is false, when both my seeing and my believing are mistaken. Reasonable beliefs may be false beliefs, quite generally. Reasonable perceptual beliefs may be false beliefs, too. Still, sense-experience delivers us evidence, particular beliefs or statements about the world against which we can test other beliefs and statements. It is 'foundational' not in the sense that 'the evidence of the senses' is infallible, but just in the sense that it is non-inferential. Or so my critical rationalism maintains. I suspect that Currie's own view of the justificatory role of experience in epistemology is not much different from this.

He saddles Popper with the view that "experience lies outside the space of reasons" (Sellars, McDowell), that "nothing can count as a reason for holding a belief other than another belief" (Davidson). What Popper actually said was that *"statements can be logically justified only by statements" (Logic of Scientific Discovery*, p. 43, italics in the original), which is quite different. Davidson's slogan "nothing can count as a reason for holding a belief other than another belief" is ambiguous. Holding a belief, like forming or acquiring a belief, is an action, something we do. So is the slogan "Nothing can count as a reason for holding a belief other than [holding] another belief"? Or is it "Nothing can count as a reason for holding a belief [with content P] other than another belief-content [different from P]". The latter is ridiculous: belief-contents or propositions are not reasons for actions. The former is implausible: it means that all foundational believings, believings that do not arise by inference from other believings, are unreasonable.

In Section 4 of his paper, Currie tells us that McDowell sought to "bring experience into the space of reasons by seeing it as possessing … conceptual content". Currie thinks this is a mistake, because 'conceptual content' is a misleading term (8). True, an experience can have exactly the same content as a judgement or belief, and be a (conclusive) reason for it. But neither content is 'conceptual' in the sense that concepts are 'constituents' of the content. The difference between perception and belief is that the subject needs no concepts to perceive that P, but does need concepts to believe that P. It seems that my cat can see that the mouse has run under the typewriter after all. It is just that, lacking the requisite concepts, she cannot bring herself to believe it!

Is sense-experience the only source of such 'foundational' or non-inferential beliefs? No, testimony is another, arguably more important, source. I have written little about testimony. I extended the pretty commonsensical critical rationalist view of sense-experience to testimony as well. I said that it is reasonable to "Trust what other folk tell you, unless you have a specific reason not to". But that was barely scratching the surface.

I am grateful to **Colin Cheyne** for digging deeper. I agree wholeheartedly with most of what he has uncovered. We agree that many, perhaps most, of our beliefs are acquired from testimony, and that if we never believed what other folk told us our belief sets would be extremely meagre. We agree that if it is reasonable to accept the testimony of others, then the problem of induction is solved. Testimony can provide you with reasonable, evidence-transcending beliefs (believings). Cheyne occasionally writes as if this is a *criticism* of my critical rationalist attitude to testimony. I regard it, rather, as vindicating that attitude, and pointing up the absurdity of the traditional empiricist doctrine that all beliefs, or at least all reasonable beliefs, arise from personal experience. All of us do have lots of reasonable evidence-transcending beliefs that we acquired from other folk, some of which will, no doubt, turn out to be wrong.

Cheyne correctly reports me as maintaining that even a contradictory belief may be reasonably acquired through testimony. And, in pursuit of a *reductio*, he points out that according to me testimony might also yield reasonable belief in the validity of affirming the consequent, or of enumerative induction! Well, as they say, one person's *reductio* is the next person's derivation of interesting conclusions. The unsuspecting logic student who has the misfortune to have a very bad teacher may well come reasonably to believe that affirming the consequent is OK. As for the widespread belief in the validity (or 'cogency') of induction, that belief is not necessarily unreasonable, either – it may have been inculcated in those who possess it by bad teachers, heirs to a bad philosophical tradition. To paraphrase Cheyne (25), people who are surrounded by inductivists, pay close attention to them, perhaps even attend their religious services (seminars and conferences on inductive logic), may well acquire a reasonable belief in the validity of induction. Cheyne thinks this absurd: "The problem of induction is not so much solved as blown away! … a belief that inductive reasoning is reasonable may not be unreasonable, from which it appears to follow that inductive reasoning may be reasonable" (23).

Once testimony is admitted as a source of reasonable belief, it must be admitted that some folk may acquire from their elders and betters a reasonable belief that inductive reasoning is valid. That does not mean, of course, that inductive reasoning *is* valid - one may reasonably believe a falsehood, according to critical rationalism. Furthermore, folk who reasonably believe that inductive arguments are valid may also reasonably act in accordance with their false belief and reason inductively.

Is the problem of induction "blown away" by this? The problem is to avoid the irrationalist conclusion of the following Humean argument (19):

> We do, and must, reason inductively.
> Inductive reasoning is logically invalid.
> To reason in a logically invalid way is unreasonable or irrational.
> Therefore, we are, and must be, unreasonable or irrational.

What Cheyne has shown is that, once we admit testimony as a source of reasonable belief, we can avoid the irrationalist conclusion by rejecting the third premise. People who reasonably yet falsely believe that induction is valid, may reasonably act on their false belief. Just as children who reasonably believe in Santa Claus, because their elders and betters told them so, reasonably put Santa's supper in the hearth on Christmas Eve in accordance with their false belief. It is crazy to deny that the children do reasonably believe in Santa Claus. Just as it is crazy to say that the countless generations who were taught by their elders and betters that the earth stood still were unreasonable to believe this, just because it is false. What goes for Santa Claus or the stationary earth now goes for inductive logic. Or so deductivists like Popper and me believe.

Cheyne's discussion reinforces the point that words like 'reasonable', 'rational' and their cognates should be reserved for believings –beliefs are true or false, rather than reasonable or unreasonable. For once we admit testimony as a source of reasonable believing in certain circumstances, it will bc impossible to say of any belief that it is unreasonable, meaning that it would be unreasonable in any circumstances for anyone to adopt that belief. That goes for belief in Santa Claus, God, a stationary earth, inductive validity, whatever.

Still, my formulation of Principle T (for testimony) was not careful enough. Cheyne objects that "as long as you refrain from criticising what you are told, your testimonial beliefs are reasonable. That cannot be right."(23). Indeed, it cannot. If you are in a position to criticise what you are told or to cast doubt on the veracity of your informant, and you refrain from doing either, you cannot be described as reasonably believing what your informant tells you. I accept Cheyne's more careful formulations of principles governing testimony.

I would add only one further point. There is, in these matters, an age of epistemic responsibility. Children reasonably believe in Santa Claus because Mum and Dad tell them so. They are in no position either to criticise what they are told or to doubt the veracity of their informants. It is different with grown-ups. And what goes for the children of today also goes for earlier generations of grown-ups. What used to be called 'the ethics of belief' is a neglected subject, chiefly because of the misguided empiricist notion that all reasonable evidence-transcending beliefs must arise by so-called 'inductive inference' from the so-called 'evidence of the senses'. Critical rationalism rejects this notion, and its 'ethics of belief' is the better for it – as well as being closer to common sense.

3. SEVERE TESTING

Sense experience and testimony are ways of getting started. They are sources of non-inferential reasonable beliefs, against which we can criticise and test other candidate beliefs. They are not infallible sources – some of the reasonable beliefs acquired from them will be false. Still, critical rationalism says that it is reasonable to believe (adopt, prefer) that evidence-transcending hypothesis, if there is one, which has best withstood serious criticism from these or other sources. One way to criticise a view is to subject it to the 'tribunal of sense experience'. (This does not just encompass our personal experience – it includes the experiences of others, transmitted to us through their testimony.) In the sciences this turns into the method of experimental testing. Clearly, critical rationalism owes us a story about what counts as a serious empirical criticism, or a severe experimental test, the result of which might genuinely confirm or corroborate a theory. John Worrall and Deborah Mayo both revisit this issue, and disagree sharply about it. I was tempted to let them fight it out – but I cannot resist entering the fray.

John Worrall and I agree that whether evidence **e** confirms theory **T** is to be assessed, not merely by considering the logical relations between **e** and **T** (as a 'purely logical' theory requires), but also by considering a third thing, 'background knowledge'. We also agree that a 'strictly temporal' view of background knowledge is no good. That leaves what I called the 'heuristic view' and the 'background (or touchstone) theory view'. Worrall favours the former, I once tentatively favoured the latter.

Worrall's first objection to the background theory view is that it yields the result that evidence that confirms the background theory **B** to a new theory **T** cannot also confirm **T**. He says "this is surely an extraordinarily counterintuitive judgement" (36). As he sees it, scientists will say that such evidence confirms *both* theories – though he concedes that scientists will be especially interested in whether the new theory is *better* supported than the old, a question to which evidence that supports both of them is irrelevant.

What this objection makes clear is that the background theory view makes evidential support an irredeemably *comparative* affair. On this view we cannot ask "Does **e** support **T**?", but only "Does **e** support **T** as against **B**?". (That evidential support is irredeemably comparative has also been argued by Larry Laudan and Elliot Sober. An early anticipation of it is 'Refutation or Comparison' by Archibald.) The background theory view is also irredeemably *historical*, because it is history that determines what the background theory **B** to **T** actually is. Worrall seems to ignore this historical dimension when he invites us to view the General Theory of Relativity as the background theory to Classical Physics, as well as viewing Classical Physics as the background theory to the General Theory of Relativity (32). You cannot do the former, if you take the *historical* character of the theory seriously.

Worrall finds it absurd to say that while there are phenomena that support Relativity but not Classical Physics, there are also phenomena that support Classical Physics but not Relativity (35). But these oddities fall away if you take the *comparative* nature of the theory seriously. There is nothing odd about saying that there are phenomena that support Classical Physics as against its rival (whatever that was), and other phenomena that support Relativity as against its rival, Classical Physics. Worrall insists that the former phenomena also support Relativity (provided that Relativity yields them in a non ad hoc way) "in the non-comparative sense of support" (38). But on the background theory view, there *is* no "non-comparative sense of support".

Is such a historico-comparative view of evidential support acceptable? It introduces a historical relativity into the issue, which seems unacceptable. I drew attention to this myself in my original 1974 paper: "… because Einstein had the misfortune to be preceded by Newton, his theory cannot be confirmed by all the evidence which it predicts, but which is also predicted by Newton's theory" (ERR, p. 246). If the 'Newtonian interlude' had never existed in the history of science, and Aristotle had been succeeded by Einstein, then Einstein's theory would have been much better supported (comparatively speaking) than it actually was! This is Worrall's chief worry, yet again.

But Worrall's own preferred theory confronted a similar worry. Worrall prefers the 'heuristic' view of novelty: evidence **e** is novel for theory **T** and supports it if **T** entails **e** but **e** was not used to construct **T**. My original worry about this was that if scientist **A** uses **e** to construct **T**, and scientist **B** constructs **T** without using **e**, then **e** supports **T** as proposed by **B** but does not support **T** as constructed by **A** (ERR, p. 241; cited by Worrall, p. 42)). Should theoreticians lie down on their couches and forget about the available data, if they want well-supported theorics? Worrall's reply to this was "Science is not like that". Two independent considerations suggest that this is right.

The first is that scientists do not typically arrive at hypotheses randomly or through mystical flashes of intuition, but rather (as Newton said) by 'deducing them from the phenomena'. Newton was right that 'deduction from the phenomena' is deduction (not induction, abduction, or any other ampliative process of inference). Newton was wrong that its premises are just observed phenomena - scientists also need general 'heuristic principles' of one kind or another as well. Scientists would be crazy not to use known facts to help construct their theories, and there is nothing wrong with doing so. But you cannot use the same fact twice, as a premise from which you deduce your hypothesis, and as support for it. Worrall argues that the real problem here is the 'prediction *versus* accommodation' problem, a.k.a. the '*adhocness* problem', a.k.a. the 'independent testability' problem. A theory is not confirmed by evidence that it entails if it merely accommodates it, or is *ad hoc* with respect to it, or if it did not result from an independent test – where 'accommodates', '*ad hoc'*, and 'not independent' are all cashed out in terms of 'used to construct'.

The second independent consideration is more philosophical. The Miracle Argument says (roughly) that the success of a theory would be miraculous if that theory were not true. The success spoken of is predictive success. But some predictive success is not miraculous at all. It is no miracle that a theory successfully predicts facts used to construct it.

Baby examples can illustrate both considerations. Suppose we do not know and want to find out what colour emeralds are. Should we lie on our couch, dream up hypotheses, and subject them to test? No, we should find an emerald, note its colour, and run through a so-called 'demonstrative induction' (actually a deduction): "Emeralds all share a colour, this one is green, so emeralds are all green". Again, suppose we do not know and want to find out what the relationship is between two measurable quantities, P and Q, and we have a hunch that it might be a linear relationship. Should we lie on our couch, dream up linear hypotheses, and subject them to test? No, we should measure two pairs of values of P and Q, and do some 'curve-fitting' (actually a deduction): "P = aQ + b, for some values of a and b; when Q = 0, P = 3; when Q = 1, P = 10; so P = 7Q + 3". It is no miracle that the hypotheses in these baby examples yield the observed facts used to construct them - neither do those facts support the hypotheses.

Of course, real science is not like these baby examples. In real scientific cases, the 'heuristic principles' that figure as premises in 'deductions from the phenomena' are specific 'hard core' principles of particular research programmes. Confirmation is not relative to persons, on the heuristic view, but relative to research programmes. And there is nothing undesirably subjectivist about that. Scientists in different programmes cannot come up with the same hypothesis (44). Scientists in the same programme can come up with the same hypothesis in different ways, but this is quite benign. Either one of them has read a parameter off the data when there was no need to do so, or one of them has mistakenly fixed a parameter from theory when it really could only be read off from the data. In the first case the data support the specific hypothesis and the programme in which it is embedded, in the second case they do not (44-47).

Can this be the whole story? There is a type of hypothesis that is constructed, not from particular observed facts, but from another hypothesis. This is the *surrealist* hypothesis T*: "The phenomena are *as if* T were true". Surrealist hypotheses are constructed, not by scientists, but by antirealist philosophers of science. They are constructed by simple deduction from T: "The phenomena are as if T were true" is a fancy way of saying "T is empirically adequate", and truth entails empirical adequacy. T* is, by design, empirically equivalent with T. But is it evidentially equivalent with T, is it equally well-supported by the evidence? On the 'background theory' view, clearly not: since T is obviously the 'background theory' to T*, there is no independent evidence at all for T*. Worrall's heuristic view is less clear about the case. But he can say, perhaps, that we implicitly use all the 'phenomena' that T entails to construct "The phenomena are as if T were true". (This would be in line with his analysis of the 'Gosse dodge'. Armed with the general principle that God created the Universe in 4004 BC as if the teachings of geology and evolutionary biology were true, the creationist finds out that geology and evolutionary biology entail fossils in the rocks, and straightway formulates the specific hypothesis that God decided to install 'fossils' in the rocks at the creation.)

Worrall considers another objection to his view, which my baby examples will make clear. *Given that* emeralds all share a colour, the observation of one green emerald *deductively entails* that all emeralds are green - what better evidence could there be? *Given that* the relationship between P and Q is linear, the results of a couple of measurements *deductively entail* a specific linear equation – what better evidence could there be? (This assumes, of course, that the observation is correct, and that the precise values of P and Q obtained from the imprecise measurements are correct as well. That is a different issue.)

In response to this objection, Worrall develops a dual view, according to which there are two types of confirmation or evidential support. First, e supports T' *relative to T* if T and e entail T' - in this case, e does not also support T. Second, e supports T' absolutely or unconditionally if T' predicts e and this prediction is experimentally verified - in this case e also supports the general T of which T' is a special case (in the baby examples the general Ts are that emeralds share a colour and that the relationship between P and Q is linear).

Worrall says he is a residual Popperian who thinks that "a test of a theory must surely be able to refute that theory" (58). I am more of a residual Popperian than Worrall. I also think that evidence for a theory must come from testing that theory. Worrall does not accept this. I would not call e 'evidence' for T', just because there is some T which together with e is deductively conclusive reason for T'. If we are not fussy, there will always be a T which together with e entails any T' – "If e then T'" will do. Nor, for similar reasons, would I grant that if e is evidence for T', then it is evidence for any T entailed by T'. Suppose that observing swans in Europe gives us evidence that all swans are white (T'). Does observing European swans give us evidence that *Australasian* swans are white (T), just because it is evidence for T' and T' entails T? (I had to get this example in – after all, having black swans in it is Australasia's chief contribution to the philosophy of science.)

I wish Worrall had stuck to his guns, and not dcveloped his dual theory of confirmation. I wish he had said instead that observation and experiment, and the data yielded by them, play two roles in science. First, they help us construct theories. Second, they enable us to test theories and, if we are lucky, confirm or support them. Worrall should grant his critics that e can be a conclusive *reason* for T' *given T*, but not grant that e is any kind of *evidence* for T'. As we will see, Worrall says something very close to this against Mayo's theory of the severity of tests.

I turn to **Deborah Mayo**'s wide-ranging and combative piece, which puzzled me at first. As explained already, critical rationalism owes us a theory about what a serious (or severe) empirical test is. Deborah Mayo is best known to the world for her resolute defence of a particular theory about this. What puzzled me initially was that there would seem to be nothing to stop the critical rationalist from adopting Mayo's theory. She formulates the general principle of critical rationalism thus: "CR: It is reasonable to adopt or believe a claim or theory P which *best survives* serious criticism" (64). Why cannot we expand this to "which *best survives* serious criticism in Mayo's sense"?

Mayo obviously sees things differently. She thinks that her theory of severity of tests is quite at odds with critical rationalism. She says repeatedly that a hypothesis can be the "best-tested" or "best survive serious empirical criticism" according to CR without having been severely tested or seriously criticised at all (e.g. 70, 71, 72). She asks "But why should it be reasonable to believe in the first hypothesis put forward …?"(71). She talks about "the critical rationalist's problem: being unable to say what is so good about the theory that (by historical accident) happens to be best-tested so far" (93). Critical rationalism has obvious answers to all this. If the 'first hypothesis put forward' has not survived serious criticism or severe tests, then CR will not say that it is reasonable to believe it. A hypothesis that has not been tested at all cannot have been tested better than any other hypothesis. What is good about the best-tested theory just *is* that it is the best-tested theory (so far).

What is going on? Mayo's own theory of severity is, in brief, that a test is a severe test of some hypothesis *h* if its outcome would be highly improbable if *h* were false. So the severity of a test with outcome **x** depends not just on p(**x**, h), but also on p(**x**, not-h). In the simplest case, where h entails **x**, the former is 1. To find out whether the test is severe we need to estimate p(**x**, not-h). How to do this?

At one point, while bashing Popper, Mayo says:

> "P is false" includes the disjunction of all possible hypotheses or claims other than P that would also "fit" or accord with **x** – the so-called 'catchall hypothesis' – including those not even thought of. Existing data **x** would be just as probable were one of the catchalls true, and P false. (71)

In what sense does "P is false" (equivalently "not-P") *include* this disjunction? It certainly does not logically include or entail it. The disjunction had better not include **x** itself, which fits **x** like a glove. If it did, "one of the catchalls" – which I take to mean one of the disjuncts in the catchall – would also entail **x**, and **x** would always be "just as probable were one of the catchalls true, and P false".

(By the way, critical rationalism does not traffic in the undreamt-of possibilities of the catch-all. It need not traffic in them, because it does not seek to justify any hypothesis. Its question is, which of the *available* competing theories is it reasonable to adopt or prefer or believe? You cannot believe an undreamt-of hypothesis.)

For Mayo, *'H is false' is Not the So-called Catchall Factor* (as she tells us in a section heading on 92). Mayo's view is that "mere accordance between **x** and P – mere survival of P - is insufficient for taking **x** as genuine evidence for P. Such survival must be something *that is very difficult to achieve* if in fact P deviates from the truth (about the phenomena in question)." (71). Well, what does P say about the phenomena in question? In the simplest case, P says (entails) **x**. So if P "deviates from the truth (about the phenomena in question)", we have **not-x**. Now "mere accordance between **x** and P – mere survival of P – is *very difficult to achieve* if in fact P deviates from the truth (about the phenomenon in question)", since p(**x**, **not-x**) = 0. On this reading, then, all tests of deductively entailed predictions have Mayo-severity 1.

This second reading is the right reading of simple cases like this, if Worrall is right. He complains that for Mayo, the process of adding up the SAT scores of the students in her class and dividing the total by the number of students to arrive at the number 1121, is a maximally severe 'test' of the 'hypothesis' that the average score is 1121. That is because the chance of the number arrived at being 1121 *if the 'hypothesis' were false* (that is, if the average score was not 1121) is zero. Like Worrall, I find it bizarre to talk of a 'test' or a 'hypothesis' in this case. The procedure is a demonstration that the average score is 1121, not a 'test' of the 'hypothesis' that it is. If we have done our sums correctly, the procedure is completely reliable 'error probe' and we can infer from its results that the 'hypothesis' is true. In this case, the hypothesis is deduced from phenomena alone (the definition of what an average score is, being true by definition, is a redundant premise). In other cases, the hypothesis is deduced from phenomena and other premises. In all cases, we gather the data not to test the hypothesis that is to be deduced from the data, but to figure out what the hypothesis is.

Suppose the hypothesis that we are interested in is the crazy hypothesis that the average SAT score in *all* classes is 1121. What does this crazy hypothesis say about the 'phenomenon in question', namely Mayo's class? It says that the average score is 1121 in that class. And the sums are a severe test of that hypothesis, too – for the chance of the sums yielding the answer 1121 if what the crazy hypothesis says about Mayo's class is false is zero.

(I said just now that the chance of arriving at the number 1121 for the average score, if the average score were not 1121, is zero. That was not strictly true: there is a small chance that I made an error in my sums. But this chance is the same whether or not the average score is 1121. The question of whether the evidence is reliable is not the same as the question of whether it represents a severe test of some hypothesis.)

As well as speaking of P deviating from the truth "about the phenomena in question", Mayo repeatedly cashes out the supposition that H is false as "a specified flaw in H is present" or "a specified discrepancy from H is present" (82). Again, she says that "H is false" refers to a "specific error that hypothesis H may be seen to be denying" (92). What 'specific error' is this? The specific error denied by H is, in the simplest case, the denial that its prediction about the case is mistaken. Mayo is quite open about this:

> What enables this account of severity to work is that the hypothesis H under test by means of data x is designed to be a specific and local claim, e.g., about parameter values, about causes, about the reliability of an effect, or about experimental assumptions. 'H is false' is not is disjunction of all possible rival explanations of x ... This is true, even if H is part of some large scale theory T: the condition 'given H is false' always means 'given H is false with respect to what it says about *this particular* effect or phenomenon'. If a hypothesis T(H) passes a severe test we can infer something positive: that the theory T gets it right about the specific claim H, that severely passes.
>
> The price of this localisation is that one is not entitled to regard full or large-scale theories as having passed severe tests as long as they contain hypotheses and predictions that have not been well probed. (92)

The upshot is that Mayo has nothing to say about which "full or large-scale theories" should be believed or accepted or preferred. Do not be fooled by the phrase "full or large-scale theories". Let T be any theory that entails but is not entailed by H, so that it might have another testable consequence H'. A Mayo-severe test of H does not allow us to say that T has been Mayo-severely tested: we are "*not* allowed to say that the entire theory is severely probed as a whole" (93). All we are allowed to say is that the particular testable consequence H has been 'severely probed'. In effect, she denies that we test a theory by testing its consequences, insisting that all we have *really* tested are the consequences! No wonder she rejects the 'comparativist' view that we should tentatively believe or accept or prefer that theory (if there is one) that has been Mayo-well-tested. No "full or large scale" theory can be Mayo-well-tested. It is not for nothing that her book is called *Error* *and the Growth of **Experimental** Knowledge.* She denies that "full or large-scale" theoretical knowledge grows.

Critical rationalism goes further than this. It proposes that it is reasonable to believe (adopt, prefer) that theory, if there is one, whose consequences have been best tested. Why is Mayo reluctant to take this further step? Basically, because this proposal about which theories we should (tentatively) believe has not been shown to be *reliable.* Can we show that following this proposal, adopting this belief-producing stratagem, will lead us to believe more truths than falsehoods? Unless we can, the proposal is to be rejected. Mayo rejects critical rationalist methods because they have not been shown to be reliable. Not that she has up her sleeve an alternative method of choosing between evidence-transcending theories that she thinks reliable. Rather, she thinks that there is no such method. That is why we should not believe any evidence-transcending theory. We should stick to reliable experimental methods and the experimental beliefs they licence.

But must a rationally adopted method be a reliable method? Consider the parallel question: must a rationally adopted belief be a true belief? Critical rationalism answers NO to the parallel question – we can rationally believe falsehoods, if we have tried and failed to show them to be false. All this is part and parcel of the rejection of justificationism. But when it comes to the meta-level or methodological level, Mayo is a justificationist. She is in good company. Other contributors to this volume, such as Sankey, Psillos, and even (in one small place) Gadenne, are meta-level justificationists, even though they all explicitly reject lower-level justificationism.

Perhaps, on this point, I may be allowed to quote myself (in English – 'BPS' stands for a belief-producing stratagem or method):

> Must a rational BPS be a reliable BPS?
>
> It might seem obvious that rationality requires reliability. After all, to believe is to think true. *If I find out* that something I believe is false, then it is no longer rational for me to believe it. Quite so. But the words 'if I find out' are crucial … After all, this is what makes room for rational beliefs which happen to be false, though I have no reason to think them so.
>
> Similarly with reliability. If I find out that BPS on which I have relied is not reliable, then it is no longer rational for me to rely upon it. But I do not need to show that a way of acquiring beliefs [is reliable] in order for it to count as rational. Of course, once I show that a way of acquiring beliefs is not [reliable], then I am epistemically at fault in persisting in that general strategy of acquiring beliefs. There is a kind of asymmetry here …
>
> Reliability is a desideratum on BPSs, just as truth is a desideratum on beliefs. As with beliefs and truth, so with BPSs and reliability. A belief does not have to be true to be reasonable … But if you find out that a belief is false then it is unreasonable to persist in it. A BPS does not have to be reliable to be reasonable … But if you find out that a BPS is unreliable then it is unreasonable to persist in it. (Musgrave 2001, pp.111-112)

If this is right, then whether it is reasonable to adopt some method or BPS depends on whether it has been criticised and shown to be unreliable.

Here I should confess an error. Part of Mayo's trouble is of my making. She says that critical rationalists "deny the reliability of the method they espouse" (64), "feel bound to deny the reliability of the methods they espouse" (64), "deny that tests which are severe in the critical rationalist's sense are reliable tools for uncovering errors"(64). When I read these statements, I wondered where Mayo got them from. Then I found that she got them from me! I did once say "Critical rationalists deny that the process they commend is reliable" (ERR, p. 346; cited by Mayo on 73). That was a mistake. What I should have said is that critical rationalists need not assert or prove that their method is reliable before they can rationally adopt it. However, if the method can be criticised by showing to be *un*reliable, then they should cease to employ it.

Never mind my mistake. What is more interesting is Mayo's claim that the critical rationalist rule CR is "demonstrably unreliable" (64). That claim, if it could be made out, should give the critical rationalist pause. But has she made it out? Has she shown that to adopt or believe claims which best survive serious criticism is to adopt false claims more often than true ones? I do not think so.

What she does instead is discuss how we might defend epistemic principles or methods such as CR. She describes an argument of mine as "subtle and interesting", and then amuses herself by trying to show that it is neither subtle nor interesting. The argument was:

> Any general epistemic principle is either acceptable by its own lights (circularity), acceptable by other lights (hence irrational by its own lights …), or not rationally acceptable at all (irrational again). So even though the rational adoption of CR involves circularity, this cannot be used to discriminate against it and in favour of some rival theory of rationality." (ERR, p. 331: cited by Mayo on p. 74)

By a 'general epistemic principle' I meant a *comprehensive* principle, which says that a belief is rational if it has some feature F and that having feature F is the *only* way that a belief can be rational. The context of my discussion was two-fold: Bartley's comprehensively critical rationalism, developed in opposition to Popper's view that belief in reason must be based on irrational commitment; and Nozick's view that "justificatory principles … deep enough to subsume themselves" are "a triumph". Mayo says that the form of a "general epistemological principle" is "EM: a claim P is acceptable if it is classified as acceptable or beliefworthy by belief-classification method M" (74). She supposes that EM is acceptable by its own lights. She then says "this does not yet entail that the *only* warrant for EM is EM" (74). Perhaps not. But what EM does entail, if it is a 'general epistemological principle' in my sense, is that this other 'warrant' for EM is unacceptable so that beliefs 'warranted' by it (including belief in EM) are not thereby shown to be acceptable. (By a general epistemic principle I mean what Mayo calls a 'self-sealing' principle. If a method is not self-sealing, then it is not the only method, and we can ask why we should accept this further method, and be off on a regress.)

This is all terribly abstract. Should we care about 'general epistemic principles'? Do we not always have a *multiplicity* of principles, or methods, or 'warrants'? Perhaps. But let the multiplicity of methods be M_1, …, M_n, form their disjunction M*, and let the general principle EM* be that a belief is rational if and only if it is classified as such by M*. (This assumes, non-trivially, that the different methods will yield consistent results.) Now ask whether belief in EM* is rational. This is the question of whether reason can be defended by reason, which Popper, Bartley, Nozick and I were discussing. (Related issues are whether scientific methods can be defended by scientific methods, whether logical principles can be defended by logic, and so on.) I stick by my argument.

Mayo discusses something different, but I am not sure what. She formulates a principle saying (in effect) that we should accept the claim that some book is in print if the book is listed in the *Handbook of Books in Print* (BIP). This principle clearly specifies a 'method' which applies only to a restricted class of claims. It is also clearly a 'meta-principle' about that restricted class of claims – it is not itself a claim that some book is in print. Mayo imagines that the principle is printed on the first page of BIP, and so is "acceptable by its own lights". No, since it is not itself a claim that some book is in print, it is not acceptable by its own lights at all. (If BIP listed BIP, the claim that BIP is itself in print would be acceptable according to the principle.) Obviously, we need to assess the acceptability of this principle in other ways, by trying to criticise the claim that BIP is a comprehensive list. As Mayo says, we reject Sloppy Joe's Books in Print because we could "adduce many reasons for regarding its listing as unreliable, out of date, and so on" (75). So, I would invoke CR to decide between Sloppy Joe and BIP. Mayo is wrong that "According to Musgrave, even "Sloppy Joe's Books in Print" would be as acceptable as the authoritative BIP!" (75).

Mayo also discusses another epistemic principle that I invented to show, contrary to Nozick, that "self-subsumption is not a virtue" (ERR, p. 331). That was "It is reasonable to believe anything said in a paper by Alan Musgrave". (Lest I be accused of hubris, I have never *said* this in any paper of mine. I merely *mentioned* it as an example of a crazy principle! It was a use-mention confusion for me to call it 'self-subsuming'.) Mayo asks me on what grounds I think this principle crazy. Critical rationalist grounds, of course: the principle could not withstand obvious criticism. Mayo seems to think that I have committed myself to the view that all self-subsuming principles are on a par and equally acceptable. This is precisely what I was concerned to deny: saying that two principles are alike in being self-subsuming is not saying that they are on a par or equally acceptable.

4. OTHER CRITICAL METHODS

Returning to the central issue, must a method for determining what to believe be shown to be reliable, before it can be reasonably adopted? As explained, critical rationalism denies this. **Volker Gadenne** is a critical rationalist, but he leaves it behind when he ascends to the meta-level, and asks about the rationality of methods themselves (including critical rationalist methods). He writes:

> But let us assume that A is the goal of science and our task is to decide whether procedure M should be recommended with respect to A or not. In this case, a rational person will recommend M if and only if he or she *believes that M contributes to A,* or that *M gives us a greater chance to achieve A.* And this belief is reasonable if there are good arguments in favour of the hypothesis that M is the best we can do to achieve A. (101)

My quarrel is only with the last sentence here, which suggests that rational belief in the meta-hyothesis MH ("M is the best we can do to achieve A") requires good arguments in favour of MH. This is meta-level justificationism. If we reject it, what is needed for rational belief in MH is that MH has withstood criticism, that there are no good arguments against it. Or so critical rationalism, applied at the meta-level or methodological level, maintains.

I suspect that Gadenne would be sympathetic with this. He goes on to say, rightly in my view, that any attempt to 'guarantee' that a method will be successful is "too strong for a methodology that is committed to fallibilism", whereas merely to hope that a method will be successful is not enough. We need something "less than guarantee but stronger than mere hope" (101). I suggest that having withstood criticism, there being no good arguments against it, is precisely what Gadenne needs here – it is more than mere hope, yet no guarantee. Here, by the way, I agree with Gadenne and others who urge that methods or methodological rules are not mere conventions, if by 'convention' we mean something that cannot be rationally assessed. As Gadenne says, "We can argue for [the adoption of] methodological rules, in the same way we argue for [the adoption of] other … assumptions" (102).

Like Gadenne, **Howard Sankey** seems to reject justificationism at the level of first-order belief, only to insist upon it at the level of method. Sankey correctly reports that the rejection of justificationism lies at the heart of critical rationalism (116-118). He concedes that:

> The critical rationalist account of theory acceptance is … of clear relevance to the problem of method and truth. For the critical rationalist asserts that survival of critical scrutiny provides the basis for rational belief in the truth of scientific theories. (118).

And yet Sankey insists that:

> An explanation is therefore required on the part of the [critical] realist of why certification by method provides warrant with respect to truth. I will refer to the need to provide such an explanation as the problem of method and truth. (109)
>
> By itself, the rejection of justificationism does not suffice to solve the problem of method and truth. If truth is non-epistemic, and the critical method is the basis of theory acceptance, the connection between method and belief in the truth is left entirely unexplained. (119)

But the connection between critical method and (rational) believing is precisely what critical rationalism explains. What is going on? What exactly is the problem of method and truth, left unsolved even if we reject justificationism? Are we being asked for an explanation of why theories certified by the critical method are true? Or are we being asked for an explanation of why theories certified by the critical method are rationally believed? Sometimes it is the latter: "Musgrave … must confront the question of why it is rational to believe theories certified by the methods of science" (109). Sometimes it is the former: "What bearing does method have on truth? Why should the use of method lead to theories that are … true …? This is the problem of method and truth." (109)

Sankey's formulations systematically conflate beliefs and believings, in line with justificationism. One example will suffice, although many could be given:

> But if it does not follow from survival of criticism that a theory is true [this concerns beliefs], then neither does it follow that the theory is to be accepted as true [this concerns believings, or 'acceptings'] (119)

In fact, Sankey does not reject justificationism, but assumes it, both at the level of particular belief and at the level of method.

Sankey has another worry about critical rationalism: "The trouble is that nothing has been done to secure belief in truth as the unique mode of theory acceptance" (119). He considers inference to the best explanation (IBE) and its meta-instance, the Miracle Argument for Realism (MA). Having displayed my deductivist formulation of IBE, he says:

> The question is why it is reasonable to accept the best explanation *as true*. Might it not be equally reasonable to accept the best explanation as empirically adequate ...? (116)

My answer to this question is NO. Suppose that H is the best explanation that we have of some phenomena. The T-scheme says: H if and only if it is true that H. Given the T-scheme, to believe that H and to believe that H is true are the same. Given the T-scheme, to accept that H and to accept that H is true are the same. So what is it to "accept H as empirically adequate"? It is not to accept H, for this is the same as accepting that H is true. Rather, it is to accept a meta-claim about H, namely the meta-claim "H is empirically adequate" or equivalently "The phenomena are as if H were true". Call this meta-claim H*. Now, and crucially, H* is no explanation at all of the phenomena that (we are assuming) H is an explanation of. "It is raining" explains why the streets are wet, but "The phenomena are as if it is raining" does not. At least, H is a better explanation than H* is. So according to IBE, H* should not be accepted as true. That is, according to IBE, H should not be accepted as empirically adequate.

I wonder what part of this argument Sankey and others who think like him will reject. Not IBE – at least, they pretend to accept IBE. Not, presumably, the T-scheme. Not, presumably, its consequence, that to accept H and to accept H as true are the same thing. Not, presumably, the equivalence of "H is empirically adequate" and "The phenomena are as if H were true". Not, presumably, the claim that H is a better explanation of the phenomena than "The phenomena are as if H were true".

This does not refute constructive empiricism. But it does refute the claim that constructive empiricists can happily accept IBE and give a constructive empiricist reading of it. If you try to recast IBE in terms of empirical adequacy rather than truth, you end up with something quite incoherent. Your major premise says that it is reasonable to accept the best explanation as empirically adequate. But this is to accept something that is no explanation at all! The fact is that realism and explanation go hand-in-hand. It is no accident that down the ages antirealists have pooh-poohed the idea that science explains things.

The same applies to the Miracle Argument (MA), a meta-instance of IBE where the fact to be explained is not a fact about the world but a fact about science, the fact that some scientific theories have (novel) predictive success. The truth of a theory T is a better explanation of T's (novel) predictive success than "T is empirically adequate" or "The phenomena are as if T were true". Or, putting the point in terms of empirical adequacy, truth explains empirical adequacy, but empirical adequacy does not explain itself. Given MA, a meta-instance of IBE, it follows that it is reasonable to accept that T is true.

Sankey comes up with a "constructive empiricist version of critical rationalism" according to which "theories which survive criticism are to be accepted as empirically adequate" (120). This is a possible view. What is not possible is to combine this view with IBE. But the trouble with IBE, as I construe it, is that the best explanation is not shown to be true (just reasonably accepted as true). And the trouble with critical rationalism is that theories which survive criticism are not thereby shown to be true either (just reasonably accepted as true). Notoriously, the same trouble arises for Sankey's critical constructive empiricism: theories which survive criticism are not thereby shown to be empirically adequate (just reasonably accepted as such). Which brings me back to Sankey's problem of method and truth.

How does Sankey solve his problem? He says that epistemology must rest on metaphysics (111), and wheels in some metaphysics. Well, resting epistemology on metaphysics is a welcome change from resting metaphysics on epistemology. That was the basic mistake of idealists down the ages – including that version of idealism known as 'internal realism', which Sankey discusses and joins me in rejecting. As he rightly sees, the internalist "closes the gap between method and truth" (113) by going for an epistemic theory of truth. But such a theory, if combined with the T-scheme, closes the gap between method and reality as well. The world comes to "depend on our methods of inquiry or our theories in idealist fashion" (114). But, Sankey rightly says, "that is evidently not something that a realist can accept" (114).

I wish that Sankey would complete the separation of epistemology from metaphysics. That is precisely what critical rationalist epistemology does, with its rejection of justificationism. As he himself puts it, when introducing his problem of method and truth:

> But matters of method and rationality are separate matters from those of reality and truth. This is especially the case from the perspective of realism. ...Reality is not subject to determination by human thought. This remains the case even if the belief that the world is a given way is a belief that is rationally justified. For one may rationally believe what is false. The point applies with equal force to scientific theories certified by the norms of scientific method. A theory that is certified by the norms of scientific method is not thereby shown to be true. A theory which satisfies methodological norms may yet be false. (108)

The critical rationalist proposal is that a theory that is 'certified by the norms of scientific method' is rationally believed, despite the fact that it is not thereby shown to be true. Of course, from the point of view of justificationism, this is an absurd proposal – any reason for believing must be a reason for what is believed. But IF justificationism is wrong (a big 'if', of course), THEN Sankey's 'problem of method and truth' is solved – or to be more precise, the problem of method and rational belief (in truth) is solved. In fact, if we reject justificationism at the meta-level, Sankey's 'problem of method and truth' actually becomes two problems. First, there is the problem of whether 'certification by critical methods' shows truth. Second, there is the problem of whether 'certification by critical methods' shows rational belief (in truth). Critical rationalism solves the second problem, but not the first. According to critical rationalism, we cannot show of any method that it yields truths infallibly, or even that it yields truths reliably (yields more truths than falsehoods). At least, we cannot show this unless we argue in a circle, or set off on an infinite regress. For what we have here is, basically, the problem of induction yet again.

We can try to solve the problem of induction by wheeling in a metaphysical principle like "Unobserved cases resemble observed cases". But where does this come from – inductive reasoning (the circle), or some other metaphysical principle (the regress)? Similarly, we can solve the problem of method and truth by wheeling in a metaphysical principle like "Scientific methods yield truths". But where does this come from – scientific methods (the circle), or some other metaphysical principle (the regress)? Nothing is changed if we weaken the metaphysics, and say "Unobserved cases resemble observed cases more often than not" or "Scientific methods yield more truths than falsehoods".

Something like the last principle is Sankey's own solution to his 'problem of method and truth'. He claims that "the rules of method are reliable means of promoting the realist aim of truth" (122). This means, I take it, that most theories certified by the rules of method are true. How do we know this? From empirical or scientific inquiry, "For it is an empirical matter whether use of a particular method reliably conduces to a given cognitive goal" (122). Now suppose we could somehow show that most theories certified by the rules of method *up until now* have been true. Would this show that the rules of method are reliable? Not without inductive reasoning, not without invoking precisely the 'rules of method' Sankey is supposed to be justifying. Would it show that it is reasonable to believe that the rules of method are reliable? Yes, provided we abandon justificationism and adopt critical rationalism.

Do not mistake me. I do not object to the empirical or scientific study of our cognitive apparatus (perception, rules of scientific method, or whatever). And I do not object to pointing out ways in which the results of such studies may cohere with or mutually support the results of studying the world using that cognitive apparatus. I once did a bit of this myself, when I made the simple-minded suggestion that if the theory of evolution is to be believed, then perception is a reliable process. This suggestion may be criticised in all sorts of ways – never mind that. The relevant point here, as I immediately pointed out when I made the suggestion, is that circularity looms. If the theory of evolution is to be believed, perception is reliable. And why is the theory of evolution to be believed? Because of the evidence in its favour, got through reliable perception.

Sankey discusses the views of Kornblith about natural kinds, which come from the same stable as my simple-minded suggestion. But what does Kornblith's suggestion come to? If science is to be believed, then there are natural kinds in nature. And if there are natural kinds in nature, then science is to be believed. More generally, given some metaphysical M, science is true or mostly true. And how do we know M? Why, science teaches us that M is so. M is not some science-transcending metaphysical 'sky-hook' (to borrow Dennett's term), more in need of justification than the science it is supposed to justify. The only escape from this justificationist circle is to free epistemology from metaphysics, and adopt critical rationalism. The problem of method and rational belief (in truth) is solved by critical rationalism. The problem of method and truth is insoluble, for much the same reason as the traditional problem of induction is insoluble.

But what if the empirical or scientific study of science were to show that most theories so far 'certified by the rules of scientific method' were false? What if it could be shown that inference to the best explanation (IBE) has so far led folk to believe more falsehoods than truths? What if it could be shown that believing the hypothesis that best withstands criticism has so far more often than not been believing falsehoods? Would such results not constitute severe criticisms of the rules of scientific method, IBE, and critical rationalism in general? Yes, they would. As I already said, if a method for forming beliefs is shown to be unreliable, it is no longer reasonable to persist with it. But the critics of critical rationalism and critical realism have got nowhere near showing any of these things, as we will see later.

If somebody can propose non-critical non-realist methods that withstand criticism better, then we should adopt them. But the principles advocated here are very general, and I can think of no better alternative to them:

> Unless you have a specific reason not to, trust your senses.
> Unless you have a specific reason not to, believe what other folk tell you.
> Prefer that evidence-transcending hypothesis that best withstands criticism.
> Accept the best available explanation of any puzzling phenomenon.

Having no beliefs is not an alternative. Having beliefs but thinking them irrational is just irrationalism. Saying that you believe things but do not think them true is just a linguistic confusion.

Stathis Psillos says that the rejection of justificationism "is exactly right" (140). Yet he joins Sankey in thinking that it is incumbent upon the scientific realist to show that scientific methods are reliable. For Psillos, these methods are "ampliative-abductive methods" (more on this in a moment). The realist must show that "the ampliative-abductive methods employed by scientists … are reliable: they tend to generate approximately true beliefs and theories" (134). How is the realist to show this? Psillos says that "The reliability of abductive reasoning is an empirical claim, and if true, it is contingently so." (135). Presumably, then, we are to employ the ampliative-abductive methods of science to show that the ampliative-abductive methods of science are reliable.

This is obviously circular. But in an attempt to disarm this objection, Psillos introduces a distinction between premise-circularity, which is vicious, and rule-circularity, which is not (135; see also his 1999, pp. 81-90). My problem with this is that non-vicious rule-circularity turns into vicious premise-circularity, given Psillos's own admission that an invalid ampliative inference can always be turned into a valid deductive inference by the addition of suitable missing premises (140). Somebody notes that all observed As were Bs, and infers that all As are Bs. We protest that this is invalid. To disarm this objection, the extra premise that unobserved cases resemble observed cases is wheeled in. If we ask what justifies this extra premise, we are told that it is inferred from the fact that unobserved cases have resembled observed cases so far. Enter Psillos. Though we *can* wheel in the extra premise, whereupon vicious premise-circularity befalls us, we *should not* wheel it in. Instead, we should say that inductive generalisation is not deduction at all, but an ampliative method, which proceeds according to an ampliative rule whose use can be vindicated by employing that very rule. This rule circularity is not vicious. Similarly with other ampliative-inductive methods of science. We use them to show that they are reliable, in a non-vicious rule-circular way. How is a problem solved by refusing to state explicitly what you admit might be stated explicitly?

In the course of his earlier discussion of these matters, Psillos wrote something with which I entirely agree:

> If one knew that a rule of inference was unreliable, one would be foolish to use it. But this does not imply that one should first be able to prove that the rule is reliable before one [non-foolishly, wisely, reasonably?] uses it. All that is required is that one should have no reason to doubt the reliability of the rule … . But we have no such reason. (1999, p. 85)

Having no reason to doubt P does not show P to be true – but it does show that it is reasonable to think that it is. Having no reason to doubt that R is reliable does not show that R is reliable – but it does show that is reasonable to think that it is. Or so critical rationalism assumes.

Chief among the 'ampliative-abductive methods of science', according to Psillos, is IBE. I regard IBE as a valid deductive scheme, instances of which might be sound as well. First, taking a leaf out of Peirce's book, I add an epistemic modifier to its conclusion. Then I add an epistemic principle to its premises so that it becomes a valid deductive scheme. This deductivist reconstruction simply makes explicit what is being implicitly assumed.

Psillos has no problem with this, but he does object to the deductivism that motivates it (132). His objections are familiar ones. He says that human reasoning is content-increasing, while deductive reasoning is not. But the undoubted fact that human thought is content-increasing, that I may think more today than I thought yesterday, does not mean that yesterday's thoughts were the premises from which I reasoned to today's thoughts. There is more to thinking than reasoning or arguing.

Again, Psillos says that human reasoning is defeasible, sensitive to new information, evidence and reasons, while deductive reasoning is not. The defeasibility of ampliative reasoning means that so-called 'ampliative logic' becomes an empirical science – new information, evidence and reasons can show you that an argument you thought valid (cogent) was not. Besides, valid deductive reasoning is also defeasible: new information, evidence and reasons may show that some explicitly-stated premise is false.

Finally, Psillos complains that deductive reasoning cannot establish the truth of its premises – for that we need ampliative or non-deductive reasoning. Deductive reasoning must stop somewhere, on pain of infinite regress – whereupon ampliative reasoning takes over. But ampliative arguments have premises, too, so there is no stopping the regress with their help. Besides, the assumption here seems to be that everything we assert is the conclusion of some argument or other. This logomaniac assumption is absurd: human reasoning requires premises that cannot themselves be conclusions of arguments, on pain of infinite regress.

The bulk of Psillos's paper contains an attempt to rebut Colin Howson's criticism of the No Miracles Argument (NMA), to the effect that it commits the 'base-rate fallacy'. The version of NMA that Howson criticises is the traditional one, where it is the truth or high probability of the best explanation that is supposed to be established. Since my version of NMA concludes neither that the best explanation is true nor that it is probably true, my version is immune to this criticism. Moreover, my version is deductively valid, and involves no fallacy.

Still, the question remains whether NMA (or IBE) are reliable methods of forming beliefs. Psillos thinks it incumbent upon realists to show that they are (which I deny), and he thinks that they can show it. He regards the following thesis as constitutive of realism:

> *The Epistemic Thesis*: Mature and predictively successful scientific theories are well-confirmed and approximately true. So entities postulated by them, or, at any rate entities very similar to those postulated, inhabit the world. (133)

This is not an epistemic thesis at all – it is a metaphysical thesis. It says (cutting out the complications) that mature science is true so that the entities it postulates exist. Next he says "What is worth stressing is that Musgrave takes NMA to aim to tell in favour of the *Epistemic Thesis* ..." (138). It is worth stressing that this is wrong. As I see it, "NMA makes it reasonable to accept that truth has been achieved" (138), but it does not show that truth has been achieved (which is what the so-called 'Epistemic Thesis' says). Justificationists assume that a reason for accepting something as true must show that it is true. That is precisely what I deny.

Nor do I claim that most theories whose acceptance is licensed by NMA are true – which is where the base-rate fallacy comes in. I do not even claim that most theories *up until now* whose acceptance was licensed by NMA were true. (To get from the latter to the former, to the reliability of NMA, would involve, of course, a simple inductive leap. How *could* a critical rationalist assert the reliability of CR, without engaging in inductive reasoning? How could a critic of critical rationalism assert the unreliability of CR, without engaging in pessimistic inductive reasoning?)

Never mind this. I agree with much of what Psillos says in defence of NMA. I read his arguments as showing, not that NMA is reliable, but that its critics have not shown it to be unreliable. More generally, have the critics of critical methods shown that they are an unreliable way of forming beliefs? I do not think so. Some critics have claimed that all scientific theories are false, usually by arguing according to the pessimistic induction – "All past theories were false, therefore all theories are false". But the premise of this argument is preposterous, and the conclusion does not follow. Nor does confining attention to 'big theories' ('global theories', 'paradigms', 'world-views', 'research programmes', etc.) help. Saying "Aristotle was wrong, Galileo was wrong, Kepler was wrong, Newton was wrong, Einstein is likely wrong" gets nowhere near to showing that all the 'little theories' (including those that follow from the big ones) are wrong. Positivists argue that all theories that postulate unobservables are false, because there are no unobservables. But why should we accept the positivist, human-chauvinistic premise of this argument? But I have discussed all this elsewhere, and will say no more here.

Save for one point. Antirealists like to point to past theories that were partially successful, but which then were falsified. Obviously, the partial success of falsified theories cannot be explained by their truth. How then to explain it? And if partial success is explained by something other than truth, why cannot total success be explained by something other then truth as well? Here Psillos goes in for truthlikeness or verisimilitude, which is (notoriously) problematic. Why not go in for partial truth instead? This is not the same as truthlikeness. Despite its falsity, "All swans are white" is predictively successful in Europe, and bird-watchers find it useful to employ it there. I do not know how close to the (whole) truth "All swans are white" is, and none of the captains of the verisimilitude industry can tell me in less than 100 pages of formulas. I do know that it has a true part, "All European swans are white", whose simple truth explains the success European bird-watchers have. Psillos qualifies NMA as follows:

> … realists should *refine* the explanatory connection between … predictive success, on the one hand, and truthlikeness, on the other. They should assert that these successes are best explained by the fact that theories which enjoyed them have had *truthlike theoretical constituents* (i.e., truthlike descriptions of causal mechanisms, entities and laws). The theoretical constituents whose truthlikeness can best explain empirical successes are precisely those that are essentially and ineliminably involved in the generation of predictions … (135)

The point is important, I agree. But why cannot we replace 'truthlikeness' with 'truth', and 'truthlike' with 'true'? Then NMA could be qualified in the following very simple way: truth is the best explanation of total empirical success; partial truth is the best explanation of partial empirical success.

Science grows out of commonsense, and scientific methods grow out of commonsense methods. We use critical methods, and IBE, all the time, and most of the time they serve us very well indeed, in everyday life and in science. Armed with these methods, we have come up with innumerable rational beliefs that are true (and some that turned out to be false).

5. THE METAPHYSICS OF CRITICAL REALISM

I have so far been resisting the idea that critical rationalist epistemology must be *underpinned* by some metaphysical assumption. Still, that leaves open the question of what metaphysics or ontology a critical rationalist/realist will adopt. Obviously, scientific realists will have no truck with the positivist idea that only things that happen to be observable by humans exist. Scientific realists will be realists about the things postulated by our best scientific theories, whether they happen to be observable by humans or not. Scientific realists believe in what **Michael Redhead** calls 'the Unseen World'. I agree with pretty well everything Redhead has to say about this. I have a few nit-picks. I wish Redhead had not floated the old, discredited idea that it is not the table we see, but "light reflected off the table, or … electrical stimulation in the retina caused by the light, or … " (135). We do not see light, or electrical stimulations somewhere in our heads caused by light. These things are part of the Unseen World. The science of vision posits them in order to explain how we see the table. To say that we observe things "by the *effects* they produce" in us (157) is not to say that we only observe the effects.

Several of my critics want me to go further. Alan Chalmers wants me to be a realist about *essences*, Robert Nola wants me to be a realist about *structures*, and Mark Colyvan wants me to be a realist about *numbers and other abstract platonic objects*. I am not so sure. My realist metaphysics is explanatory or more generally problem-solving. I believe in things that do explanatory or problem-solving work for us. So my question is, what explanatory or problem-solving work is done with essences, structures, and platonic objects?

Alan Chalmers wants me to become an essentialist. He says there are two problems I cannot solve that essentialism can solve – "straightforwardly" (163). First, there is the problem of explanatory asymmetries: some scientific deductions are explanatory and some not. Second, there is the problem of distinguishing genuine laws from accidentally true generalisations. The problems seem connected. One appealing thought is that a deduction is explanatory if a genuine law figures in its premises, non-explanatory if an accidentally true generalisation figures in its premises. Thus, if we can solve the second problem, we can solve the first problem as well. This appealing idea is hinted at by Chalmers when he says:

> Appeal to the law governing the expansion of metals can help explain why the bottle top is loosened when held under the hot water tap, whereas 'all the coins in my pocket are silver' cannot help to explain why any one of them is silver. (164)

But this will not do – laws figure in non-explanatory deductions as well as explanatory ones. Chalmers' own example proves the point:

> (The deduction of the range of a projectile from Galileo's laws of motion plus initial conditions explains why the projectile has that range, but the deduction of the height of a cliff from those laws plus the time of fall of a stone from top to bottom does not explain why the cliff has that height.) (163)

The example is quite typical. Scientific laws are typically functional dependencies, rather than universal generalisations of the form "All As are Bs". And a functional dependency can always be manipulated to yield a deduction that is intuitively non-explanatory. The loosening of the bottle top does not explain why it was held under the hot water tap.

That is why I suggested a different answer. A deduction is explanatory if its initial conditions specify the cause of the event to be explained, non-explanatory if they do not. But to make this answer good, we need an independent account of causality, one that will make good our intuitions that the time it took the stone to fall did not cause the cliff to have a certain height, that the loosening of the bottle top did not cause it to be held under the hot water tap (though the antecedent desire to loosen it might have), and that being in my pocket does not cause a coin to be silver. The principle that an effect cannot temporally precede its cause suffices in these cases.

What about the second problem, distinguishing genuine laws from accidentally true generalisations? Chalmers says that laws are "necessarily true, as opposed to accidentally true" (164). Now the 'necessity' spoken of here is not logical necessity, truth in all possible worlds, but rather physical necessity, truth in all physically possible worlds. And what are physically possible worlds? They are worlds in which the laws of nature obtain. The circle is complete: laws as opposed to accidental generalisations are physically necessary, true in all physically possible worlds, and a physically possible world is a world in which the laws are (accidentally?) true. Possible worlds are no help, as Chalmers says (164).

Can counterfactuals help? What about the idea that genuine laws support counterfactuals, whereas accidental generalisations do not? To make this idea work, we need accounts of the truth-conditions of counterfactuals and of the 'support' relation. The favourite account of the former is David Lewis's. To find out whether "If it were the case that A, then it would be the case that B" is true, inspect the closest possible world(s) to ours in which A is true and see if B is true in that world as well. And what world(s) are closest to ours? Why, worlds in which our laws of nature are true. The circle has got bigger, but it is still a circle. Possible worlds are no help once more.

I favour the idea that counterfactuals are elliptical statements of logical consequence: "If it were the case that A, then it would be the case that B" is true just in case A, together with some unstated premise(s) C that are true, entails B. Given this, to say that a generalisation G supports the counterfactual "If it were the case that A, then it would be the case that B" is just to say that G and A entail B. But this simple (simple-minded?) view means that accidentally true generalisations support counterfactuals just as well as laws do. We are still stuck.

Still, this theory of counterfactuals can illuminate the status of 'ideal laws', laws about 'ideal entities' that do not exist. These are a problem for realists, who seem forced to say that any such 'law' is false if construed as having existential import or vacuously true if construed as having no existential import. Neither option is satisfactory: the ideal gas laws are important, non-vacuous (and non-accidental) truths. I suggested that the ideal gas laws are true counterfactuals, thought neither vacuously nor accidentally true because they are supported by the kinetic theory of gases. Chalmers objects (167) that the ideal gas laws were empirically supported, independently of their derivation from the kinetic theory, by regarding ideal gases as the limit to which real gases tend as pressure is reduced. Quite so. I agree that experiments convinced physicists that the ideal gas laws were not mere vacuous truths. But it took the derivation from kinetic theory to convince physicists that they were not just accidentally true. The kinetic theory told them that if any gas were to be ideal, then it would obey the ideal gas laws. But what about the laws of the kinetic theory itself? If these are only true by accident, so are the counterfactual ideal laws that they support.

Can essences help us to break out of the circle? Chalmers says that "there is something about the nature of metals that makes it physically necessary that they expand when heated" (164). Metals by their nature have ontologically basic dispositional properties, powers and capacities, and precise statements of these dispositional properties or modes of acting and interacting are genuine laws of nature. If you ask why metals *must* expand when heated, the answer is that it is in their nature to do so. They would not be metals if they did not do so.

Chalmers correctly reports me as having two reservations about essences. First, are explanations in terms of essences to be regarded as *ultimate*? Is there no explaining why metals expand when heated, except to say they would not be metals if they did not? Second, is knowledge of the essences of things a priori, as is suggested by talk of *defining* things by their essences? Chalmers says that neither reservation applies to scientific essentialism, as defended by Brian Ellis. I am not so sure.

Chalmers insists that essences "needed to be discovered, not merely defined" (170), and that the "adequacy of our essentialist definitions needs to be established empirically" (170). But how do we find out empirically that a metal *must* expand when heated, because this is part of its essence, as opposed to finding out that all metals *do* expand when heated? "Metals must expand when heated" entails, but is not entailed by, "Metals expand when heated". There is no independent *evidence* for the stronger statement. Nor is the stronger statement more refutable than the weaker one – indeed, it does not seem refutable at all! Chalmers says that our characterisation of the essential properties of metals "may or may not correspond to what they actually are" (170). But he also says that anything that lacks an essential property of Xs is not an X (168). So we cannot discover a *metal* that fails to expand when heated – anything we find that fails to expand when heated was not a metal in the first place. Empirical methods seem powerless here. The most they can establish is that all metals do expand when heated, not that they must do so as a matter of physical necessity. Nor could we ever empirically refute "It is an essential property of metals that they expand when heated". Essentialism is unconfirmable and irrefutable metaphysics, not physics.

Something has gone wrong here. On the one hand, "Metals must expand when heated" seems irrefutable. On the other hand, it entails "Metals expand when heated" which is refutable. But if the latter is refutable, so is the former. Here is a diagnosis of what has gone wrong. If we stipulate that the term 'bachelor' means unmarried man, this cannot be refuted by finding a married bachelor. Similarly, if scientists stipulate that the term 'metal' is to be (partially) defined as something that expands when heated, this cannot be refuted by finding a metal that does not expand when heated. Statements of the essences of things (real or essentialist definitions) are irrefutable because they are really verbal stipulations (nominal definitions).

How plausible is this diagnosis? Scientists found out empirically that all metals expand when heated. They came to take this for granted in their future researches, and they sought to explain it. Suppose they went further, and decided that henceforth the term 'metal' was to be (partially) defined as something that expands when heated. This verbal stipulation marked a change in the meaning of the term 'metal'. To mark that change pedantically, we might say that tin was once called a 'metal', now it is called a 'metal*'. The new stipulation makes nonsense of the old empirical researches that established that metals expand when heated – those researches must be described using the old term 'metal', not the new term 'metal*'. The stipulation seems to turn a contingent truth into a necessary one. But this is an illusion that stems from overlooking the change in meaning. There is not one truth here, once thought merely contingent, now discovered to be 'physically necessary'. "Metals expand when heated" was and remains contingent. "Metal*s expand when heated" was and remains necessary, but only verbally or conceptually so.

Now I do not object to changes in language of this kind, brought about by new verbal stipulations. I do object to reading essentialist metaphysics off them. Words are one thing, things another. Our stipulations about what our words shall mean yield necessary truths – but the necessities they yield are logical or conceptual or verbal necessities, not real or natural or physical necessities. Nor can these verbal necessities do any explanatory work for us. The necessary truth "Metal*s expand when heated" does not entail the contingent truth "Metals expand when heated", let alone explain it or bestow necessary status upon it.

I said that I do not object to (partially) defining a metal as something that expands when heated. Or more precisely, since we do not define things but rather words, I do not object to (partially) defining the word 'metal*' to stand for things that expand when heated. This is just to make the word 'metal*' a *dispositional* word. Dispositional words have necessities built into them from the start – but the necessities are logical or conceptual. The meaning of an ordinary dispositional word involves from the outset a generalisation about behaviour. Something is brittle if it will break when struck by something else, something is soluble in water if it will dissolve when placed in water, and so on. The generalisations are, of course, vague and rough-and-ready bits of 'folk science', which need specification and refinement. No matter. The key point here is that such statements as "Something is brittle if it will break when struck by something else" are conceptual truths, matters of logical or conceptual necessity, not matters of fact. We do not discover empirically that they are true, by inspecting brittle things and checking whether they break. Just as we do not discover empirically that "Bachelors are unmarried men" is true, by inspecting bachelors and checking whether they are unmarried. Nor can we explain why something broke by saying that it was brittle. Just as we do not explain why some bloke is unmarried by saying that he is a bachelor.

Chalmers wants me to adopt Brian Ellis's version of essentialism. This claims that the essences of things need not be monadic properties (how the things are in themselves, without relation to other things), but may also be dispositions, powers, capacities, "how they are disposed to act and interact with other objects" (168). This is a welcome change from the usual essentialism, which insists that essential properties must be intrinsic or monadic properties. Now suppose that essences are dispositions, and consider the following:

> A charged body will attract or repel other charged bodies, give rise to a magnetic field when moving and radiate when accelerating because it is in the nature of charged bodies to do such things. Precise statements of these modes of acting, such as Coulomb's law or the Lorentz force law, describe the laws of nature. They are not something imposed on charged bodies because they are already implied in what it is to be a charged body. So charged bodies necessarily obey the laws that they do. (168-9)

If 'charged' is a dispositional term (which I will not dispute), then the laws here specified are *logically or conceptually* implied by describing something as 'charged'. It is logically or conceptually necessary that charged bodies obey those laws. (Which is not the same as saying that "charged bodies necessarily obey the laws" – see below). Nor do we explain why some body obeys the laws by saying that it is charged.

Why do we think that "Metals expand when heated" is not just 'accidentally true', like "All the coins in my pocket are silver"? The answer lies, I believe, in their causal explanations. We can explain why metals expand when heated in terms of deeper regularities, which reveal to us how heating up a piece of metal causes it to expand. There is no such explanation in the case of the coins in my pocket – we know that being put in my pocket does not cause a coin to be silver. But given what we know about metals, they *must* expand when heated.

But there is a fallacy here. Suppose there is an explanation, and a deductive explanation to boot, of why metals expand when heated. Let its explanans (whatever it is) be E, and its explanandum G. If the deduction of G from E is valid, we can say "Necessarily, if E then G". Here, of course, the necessity spoken of is logical. We can also misplace the word 'necessarily' and say "If E, then necessarily G". Now the necessity spoken of cannot be logical, since G is not a logical or conceptual truth. So is the necessity of G non-logical or 'natural' or 'physical'? No: this route to physical necessity is just misplaced talk of logical necessity, where the misplacing stems from Stove's fallacy of misconditionalisation. "Given what we know about metals, they must expand when heated" misconditionalises on logical necessity (Necessarily, if E then G) to yield physical necessity (If E, then necessarily G).

What about the coins in my pocket? I said that putting a coin in my pocket does not cause it to be silver, which is why we think "All the coins in my pocket are silver" is just 'accidentally true'. It is not that there is no explaining why all the coins in my pocket happen to be silver. But the explanation of it involves a rare confluence of independent causal chains. It happens 'by accident' in Aristotle's sense. We pronounce a verdict of 'accidental death' on a person killed by a loose brick falling from a building as he walks beneath. We can explain the death, by explaining why he walked there when he did and why the brick worked loose just then and fell. The accidental death is not uncaused or random, it arises out of a rare confluence of independent causal chains. Similarly with the coins in my pocket all being silver.

Of course, no misconditionalisation is involved in saying "Metals must expand when heated" if the deeper explanatory principles E are themselves natural necessities. Then the explanatory deduction does establish (as a matter of logical necessity) that "Metals expand when heated" is also a natural necessity. Now we are saying that what distinguishes genuine laws from accidental generalisations is that genuine laws follow from genuine laws, whereas accidental generalisations do not. This is hardly illuminating!

Does it become more illuminating if we can make out independently that the *ultimate* explanatory principles, the places where the explanatory buck stops, are not just accidentally true? There might, I grant, be ultimate laws – the 'Russian Doll' model of the universe might be false. But why cannot the ultimate laws be ultimate contingencies? There might, I grant, be fundamental particles. But why must fundamental particles have all their properties essentially?

I would quite like to be an essentialist and to believe in necessities in nature. My problem is to find a version of these doctrines that does not stem from two sources – the unholy alliance with definition, and the fallacy of misconditionalisation. My problem is to find a version of these doctrines that does not project necessities of language and/or logic onto the world. My problem is to get over the 'positivist' principle (prejudice?) that the only necessity is logical necessity.

I confess that I cannot understand Ellis's claim that "physical necessity is a species of logical necessity" (177). Physical necessity is in the world (if it is anywhere), logical necessity in our world-representations. To say that the former is a species of the latter just seems to be projecting necessities of language and/or logic onto the world, in the ways I have tediously explained. I do not object to a 'modest essentialism' whereby an essential property of an object is not one that it must possess, so that it would not be the object that it is without that property, but rather a particularly important property possession of which determines what the object does (and identification of which helps us to give causal explanations of its actions and interactions). 'Essentialness' in this modest sense obviously comes in degrees. And it has more to do with dispositional properties than monadic ones. Why cannot objects be 'bare particulars', not in the sense that they have no properties, but just in the sense that they have none of their properties essentially? Why cannot we stop asking "What is X?" (Popper called these 'essentialist questions') and ask instead "What does X do?"?

Which brings me to structures. What are they? Should realists believe in them? And what about *structural realism*, which is (roughly) the view that structures are the only things that realist should believe in? **Robert Nola** discusses these questions, and I agree with pretty well everything he says. I would say some of it more bluntly. Like Redhead, I think it bizarre to say that "it is only structure which really exists" (157). Nola calls this loopy view *Platonistic ontological structural realism* – "all that exists are mathematical structures … there do not exist placeholders, such as objects, within the structures" (182). To my mind, a structure must be a structure *of* something. My house is a structure made of house-bricks, and it is loopy to say that the structure exists but the bricks don't. Similarly, the bricks are made of molecules, and it is loopy to say that the bricks exist but the molecules don't. Reverse loopiness is just as bad. It is also loopy to say that the house-bricks exist but the house doesn't, or that the molecules exist but the bricks don't. Eddington started all this with his tale about the 'table-of-physics' and the 'table-of-common-sense', and his silly question "Which table is the real table?" I wish Redhead had not mentioned Eddington with approval (155). I have discussed all this elsewhere, and will say no more here.

Redhead and Nola think it less easy to dispose of what Nola calls (182) *epistemic structural realism* – "we can have knowledge of structures but we cannot know the items that are placeholders in such structures (such as objects); they are a "something-we-know-not-what"" (182; also Redhead, 157). The obvious response to unknowable Kantian *ding-a-ling-an-sich* is that if we know a lot about a structure, then we *ipso facto* know a lot about how the placeholders in that structure relate to one another. Yes, comes the stock reply, but knowing how the placeholders relate to one another is not knowing what their *true natures* are (Redhead, 157), what they are *in themselves*, what they are *intrinsically*, what they are *essentially*. And we are deluged with a heady mixture of bad essentialist metaphysics and/or bad philosophy of language, much of the former fuelled by the latter.

There is supposed to be a difference between knowing truths about Xs, about what they do, about how they relate to or interact with other things, and knowing about the *nature* of Xs. But what is the difference? Must truths about the nature of Xs concern only intrinsic or monadic properties of Xs? That seems wrong – truths about how beams of light behave are truths about the nature of beams of light. Perhaps the thought is that truths about relational properties of Xs cannot tell us the essential nature of Xs. At which point the so-called 'dispositional essentialist' wonders why relational or dispositional properties of Xs cannot be 'essential' to them – whatever light is, it would not be light unless beams of it obeyed this or that law.

I think that Nola shows convincingly in his paper that the problem which structural realism is supposed to solve is actually a pseudo-problem. The problem is so-called 'ontological discontinuity': as theories change, ontology changes. Ontological discontinuity cannot mean that the contents of the world change as our theories about the world change. To credit thinking or talking with such transformative powers would be a "virulent form of human chauvinism" (184 – I wish I had coined the phrase!). Nola asks whether "there are two different objects, the Bohr-Rutherford electron and the mature-Bohr electron" (205). There are certainly two phrases, two hyphenated names. But they are both empty names, no 'hyphenated entities' are picked out by either of them. They are out of the same (Kantian) stable as Eddington's 'table-of-common-sense' and 'table-of-physics'. "The Moon-as-conceived-by-Aristotle was perfectly spherical, whereas the Moon-as-conceived-by-Galileo had mountains and oceans on it" is just philosopher's gobbledy-gook for "Aristotle thought that the Moon was perfectly spherical, whereas Galileo thought it had mountains and oceans on it".

If ontological discontinuity is not in the world, where is it? It is in our theories. It is discontinuity in what our theories *say* is in the world, in the 'ontological commitments' of our theories. Difference of theory yields, we are told, difference of ontological commitment. The false theories of our ancestors did not succeed in referring to objects at all. It is not that the world changed when Galileo got a different theory about the Moon than Aristotle's. Rather, we found out that Aristotle was not talking about any object in the world at all. The same applies to Galileo, of course, since his theory about the Moon was not quite right, either. And the same will apply to us, if our theory is not quite right.

But of course, this is just bad philosophy of language. Difference of theory does not imply difference of ontological commitment. Aristotle's theory, and Galileo's, and ours are different theories about the same object, the Moon. The expression 'the Moon' has the same referent in all these theories. People say that these simple-minded views are all very well for observational terms like 'the Moon', where we can point to the object to fix the referent, but will not work for theoretical terms. Nola shows convincingly that continuity of reference can be established for theoretical terms as well. As he says, of the example involving competing theories of light:

> Given the above account of reference determination for 'light' in the two theories, we can dispense with the claim that there is object discontinuity from Fresnel's theory to that of Maxwell. There is a "something" that both theories are about. And it is not just structure ... We have not said very much about the intrinsic properties of light or even its nature or essence ... All we have is a "something" which ... obeys F- and M-equations. None of this forms the basis for an objection to the above account of referential continuity. If the Fresnel and Maxwell equations are correct, then there will be the same "something" that satisfies them ... There is no need for extreme structuralists to deny the existence of "objects" ... that stand in the structural relations. So there is no need for ontological flight from objects to structure. (219)

What Nola may not have noticed is that essentialism makes a mess of all this. He writes as if his story about referential continuity is compatible with any kind of metaphysical view. But essentialist metaphysics is incompatible with that story. As Nola well knows, Fresnel and Maxwell also had different views about the intrinsic nature of light. Fresnel thought light was intrinsically F, Maxwell thought it was intrinsically M. Suppose Fresnel had been an essentialist, who thought light not just intrinsically F but essentially so – light would not *be* light if it were not F. Suppose Maxwell had been an essentialist, too, who thought that light would not *be* light if it were not M. Then, despite the continuity of F-equations and M-equations, there is no one "something" that both theories are about. According to Maxwell, nothing answers to Fresnel's 'essentialist definition' of light – and vice versa. According to us, who think light is neither F nor M, neither Fresnel's nor Maxwell's theory is about anything at all. All that is left are common F-equations and M-equations, neither of them about anything. The way out of this morass is obvious enough – give up essentialism.

A few paragraphs back I uncharitably described the view that mathematical structures are *all* that exists as 'loopy'. It is not, of course, loopy to think that mathematical structures are *some* of the things that exist – as also, perhaps, are the mathematical objects that mathematical structures are structures of. Which brings me to **Mark Colyvan**'s paper. Colyvan correctly identifies an apparent tension in my views. I have argued, *contra* philosophical idealists of various kinds, for realism about (some of) the observable entities of commonsense. I have also argued, *contra* scientific antirealists of various kinds, for realism about (some of) the unobservable entities of science. But I am reluctant to extend realism to platonic entities, including the entities of mathematics. The reason for my reluctance is simple. Platonic entities are *queer* entities: not only are they unobservable, like the theoretical entities of science, but also they do not exist in space, or time, or space-time, and they have no causal powers. That being the case, why should we believe that such entities exist?

Colyvan and I agree that the only decent answer to this question is the Indispensability Argument. Mathematics is indispensable to our best theories about both the observable and the unobservable world. Furthermore, mathematics is to be taken at face-value, and the usual semantics applied to it. From which it follows that we should believe in numbers for the same kinds of reason as we believe in electrons. Numbers and electrons are in the same boat, epistemologically speaking. I agree with Colyvan and many others that this is the best, indeed the only decent, argument for numbers and other platonic entities. And I reject the argument.

It is important to see that there are *two* premises in the indispensability argument, one concerning the indispensability of mathematics, the other about the indispensability of numbers. Here a use/mention confusion pervades Colyvan's writings. He says "we ought to count as real any entity that plays an indispensable role in our best scientific theories" (225). But entities such as rocks, or electrons, or numbers, do not play any role in our *theories*. What play a role in our theories are *expressions* which, if they refer to anything at all, refer to entities such as rocks, or electrons, or numbers. In a telling footnote in his book, Colyvan writes:

> I often speak of certain entities being dispensable or indispensable to a given theory. Strictly speaking it's not the entities themselves that are dispensable or indispensable, but rather it's the *postulation of* or *reference to* the entities in question that may be so described. Having said this, though, for the most part I'll continue to talk about *entities* being dispensable or indispensable, eliminable or non-eliminable and occurring or not occurring. I do this for stylistic reasons, but I apologise in advance to any reader who is irritated by this. (Colyvan 2001, p. 10, fn. 18)

But use/mention confusion is not a matter of style. Consider Santa Claus theory, the stories we tell to small children at Christmas time. The name 'Santa Claus' occurs in these stories, and (let us grant) is not eliminable from them, and is indispensable to them. But Santa Claus does not occur in the stories – how could he, he does not exist? In this case, we all agree that the indispensability of the name to the theory does not carry with it commitment to the existence of the entity. We take the name at face-value, and say that it fails to name anything.

Acausal platonic entities are odd because they play no causal role in the world, unlike the unobservable yet causal entities that scientific realists are happy to believe in. How can they be indispensable if it makes no difference whether they exist or not? Armed with use/mention confusions, Colyvan disarms this worry: mathematical entities are indispensable to our theories, and "do not need to play causal roles in those theories (indeed, it is generally agreed that they do not play such roles)" (230). But mathematical entities play no role in theories, expressions for them do. The role played by mathematical expressions in our theories is, like the role played by all expressions, not causal but semantic.

Are acausal mathematical objects any more mysterious than stars and planets outside our light cone that do not causally interact with us? And do we not "accept the existence of stars and planets outside our light cone because they play an indispensable role in our best cosmological theories"? (232). Well, no. What Colyvan should say is that we "accept the existence of stars and planets outside our light cone because our best cosmological theories say that they play an indispensable causal role in the world". Acausal mathematical entities play no causal role in the world, and you cannot argue for their indispensability as we argue for that of stars and planets outside our light cone.

Use/mention confusions ease the transition from words to things. But in the case of numbers, they are not the only thing that does this. The transition proceeds without confusion if we insist (a) that the words are to be taken at face-value, and (b) that appropriate sentences containing them are true. Given (a) and (b), the indispensability of mathematics, of number-talk, gets us to numbers as well. Antiplatonists or nominalists about numbers must resist the transition by rejecting (a), or (b), or both – they must either refuse to take the talk about numbers at face-value and give it some sort of antiplatonist construal, or take the talk at face-value and say that it is false, though indispensable. The indispensability argument, then, only shows the indispensability of thought or talk about numbers (and other mathematical entities). It does not, by itself, show the indispensability of numbers, when these are viewed as platonic entities. To get the entities out of the indispensability of the talk, more is required, and the more that is required can be disputed.

Colyvan evidently finds it absurd to dispute (a) and (b). He complains that "the nominalist cannot employ the usual semantics to account for the truth of sentences such as 'there is a number smaller than 2'" (224, note 2). We might as well say that a nominalist about the creeps cannot employ the usual semantics to account for the truth of sentences such as 'Osama Bin Laden gives me the creeps'. We are all nominalists or antirealists about the creeps. We all say that if the 'usual semantics' is applied to this sentence, then it is false. Or, more plausibly perhaps, we say that if it is true, then the 'usual semantics' is not to be applied to it, either because it is just a colourful way of saying that Osama Bin Laden makes me nervous, or because what makes it true is just the fact that Osama Bin Laden makes me nervous. (These two alternatives are not quite the same: the first seems to involve some claim of 'sameness of meaning', the latter does not. But I ignore this complication here.)

The creeps is one thing, numbers another. The nominalist about numbers can simply deny that 'there is a number smaller than 2' is true, and set to work to explain the indispensability of number-talk. That is Hartry Field's programme. It is consistent with acknowledging the *logical* truth of "If Peano's axioms are true, then it is also true that there is a number smaller than 2". As for the easy move from the undeniable truth "I have five fingers on my right-hand" to "The number of fingers on my right-hand is five", the nominalist can say that the latter is just a long-winded way of saying the former.

Colyvan discusses my argument that "If we view [the indispensability argument] from a Popperian perspective, it begins to lose its charm" (1986: p. 90). By "a Popperian perspective" I simply meant a falsificationist perspective:

> Imagine that all the evidence that induces scientists to believe (tentatively) in the existence of electrons had turned out differently. Imagine that electron-theory turned out to be wrong and electrons went the way of phlogiston or the heavenly spheres. Popperians think that this *might* happen to any of the theoretical posits of science. But can we imagine natural numbers going the way of phlogiston, can we imagine evidence piling up to the effect that there are no natural numbers? This must be possible, if the indispensability argument is right and natural numbers are a theoretical posit in the same epistemological boat as electrons.
>
> But surely, if natural numbers do exist, they exist of necessity, in all possible worlds. If so, no empirical evidence concerning the nature of the actual world can tell against them. If so, no empirical evidence can tell in favour of them either. The indispensability argument for natural numbers is mistaken. (1986: pp. 90-91)

One thing is sure – one should never, in philosophy, say "Surely". Colyvan grants that this objection "presents serious difficulties for any defender of the indispensability argument who takes mathematical entities to exist of necessity" (230). His way out is contingent platonism. It is a contingent matter whether mathematical entities exist. So, for example, the number five exists in some possible worlds but not in others, and only empirical inquiry can tell us whether the actual world is one of the worlds in which the number five exists!

I confess that I overlooked this possibility. Before I turn to discuss it, it is worth dwelling on what Colyvan loses by advocating it. The charms of necessary Platonism are considerable. Its chief charm is that it enables us to disarm Benacerraf's epistemic worry about platonism. How can we find out that the number five exists? No problem, says the necessary Platonist. "The number five exists" is a necessary truth, true in all possible worlds. Necessary truths are knowable a priori. So we know a priori that the number five exists. This is the line of thought that many Platonists, early and late, have pursued. It has difficulties of its own, of course, concerning the very idea of necessary existence. Like many philosophers, I can make little sense of that idea.

What about contingent platonism? According to this view, there is a possible world in which the number five exists, and another one in which it does not. Now, of course, we cannot identify the 'possible worlds' spoken of here with standard set-theoretical interpretations of formulas. Nothing is easier than to produce an interpretation in which the sentence "The number 5 exists" is false – let the domain of that interpretation contain just Mark and me, viz. {Alan Musgrave, Mark Colyvan}. It is equally easy to produce another interpretation in which that sentence is true – let the domain be {5}. 'Possible worlds' are not such model-theoretic trivialities. 'Possible worlds' are heavy-duty metaphysics of some kind.

So the question recurs – is there a heavy-duty possible world in which the number five does not exist? A contingent Platonist like Colyvan evidently thinks that there is. So what will he make of the equivalence "The number of fingers on my right-hand is five if and only if I have five fingers on my right-hand"? The left-hand side is false in a world bereft of the number 5. The right-hand side might be true in a world bereft of the number 5. If the right-hand side *is* true in that world, and the left-hand side false, then the equivalence is false in that world. The equivalence is contingent as well, true in some heavy-duty worlds and false in others. (One can hardly avoid this by saying that in a world bereft of the number five, there cannot be five of anything, not even numerals to count up to five with!) I think that if the equivalence is true, then it is necessarily true. And that if numbers exist, then they exist necessarily.

Which brings us back to the indispensability argument. The chief burden of Colyvan's paper is that if you go in for what he calls 'confirmational holism' or 'justificatory holism', you will be led to realism about mathematical entities. 'Confirmational or justificatory holism' is just the Duhem-Quine thesis that whole systems of theory, and not isolated hypotheses, are required to obtain testable predictions about the world. If these predictions turn out to be correct, then it is the whole system that gets confirmed. If the whole system includes mathematical theories, that postulate mathematical entities, then confirmation of the whole system gives us evidence that the mathematical entities exist.

I am afraid that this misses exactly the 'Popperian perspective' from which my objection to the indispensability argument proceeded. Never mind 'confirmational holism' – what about 'refutational holism'? When a theoretical system gets refuted, scientists know that something is wrong somewhere, but they do not rest content with that. They try to pin the blame more narrowly, and there is an enormous literature on how they do that, beginning with Duhem and Quine themselves. My question was: can we imagine pinning the blame for a refutation of a theoretical system that contained arithmetic on the non-existence of the natural numbers? Or, to take a silly example, when we refute a system by finding out that one drop of water put together with another drop of water does not yield two drops of water, but rather one big drop, does it make sense to pin the blame on "1 + 1 = 2" and say that in our world the number two does not exist?

Never mind silly examples. When physicists found out that space is not Euclidean but rather non-Euclidean, did they find out that Euclidean geometry is *false*? A hard-won logical empiricist distinction is important here, the distinction between pure and applied geometry (or more generally, between pure and applied mathematics). What physicists found out is that real space-time is not adequately represented by Euclidean laws, so that applied Euclidean geometry did not work. But Euclidean geometry considered as a theory of pure mathematics remained intact.

Colyvan tries to convict me of a Popperian 'separatist' theory of confirmation or justification. He says that "The separatist wants a crucial experiment that identifies the causal roles of the entities in question" (231). He asks "What are the crucial experiments that establish the existence of [stars and planets outside our own light cone]?" (231). Of course, there are no such crucial experiments. No 'crucial experiment' can *establish the existence* of any entity. Since experiments test whole bodies of theory, one cannot say that any successful experiment confirms (let alone establishes the truth of) some particular existential assertion in that body of theory. Similarly, if an experiment contradicts a prediction drawn from a body of theory, it is muddle-headed to say that it contradicts some particular hypothesis in that body of theory. Still, scientists want to try to figure out which particular hypothesis is responsible for the failed prediction. And the question recurs: would it ever make sense to pin the blame on the hypothesis that some abstract mathematical entity happens to exist in our world?

I confess that I can make little sense of this. The reason is simple. Existence claims regarding acausal abstract entities form no part of testable theories about the (actual) world. As Cheyne has shown elsewhere, when Colyvan is defending his contingent Platonism he points out that it is neither obvious nor necessary nor knowable a priori that there are odd numbers greater than five. Rather, what is obvious, necessary and knowable a priori is the logical truth "If the axioms of arithmetic are true, then there are odd numbers greater than five". Conflating the two is committing the 'conditional fallacy'. But when Colyvan is trying to convince us that acausal objects can help explain things, he commits that same fallacy. We are to explain why a square peg of side length l will not fit into a round hole of diameter l. Colyvan says there is a non-causal explanation involving acausal squares and circles. Cheyne points out that there is only a causal explanation involving square pegs and circular holes, and the claim that these concrete objects satisfy the antecedent and consequent of the conditional geometrical claim. The conditional claim may be indispensable to the explanation, but it does not assert the existence of acausal objects. (For further details and examples, see Cheyne (unpub.).)

6. ANTIREALISMS

I turn, finally, to those contributions that discuss antirealist views of one kind or another. One currently fashionable antirealist view is the *semantic conception of theories* (hereafter SC). According to SC, the traditional 'statement view of theories' is wrong: theories are not true or false statements, but sets of models. Laws or generalisations are clauses in definitions of those sets of models. Now if anything is 'true by definition', a definition is. But definitions say nothing about the world, and neither do scientific theories construed as clauses in such definitions. Claims about the world only enter the picture when it is claimed that some definition is not empty, that it applies to some part of the world, that this or that set of data can be 'fitted' into one of the models defined. Antirealism is already to the fore. The aim of scientific theorising is to provide a toolkit for building models of the phenomena. The name of the game is 'saving the phenomena', as van Fraassen put it. The aim is empirical adequacy, not truth.

Noretta Koertge criticises SC on methodological grounds: its associated methodology is "either antithetical to commonly accepted norms of scientific inquiry or hopelessly ad hoc" (237). It is an appealing line of criticism. Popper launched a similar criticism of what he called 'instrumentalism', the view that theories are more or less useful instruments or tools for saving phenomena:

> "Instruments ... cannot be refuted ... . The instrumentalist interpretation will therefore be unable to account for real tests, which are attempted refutations, and will not go beyond the assertion that *different theories have different ranges of application.* But then it cannot possibly account for scientific progress. Instead of saying (as I should) that Newton's theory was falsified by crucial experiments which failed to falsify Einstein's, and that Einstein's theory is therefore better than Newton's, the consistent instrumentalist will have to say ... 'Classical mechanics ... is everywhere exactly "right" where its concepts can be applied'" [*Conjectures and Refutations*, p. 113. The quoted sentence is from Heisenberg.]

It is no wonder that Koertge's arguments against SC resemble Popper's. As she rightly sees, SC is yet another version of the instrumentalism that Popper was attacking. According to SC, recalcitrant phenomena do not refute a theory, they merely show that a definition fails to apply. Newton's theory was not refuted by data about Mercury's perihelion – rather, "scientists failed to find a Newtonian model for the trajectory of Mercury" (239). Similarly, Lavoisier's experiments with mercury merely showed that "the phlogiston model fails to represent the controlled combustion of mercury as carried out in this experiment" (Koertge, 239, quoting Giere, 1991, p. 76).

Koertge asks why the failure of scientists to find a Newtonian model for the trajectory of Mercury loomed so large in the history of science, and led to the replacement of Newton's theory by Einstein's. She suggests that SC adherents "must also posit that scientists value theories whose models fit more phenomena" (4). Well, they might posit that, but what methodological justification of such a 'posit' might they give? If we have a tool that saves many phenomena, why not keep it for those purposes, and only use a more refined tool when we need to? Hanson saw the point. (Hanson was defending yet another antirealist view of theories, the inference-licence view. No matter.) Writing about 'The Scientist's Toolbox', he warmed with characteristic verve to the analogy between theories and tools:

> There are those who, knowing something of modern physics, dismiss Newtonian mechanics with a snap of the fingers. I suggest that such people ... are confusing the *purposes* of what are now seen as two distinct ... methods of representation. It is no longer a question of Newton's laws being wrong and Einstein's laws being right. ... because of its greater simplicity, the Newtonian formulation is greatly preferable to relativistic quantum laws ... Only when our experiments absolutely require more refined representation do we place the Newtonian formulation to one side. But then there is no question of appealing to ... the truer method of representation. Newtonian mechanics is simply *inappropriate* to the representation of relativistic and quantum phenomena. By exactly the same token, relativity theory and quantum mechanics are inappropriate to the representation of a good deal of the macro-physical world. (We do not distemper walls with water-colour brushes, nor do we repair watches with sledge-hammers.) [Hanson, 1969, 315-7]

This fits the semantic conception like a glove. As Hanson makes clear, if we take seriously the view that theories are tools for representing bodies of data, then failure to represent some data-set does not refute a theory, it merely shows that it is not a good tool for that particular job. The methodological demand for theories of broad scope makes as much, or as little, sense as the demand for multi-purpose tools. Nor does the methodological demand for theoretical unity make much sense. If theory A saves all its phenomena, and theory B saves all its phenomena, then all the A and B phenomena have been saved. Why search for a unified theory that will save A-phenomena and B-phenomena all at once? Scientists seek comprehensive and unified theories – but toolmakers do not seek multi-purpose or all-purpose tools. (Swiss Army knives are fun, but not many professional carpenters jettison all their special purpose tools in favour of them!)

Antirealists have, down the ages, invoked simplicity or 'economy of thought' at this point. But if simplicity is viewed, as it must be by instrumentalists, merely as a 'pragmatic virtue' (van Fraassen), then it is not clear that this will work. Modern physics is not pragmatically simpler than classical physics. Hanson took it for granted that the 'Newtonian formulation' is simpler and hence pragmatically preferable to 'relativistic quantum laws'. Relativistic mechanics is not 'economical of thought', as any physics student knows. The great Duhem saw the problem better than most:

> Why should we forbid the worker the successive employment of disparate instruments when he finds that each one of them is well adapted to a certain task and not well adapted to another job? [P. Duhem, *The Aim and Structure of Physical Theory*, p. 294.]

Duhem's answer to his question was not pragmatic simplicity or economy of thought. Duhem's answer was, in a word, metaphysical faith (*op. cit.,* pp. 296-7). When Giere talks about the world having "a single structure", he merely echoes Duhem. But professions of metaphysical faith are anathema to Duhem's latter-day incarnation, van Fraassen, who has managed to convince the world that it is realists who must invoke metaphysics, not constructive empiricists. Van Fraassen's constructive empiricist demands for comprehensive and unified theories are quite ad hoc.

Koertge is right. One can graft demands for comprehensiveness and unity onto SC, and make it seem that there is no methodological difference between realism and instrumentalism. But such grafts are ad hoc. They should be rejected by an SC that takes seriously the idea that theories are just tools for representing data-sets.

Another route to antirealism is to jettison the commonsense realist theory of truth in favour of an epistemic theory of truth. Are you obsessed with the sceptical nightmare that our best theories might be false and the entities they postulate non-existent? Then *define* truth as what ideal science, pursued to its limit, will throw up. Earlier I used a Fitch-style argument, made famous by Williamson, against epistemic theories of truth in general, more precisely, against the idea that all truths are knowable which is a common presupposition of such theories. As **Graham Oddie** explains, that argument can be disputed, and it would be nice to have a direct argument against Peircean idealism that does not get bogged down in those disputes. This Oddie has provided. He modestly confesses that there may be an error lurking in his knock-down-drag-out argument. I can only report that I have not been able to find one.

Nor can I find anything to disagree with in **Hans Albert**'s critique of what he calls 'methodological historism'. Like Albert, I find this doctrine unclear, to say the least. Is it the view that naturalistic methods cannot be applied in history, that history *qua* history is not amenable to them? That view seems to be refuted by the historical natural sciences, or the historical parts of natural science, such as the attempts by astronomers to explain the formation of the solar system, of geologists to explain current formations on the surface of the earth, or of evolutionary biologists to explain the current structure and distribution of animals and plants on the surface of the earth. One does not find astronomers or geologists or evolutionary biologists saying that they employ some special method of *verstehen*, that they have no need of the concept of cause, and such things. Or is it, more likely, that it is not history as such that is non-naturalistic, but rather the explanation of human action, past and present? That view seems to be refuted by the theoretical social sciences, such as psychology or economics or sociology, which try to formulate general hypotheses and give causal explanations of human behaviour and human society using them. All that remains, when confusion is set aside, is the distinctive focus of interest of historical as opposed to generalising scientists, a point that Weber stressed long ago. But we can see an historical event in all its particularity and uniqueness without thinking it uncaused or not governed by law. It may just arise from a unique confluence of independent causal chains, each of them law-governed.

And so, finally, to one such event, back at the beginnings of science. I can say little about **Andrew Barker**'s fascinating study of Ptolemy's *Harmonics,* except to thank him for it. In my amateur excursion into these matters, I criticised Duhem's instrumentalist interpretation of Ptolemy's geometrical models in astronomy, an interpretation that had become something of an orthodoxy amongst subsequent historians. Although the realism/instrumentalism issue is not the main focus of Barker's paper, I take comfort from what he says there. For what he says, as I read it, is that Ptolemy was as much a realist in his *Harmonics* as he was in astronomy. Ptolemy's harmonic ratios were a guide, not just to real musical consonances, but also to real psychological phenomena and to real astrological phenomena. In saying this, Ptolemy tapped into traditions foreign to our ears. But it is unhistorical to suppose that speculations which seem bizarre to us cannot have been seriously (that is, realistically) intended by their proponents. Besides, I am tempted to add, Ptolemy's harmonic speculations, while they may seem bizarre to us in their details, are not so bizarre in their general orientation. That reason (mathematical reason) is the key to understanding nature is a Greek legacy passed down to us through the scientific revolution, and still going strong today, as Redhead points out (pp. 8-11).

REFERENCES

Archibald, G. C. (1967) 'Refutation or Comparison?' *British Journal for the Philosophy of Science* 17: 279-296.

Cheyne, C. (unpub.) 'Getting an "Is" from an "If": Mathematical Realism and the Conditional Fallacy'.

Colyvan, M. (2001) *Indispensability of Mathematics.* New York: Oxford University Press.

Duhem, P. (1954) *The Aim and Structure of Physical Theory* (transl. by P. Wiener). Princeton: Princeton University Press.

Hanson, N.R. (1969) *Perception and Discovery.* San Francisco: Freeman, Cooper and Co.

Musgrave, A. (1986) 'Arithmetical Platonism: Is Wright Wrong or should Field Yield?', *Essays in Honour of Bob Durrant.* Otago University Philosophy Department.

Musgrave, A. (1999) *Essays on Realism and Rationalism.* Amsterdam; Atlanta: Rodopi.

Musgrave, A. (2001) 'Rationalität und Zuverlässigkeit.' ['Rationality and Reliability'] *Logos* 7: 94-114.

Popper, K. R. (1968) *The Logic of Scientific Discovery,* 2nd edn. London: Hutchinson.

Popper, K. R. (1969) *Conjectures and Refutations.* London: Routledge and Kegan Paul.

Psillos, S. (1999) *Scientific Realism: How Science tracks Truth.* London, New York: Routledge.

INDEX OF NAMES